EXS 73

Interface between Chemistry and Biochemistry

Edited by P. Jollès
H. Jörnvall

Birkhäuser Verlag
Basel · Boston · Berlin

Editors

Prof. Dr. P. Jollès
Laboratoire des Protéines
Université René Descartes
45, rue des Saints-Pères
F-75270 Paris Cedex 06
France

Prof. Dr. H. Jörnvall
Department of Medical Biochemistry and
Biophysics
Karolinska Institutet
S-17177 Stockholm
Sweden

Library of Congress Cataloging-in-Publication Data

Interface between chemistry and biochemistry / edited by P. Jollès, H. Jörnvall.
(EXS; 73)
Includes bibliographical references and index.
ISBN 3-7643-5081-4 hc: acid-free. – – ISBN 0-8176-5081-4 hc: acid-free
1. Biochemistry. 2. Analytical biochemistry. I. Jollès, Pierre, 1927– . II. Jörnvall, Hans. III. Series.
QP514.2.I57 1995
574.19'2 – – dc20

Deutsche Bibliothek Cataloging-in-Publication Data

EXS. - Basel; Boston; Berlin: Birkhäuser.
Früher Schriftenreihe
Fortlaufende Beil. zu: Experientia
73. Interface between chemistry and biochemistry. – 1995
Interface between chemistry and biochemistry / ed. by P. Jollès
; H. Jörnvall.–Basel; Boston; Berlin: Birkhäuser, 1995
(EXS; 73)
ISBN 3-7643-5081-4 (Basel ...)
ISBN 0-8176-5081-4 (Boston)
NE: Jollès, Pierre [Hrsg.]

The publisher and editor can give no guarantee for the information on drug dosage and administration contained in this publication. The respective user must check its accuracy by consulting other sources of reference in each individual case.

The use of registered names, trademarks etc. in this publication, even if not identified as such, does not imply that they are exempt from the relevant protective laws and regulations or free for general use.

This work is subject to copyright. All rights are reserved, whether the whole or part of the material is concerned, specifically the rights of translation, reprinting, re-use of illustrations, recitation, broadcasting, reproduction on microfilms or in other ways, and storage in data banks. For any kind of use permission of the copyright owner must be obtained.

© 1995 Birkhäuser Verlag, PO Box 133, CH-4010 Basel, Switzerland
Printed on acid-free paper produced from chlorine-free pulp. ∞
Printed in Germany
ISBN 3-7643-5081-4
ISBN 0-8176-5081-4
9 8 7 6 5 4 3 2 1

Contents

Foreword
P. Jollès and H. Jörnvall VII

Chemistry at interfaces and in transport

F. Carrière, R. Verger, A. Lookene and G. Olivecrona
Lipase structures at the interface between chemistry and biochemistry 3

G. Vandenbussche, J. Johansson, A. Clercx, T. Curstedt and J.-M. Ruysschaert
Structure and orientation of hydrophobic surfactant-associated proteins in a lipid environment 27

M.N. Gupta
Enzyme function in organic solvents 49

G. von Heijne
Protein sorting signals: Simple peptides with complex functions 67

Chemistry and biochemistry

J. Jeffery
Enzymes: Chemistry and biochemistry 79

E. Appella, E.A. Padlan and D.F. Hunt
Analysis of the structure of naturally processed peptides bound by Class I and Class II major histocompatibility complex molecules 105

R.A. Lerner and K.D. Janda
Catalytic antibodies: Evolution of protein function in real time 121

Analysis of proteins and nucleic acids

A.S. Inglis, G.E. Reid and R.J. Simpson
Chemical techniques employed for the primary structural analysis of proteins and peptides 141

T. Douki and J. Cadet
UV and nucleic acids 173

Synthesis of active compounds

M. Schultz and H. Kunz
Chemical and enzymatic synthesis of glycopeptides 201

A. Undén and T. Bartfai
Peptides as active probes................................. 229

Metalloproteins

B.L. Vallee and D.S. Auld
Zinc metallochemistry in biochemistry 259

R.M.A. Knegtel, M.A.A. van Tilborg, R. Boelens and R. Kaptein
NMR Structural studies on the zinc finger domains of nuclear
  hormone receptors 279

Subject index 297

Foreword

The developing importance of the interface between chemistry and biology is probably the largest change to have occurred in chemistry in the past 15 years. Increasingly more chemists work on problems dealing with biology, and interfacial research is poised to move into the mainstream of both disciplines. This merging of two types of approach has resulted in a vigorous research discipline with unprecedented potential to address important biological and chemical problems. A series of examples is developed in this book.

Analytical aspects are discussed in several chapters. Fundamental concepts do not only derive from chemistry, but chemistry has provided biochemistry with powerful tools of analysis. Equally important, physicochemical methods allow studies of nucleic acids and lipids, lipases, receptors and other membrane proteins. Several chapters deal with enzymes in different contexts. The part devoted to metalloproteins is directed toward zinc metallochemistry and NMR structural work on zinc proteins.

Chemists have been able to bring to biology their characteristic approach of synthesizing new molecules. These aspects are treated in chapters devoted to glycopeptides and uses of peptides as probes. Further fields of interest in combining different disciplines concern novel active compounds, such as surfactant peptides and catalytic antibodies in studies resulting from close collaboration between chemists and biochemists.

Combined, all chapters illustrate the broad approaches used in modern research and the frequent use of both methods and subjects at the interface between different fields, be it chemistry and biochemistry, antibodies and enzymes, metals and proteins, transport and static molecules, or hydrophobicity and hydrophilicity.

Pierre Jollès and Hans Jörnvall

Université de Paris V and Karolinska Institutet
Paris and Stockholm, January 1995

Chemistry at interfaces and in transport

Lipase structures at the interface between chemistry and biochemistry

F. Carrière[1], R. Verger[1], A. Lookene[2] and G. Olivecrona[2]

[1]*Laboratoire de Lipolyse Enzymatique, CNRS, B.P. 71, F-13402 Marseille cedex 9, France*
[2]*Department of Medical Biochemistry and Biophysics, Umeå University, S-90187 Umeå, Sweden*

Summary. In this chapter we review recent molecular knowledge on two structurally related mammalian triglyceride lipases which have evolved from a common ancestral gene. The common property of the lipase family members is that they interact with non-polar substances. Pancreatic lipase hydrolyzes triglycerides in the small intestine in the presence of many dietary components, other digestive enzymes and high concentrations of detergents (bile salts). Lipoprotein lipase acts at the vascular side of the blood vessels where it hydrolyses triglycerides and some phospholipids of the circulating plasma lipoproteins. A third member of the gene family, hepatic lipase, is found in the liver of mammals. Also, this lipase is involved in lipoprotein metabolism. The three lipases are distantly related to some non-catalytic yolk proteins from *Drosophila* (Persson et al., 1989; Kirchgessner et al., 1989; Hide et al., 1992) and to a phospholipase A1 from hornet venom (Soldatova et al., 1993).

Introduction

Exocrine lipases are soluble enzymes that bind to lipid/water interfaces where they catalyze hydrolysis of the water-insoluble lipids to more polar products which can then be incorporated into cell membranes and be used in further metabolic processes. Lipases differ from other esterases in that their substrates are insoluble and therefore aggregated. It follows that lipases carry out catalysis in a heterogeneous system. Their activity is usually maximal only when the enzyme is adsorbed to a lipid/water interface. This catalytic property, known as "interfacial activation", was first described by Sarda and Desnuelle, who proposed a conformational change of the enzyme at the interface (Sarda and Desnuelle, 1958). The theoretical treatment of lipase kinetics at the interface was previously reviewed (Verger and deHaas, 1976).

In recent years a structural basis for the action of lipases was revealed through high-resolution X-ray crystallography of some microbial lipases (*Rhizomucor miehei* (Brady et al., 1990; Brzozowski et al., 1991), *Geotrichum candidum* (Schrag et al., 1991), *Candida rugosa* (Grochulski et al., 1993), *Pseudomonas glumae* (Noble et al., 1993), and *Candida antartica* (Uppenberg et al., 1994), of cutinase (Martinez et al., 1992), as well as of a mammalian triglyceride lipase (human pancreatic lipase (Winkler et al., 1990)). The active sites contain a charge-relay triad of Ser, His and Asp/Glu, similar to what was known from serine proteases. An additional feature of the lipases is that their active sites are hidden

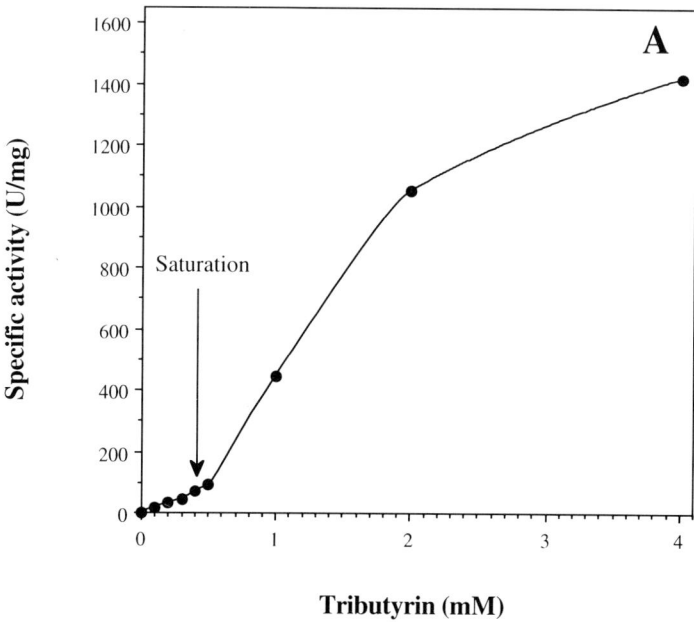
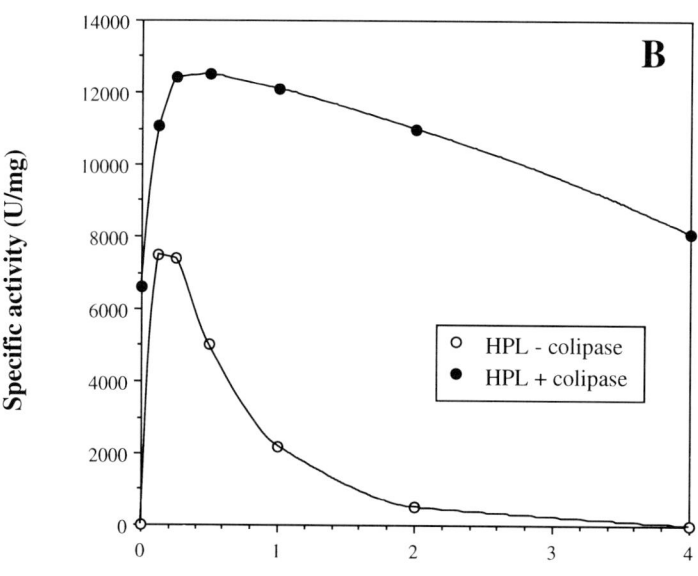

from water, and also from substrate molecules, by a surface loop or lid of about 20 amino acid residues. Crystallographic studies of the lipase from *Rizomucor miehei* in complex with active-site inhibitors elegantly showed that the lid opens on interaction with substrate molecules (Lawson et al., 1992). It was suggested that this might be the molecular correlate to the previously observed interfacial activation. The feature of a surface lid appears to be common in the families of triglyceride lipases, but is different from what had been previously encountered in phospholipases (Scott et al., 1990). Interesting variations to the theme are now emerging, as will be discussed below.

Pancreatic lipase

Pancreatic lipase is the major lipolytic enzyme involved in the digestion of dietary triglycerides. It hydrolyzes primary ester bonds of tri- and diglycerides thus generating 2-monoglycerides and fatty acids. In contrast to most other pancreatic enzymes, which are secreted as proenzymes and further activated by proteolytic cleavage in the small intestine, pancreatic lipase is directly secreted as a 50 kDa active enzyme consisting of 449 amino acid residues (De Caro et al., 1981; Lowe et al., 1989). The structural and catalytic properties of classical human pancreatic lipase (HPL) will be described below based on X-ray structures.

Main kinetic properties of pancreatic lipase

Entressangles and Desnuelle clearly showed that the rate of breakdown of a dilute solution of triacetin by porcine pancreatic lipase is very slow, and that there is a sharp increase in the enzymatic activity once the substrate solubility is exceeded and a lipid/water interface appears (Entressangles and Desnuelle, 1968). Interfacial activation of human pancreatic lipase (HPL) is demonstrated in Figure 1A with tributyrin as substrate (Thirstrup et al., 1993)).

Figure 1. (A) Interfacial activation of HPL. The influence of tributyrin concentration on HPL activity was measured by the pH stat technique. Under the present experimental conditions (37°C, pH 8.0), tributyrin saturation occurred at 0.4 mM, as indicated by the arrow. The assay contained 1 μg and 50 μg enzyme above and below tributyrin saturation, respectively, in 15 ml NaCl solution (150 mM) with various concentrations of tributyrin in the presence of a two-fold molar excess of colipase over lipase. (B) Sensitivity of HPL to bile salts and colipase. In the absence of colipase (open symbols) HPL was inhibited by bile salt concentrations above the critical micellar concentration (around 1 mM NaTDC). In the presence of colipase (closed symbols), the inhibition of HPL by bile salts was suppressed. The final assay (15 ml) contained 0.5 ml tributryrin, 14.5 ml of 0.28 mM Tris with 150 mM NaCl, 1.4 mM $CaCl_2$ and the indicated concentrations of NaTDC, in the presence of 1 μg to 5 μg HPL and a molar excess of colipase over lipase of about 2.

In addition to interfacial activation, pancreatic lipases are characterized by their behavior in the presence of bile salts (Borgström and Erlanson, 1973). They are inactive on an emulsified triglyceride substrate in the presence of micellar concentrations of bile salts, such as those found in the small intestine during digestion. It was proposed that the bile salt coating of triglyceride globules presents a negatively charged surface that inhibits pancreatic lipase adsorption and activation. To counteract this effect a specific lipase-anchoring protein, colipase, is present in the exocrine pancreatic juice. It forms a 1:1 complex with the lipase that facilitates its adsorption to bile salt-covered lipid/water interfaces (Maylié et al., 1971; Borgström and Erlanson, 1971; van Tilbeurgh et al., 1992). The inhibition of HPL by bile salts and the reactivation by colipase is shown using tributyrin as substrate and various concentrations of sodium taurodeoxycholate (Fig. 1B). When the bile salt concentration is increased above the critical micellar concentration (CMC) (1-2 mM) in the absence of colipase, HPL activity towards tributyrin is lost. HPL is desorbed from the lipid/water interface and no significant activity towards tributyrin monomers present in solution is observed, probably because the enzyme is in an inactive conformation in solution. Reactivation by addition of colipase in the assay is observed (Fig. 1B). An additional effect of colipase, and of bile salts, is to prevent inactivation of the enzyme at interfaces with high surface tension (Gargouri et al., 1986a).

The previous properties demonstrate adaptation of HPL to physiological conditions found in small intestine, where pancreatic and biliary secretions are mixed with dietary lipids. Through the concerted action of colipase and interfacial activation, HPL displays a high specificity towards insoluble triglycerides.

Functional organization of pancreatic lipase into several structural domains

The resolution of the three-dimensional (3D) structure of HPL (Winkler et al., 1990) showed the existence of two distinct domains in pancreatic lipase: a larger N-terminal domain comprising residues 1–336, and a smaller C-terminal domain made up of residues 337–449 (Fig. 2). This arrangement had previously been suggested based on biochemical studies (Bousset-Risso et al., 1985). The high degree of amino acid sequence homology observed within the lipase gene family supports the view that this folding is also common to lipoprotein lipase and to hepatic lipase, as it will be discussed below (Kirchgessner et al., 1989; Persson et al., 1989; Derewenda and Cambillau, 1991; Hide et al., 1992; van Tilbeurgh et al., 1994).

The large N-terminal domain of HPL is a typical α/β-structure dominated by a central paralled β-sheet (Ollis et al., 1992). It contains

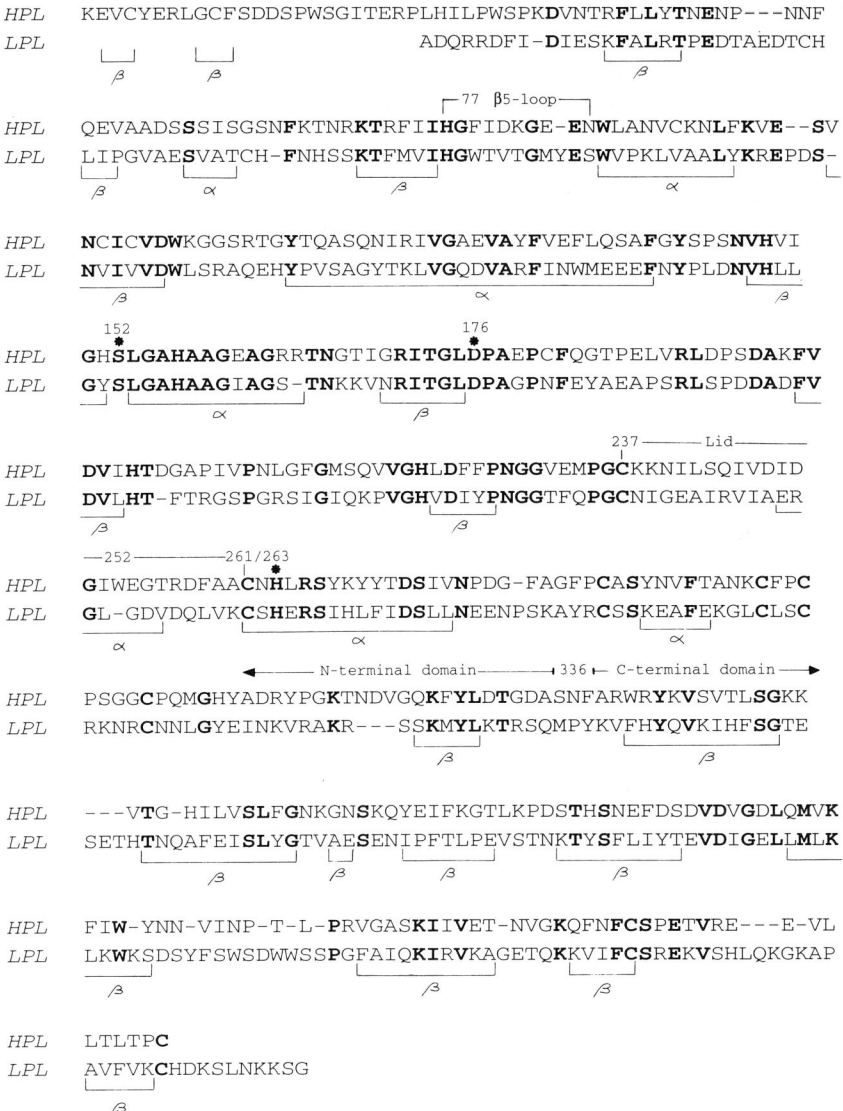

Figure 2. Sequence alignment of HPL and human LPL. This alignment results from modeling of LPL based on the 3D structure of HPL (van Tilbeurgh et al., 1994). Residues in bold are identical between the two sequences. The elements of secondary structure (α-helices and β-strands) are based on the HPL structure (Winkler et al., 1990). Stars indicate the catalytic triad residues. The important structural domains of HPL are indicated.

the active site with a catalytic triad formed by Ser 152, Asp 176, His 263. This catalytic triad is chemically analogous to that originally described in serine proteases such as chymotrypsin, but it is conformationally different. The β-strand/εSer/α-helix structural motif including the Gly-X-Ser-X-Gly consensus sequence has so far only been found in lipases and esterases (Derewenda and Cambillau, 1991; Derewenda and Sharp, 1993). The structure of HPL clearly demonstrated that Ser 152 is the nucleophilic residue essential for catalysis. This was in agreement with chemical modification of Ser 152 in porcine pancreatic lipase (Guidoni et al., 1981).

In HPL the active site is covered by a surface loop (lid) between the disulphide-bridged Cys 237 and 261 (Fig. 2). This surface loop includes a short α-helix with a tryptophan residue (Trp 252) completely buried directly on top of the active site Ser 152. Under this closed conformation, the lid prevents access of substrate to the active site. Spectroscopic studies of typtophan fluorescence had shown large spectral changes induced by acylation of pancreatic lipase with the active-site inhibitor tetrahydrolipstatin in the presence of bile salt micelles (Luthi-Peng and Winkler, 1992). By crystallization of the pancreatic lipase-procolipase complex, in the presence of mixed lipid micelles, it was shown that the lid was displaced to one side, exposing both the active site and a larger hydrophobic surface (van Tilbeurgh et al., 1993). This motion is induced on binding to lipid and is most likely the structural basis for "interfacial activation" of pancreatic lipase.

The β-sandwich C-terminal domain of pancreatic lipase is required for colipase binding as shown in the 3D structure of the HPL-porcine procolipase complex (van Tilbeurgh et al., 1992). Procolipase is an amphipatic "three-finger" protein which is topologically comparable to snake toxins, even though these proteins do not share any sequence homology. The tips of the fingers contain hydrophobic amino acid residues which form a lipid binding site. Lipase binding occurs at the opposite side of the molecule and involves polar interactions. In the absence of an interface, no conformational change in the lipase molecule is induced by binding of procolipase (van Tilbeurgh et al., 1992). The hydrophobic surface of colipase, in addition to the hydrophobic back of the lid, helps to bring the catalytic N-terminal domain of HPL into close contact with the lipid/water interface.

Details in the interfacial activation of human pancreatic lipase

In solution (absence of interface) HPL is in closed conformation as shown in Figure 3 (Winkler et al., 1990; van Tilbeurgh et al., 1992). The active site is completely inaccessible to solvent due to the conformation of the lid, which also interacts with another surface loop, the β5-loop according to Winkler's nomenclature (residues 76 to 85). A residue in the lid region, Trp 252, is structurally located directly on top of the active Ser 152, with the indole ring packed against Phe 77, a residue which belongs to the β5-loop. In the presence of an interface, the lid domain

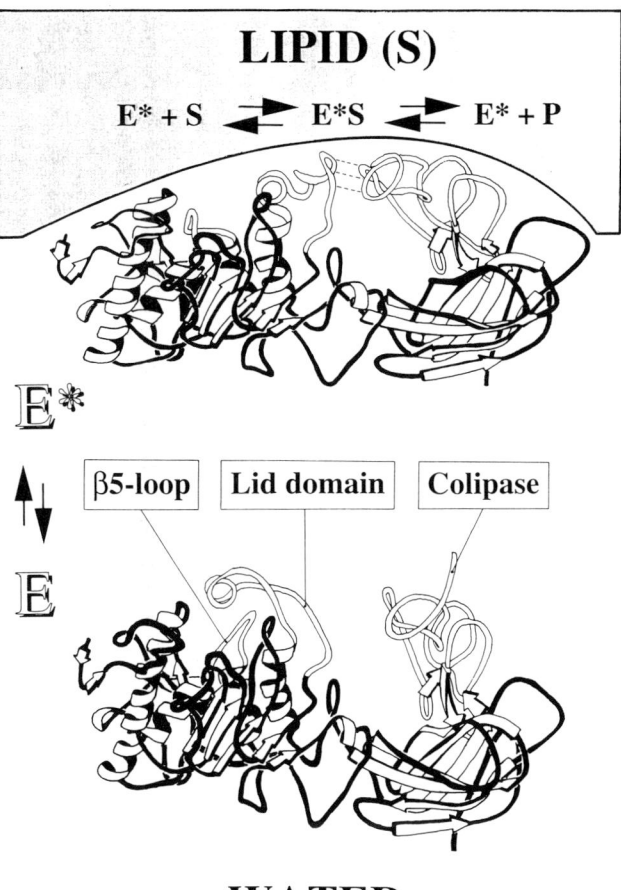

Figure 3. Interfacial activation of the HPL-colipase complex as revealed by X-ray crystallography. In solution, HPL (E) exists in a closed conformation and the active site is completely inaccessible to solvent due to the conformation of a surface loop, "the lid domain", which also interacts with another surface loop, the β5-loop. The activation of the pancreatic lipase-colipase complex at an interface (E*) results from the simultaneous conformational changes of these two surface loops. At the same time, colipase interacts with both the C-terminal domain and the lid domain of HPL. E* is the lipase adsorbed at the lipid/water interface (Verger and deHaas, 1976) where it carries out heterogeneous catalysis by hydrolysis of the insoluble triglyceride substrate (S).

as well as the β5-loop undergo large conformational changes, creating a hydrophobic active-site groove and adjusting the hydrolytic machinery (van Tilbeurgh et al., 1993), as shown in Figure 3.

The opening of the active site is caused by a complicated reorganization of the lid. As mentioned above, this surface loop contains a short α-helix (residues 248 to 255) covering the active site in the closed conformation of HPL. In the presence of lipids, this helix partially unwinds and two new helices are formed (residues 241 to 246, and residues 251 to 259). The conformational change of the lid results in a maximal main-chain movement of 29 Å for Ile 248. The substrate now has access to the active site. Trp 252, which fills the active-site pocket in the closed conformation, is conveyed to the surface of the molecule and is involved in a new interaction with the core of the protein.

Another consequence of the lid reorganization is the conformational change of the β5-loop. In the closed conformation of the HPL-procolipase complex, this loop makes van der Waals contacts exclusively with the lid. In the open conformation, these interactions are lost and the β5-loop lofts back onto the core of the protein. This movement creates an electrophilic region close to the active site Ser 152. This is probably the so-called oxyanion hole which stabilizes the transition-state intermediately formed during catalysis. The main-chain nitrogen of Phe 77 from the β5-loop moves to an ideal position to stabilize the negative charge developed during ester hydrolysis. The main-chain nitrogen of Leu 153 is also involved in the formation of the oxyanion hole.

In the open conformation procolipase also binds to the lid domain. This interaction is mediated by three direct hydrogen bonds between residues of the lid (Val 246, Ser 243, Asn 24) and residues of the procolipase N-terminal domain (Arg 38, Leu 16, Glu 15), respectively. This later domain undergoes a considerable conformational change from the closed to the open conformation of the complex.

As a result of all these conformational changes, the open lid and the tips of the procolipase fingers form an impressive continuous hydrophobic plateau, extending over more than 50 Å. This surface might be able to interact strongly with a lipid/water interface and could explain the colipase effect in the presence of bile salts.

A subfamily of pancreatic lipases with new kinetic properties

Until recently, pancreatic lipase was considered to be well characterized with respect to structural and kinetic properties, and distinct in the lipase superfamily. However, several new members of the pancreatic lipase family have now been cloned, sequenced, and partly characterized (Grusby et al., 1990; Giller et al., 1992; Hjorth et al., 1993; Wishart et al., 1993; Thirstrup et al., 1994). Primary structure analyses of these new lipases reveal that the pancreatic lipase family can be divided into three subgroups: (i) classical pancreatic lipases, (ii) pancreatic lipase-related protein 1 (PL-RP1), and (iii) pancreatic lipase-related proteins 2 (PL-RP2). Among the RP1 subfamily, only HPL-RP1 has been expressed *in vitro*, but it has not been found to display any enzymatic activity under the assay conditions reported (Giller et al., 1992). In the case of the RP2 subfamily however, the kinetic properties of two lipases found in the

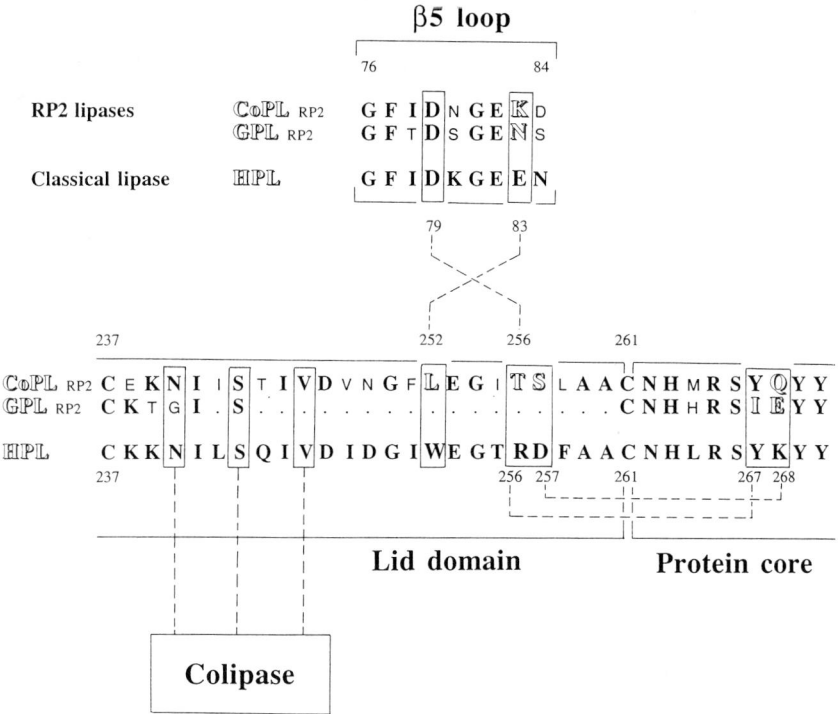

Figure 4. Sequence alignment of HPL, GPL-RP2 and CoPL-RP2 in their lid domain and β5-loop. Residues involved in interactions of the HPL lid with the protein core, the β5-loop and colipase are boxed. The interactions are shown by dotted lines. Residues in bold indicate identity with residues in HPL. Highlighted residues indicate striking differences in the RP2 lipases compared to HPL.

guinea pig (GPL-RP2 (Hjorth et al., 1993)) and in the coypu (CoPL-RP2 (Thirstrup et al., 1994)) have been studied. Both enzymes share an atypical enzymatic behavior and challenge the classical distinctions among lipases, esterases, and phospholipases.

In contrast to the classical HPL, CoPL-RP2 and GPL-RP2 display no interfacial activation, i.e., both enzymes are fully active already on monomers of partly soluble triglycerides, and their active site is accessible to a water-soluble substrate. In GPL-RP2, the lid domain is shortened to a "mini-lid" (Fig. 4) and the active center is envisioned to be freely accessible (Hjorth et al., 1993). In CoPL-RP2, the activity in solution is probably due to an open conformation of the lid domain. The lid domain sequence is poorly conserved between CoPL-RP2 and HPL (Fig. 4). Particularly the replacement of Trp 252 by a Leu in CoPL-RP2 could weaken the interactions of the lid with the β5-loop, thus facilitating the spontaneous opening of the active site.

It has also been demonstrated that colipase does not reactivate GPL-RP2 and CoPL-RP2 at high bile salt concentrations (Thristrup et al., 1994). Whereas there is absolutely no effect of colipase on GPL-RP2 activity whatever the bile salt concentration, the colipase binding to CoPL-RP2 is not completely abolished since a clear effect of colipase is observed below CMC of bile salts, at 1 mM sodium taurodeoxycholate. The behavior of GPL-RP2 can easily be explained: among the 12 residues of the HPL C-terminal domain involved in colipase binding (van Tilbeurgh et al., 1992), nine are different in GPL-RP2, including the important Lys 399. Moreover, colipase also interacts with the open lid of HPL (van Tilbeurgh et al., 1993). These interactions cannot exist in a lipase with a "mini-lid" such as GPL-RP2. In CoPL-RP2, the absence of colipase effect at bile salt concentrations above 2 mM is not easily interpretable. Almost all residues involved in the interaction with colipase are conserved. One striking exception is residue 403 which is conserved as tyrosine in all classical lipases, but is different in all RP2 lipases (Carrière et al., 1994). Tyr 403 interacts strongly with colipase through several van der Waals contacts in the HPL-colipase complex (van Tilbeurgh et al., 1992). The stacking of Tyr 403 and Arg 65 of colipase mainly confers the apolar component of binding energy. The point mutation of residue 403 in CoPL-RP2 could induce a low affinity for colipase.

Also, the lid/colipase interactions might be suppressed in CoPL-RP2 due to an open conformation of the lid-domain different from the one observed in HPL. The open conformation of the HPL lid is stabilized by interactions with the core of the protein, the colipase, and the β5-loop (van Tilbeurgh et al., 1993) (Fig. 4). The lid domain interacts with the core of the protein through a salt bridge (Asp 257 ... Lys 268) and a hydrogen bond (Arg 256 ... Tyr 267). In the RP2 subfamily, these residues are almost all different whereas they are totally conserved in the two other pancreatic lipase subfamilies (Thirstrup et al. 1994; Carrière et al. 1994). The lid domain also interacts with the β5-loop through a salt bridge (Arg 256 ... Asp 79), and a hydrogen bond (Trp 252 ... Glu 83). Here again, most of these residues are different in the RP2 subfamily. In CoPL-RP2, all the stabilizing interactions observed in HPL between the lid domain, the protein core, and the β5-loop are missing. One can assume that the lid domain in CoPL-RP2 possesses a higher degree of freedom and that the interactions with colipase are weakened if not completely absent.

Finally, both GPL-RP2 and CoPL-RP2 display phospholipase activities of 500 U/mg and 180 U/mg, respectively (1 unit = 1 μmole of fatty acid released per minute), using egg yolk lecithin as substrate (Thirstrup et al., 1994). For comparison, the classical porcine pancreatic phospholipase A2 displays a specific activity of 700 U/mg under similar conditions. In contrast, HPL showed no measurable activity on this substrate. From a structural point of view, the reason for the high phospholipase activity remains unclear. Both enzymes display a high phospholipase activity towards lecithins but also towards other phospholipid classes (Hjorth et al., 1993; Thirstrup et al., 1994), whereas classical pancreatic lipases can only hydrolyze negatively charged phospholipids at very low specific activity (Verger et al., 1977). From the GPL-RP2 modeling based on the HPL 3D structure, the core of the N-terminal domain appears to be conserved overall, with the exception of the lid-domain. Furthermore, within a 10 Å radius sphere of the active site serine, there are no insertions/deletions and only four minor residue changes pointing away from the active site. Thus, it was speculated previously that the lid domain was also involved in substrate selectivity towards phospholipids (Hjorth et al., 1993). The lack of high phospholipase activity in classical pancreatic lipases might be viewed as a depressed action on phospholipids due to the presence of the lid-domain, rather than as a total absence of catalysis (Hjorth et al., 1993). Now that CoPL-RP2 has been characterized, it is clear there is no direct relationship between the nature of the lid domain and phospholipase activity within the RP2 subfamily. One can speculate that the low affinity of classical pancreatic lipases for phospholipids is due to the peculiar open conformation of the lid domain in HPL and the large hydrophobic area exposed around the active site entrance. This hydrophobic environment, only stabilized in the presence of a lipid/water interface, might induce a repulsion towards phospholipid polar heads. The fact that both GPL-RP2 and CoPL-RP2 are active on a monomeric substrate in solution, reveals a stable active conformation of these enzymes within an hydrophilic environment. Although they can bind to the lipid/water interface, phospholipids are found dispersed in the water phase as micelles or liposomes. The affinity of GPL-RP2 and CoPL-RP2 for such amphipathic molecules might be higher than that of classical pancreatic lipases.

Lipoprotein lipase: An adaption to more complex functions

Lipoprotein lipase catalyzes the same chemical reaction as classical pancreatic lipases, hydrolysis of the primary ester bonds in tri- and diglycerides to form free fatty acids and 2-monoglycerides, but there are at least two major differences. Lipoprotein lipase hydrolyzes phospholipids to about 10% of the rate of triglyceride hydrolysis. Furthermore, lipoprotein lipase is active on simple, partly soluble esters like p-nitrophenyl acetate. As discussed above, classical pancreatic lipases have little if any activity on soluble substrates, but need interfacial activation to be fully active. Thus, lipoprotein lipase has more in common with the kinetic properties of lipases from the PL-RP2 subfamily than with classical pancreatic lipases.

Role of lipoprotein lipase in lipoprotein metabolism

The site of action of lipoprotein is at the vessel wall in all parts of the organism (Bensadoun 1991; Olivecrona and Bengtsson-Olivecrona, 1993). The lipase acts as a membrane-bound enzyme on the surface of endothelial cells. Its mode of attachment is of an unusual kind. Due to its strong affinity for polyanions, the lipase binds to heparan sulphate glycan chains that in turn are anchored in the cell membrane by their core proteins. Several studies have shown that binding of lipoprotein lipase to cells can be prevented by treatment of the cells with heparan sulphate-degrading enzymes (Bensadoun, 1991; Eisenberg et al., 1992; Mulder et al., 1992).

The physiological substrates for lipoprotein lipase are the triglyceride-rich plasma lipoproteins. These are two kinds: chylomicrons for transport of dietary lipids from the intestine and very low density lipoproteins (VLDL) for transport of endogenously produced lipids from the liver. The lipoprotein have an organized structure with a core of non-polar lipids (triglycerides and cholesteryl esters) covered by a surface layer of more polar constituents (phospholipids, cholesterol, and some specific protein called apolipoproteins) (Segrest et al., 1994). When the chylomicrons enter the blood they encounter lipoprotein lipase resulting in rapid decrease in their content of triglycerides. In addition, some phospholipids are hydrolyzed, but this cannot keep pace with the shrinking surface area of the particles (Deckelbaum et al., 1992). Therefore, surface constituents are shed from the particles to be assimilated with high-density lipoproteins (HDL) (Schaefer et al., 1982). By these actions the particles are converted to so-called chylomicron remnants, which are rapidly cleared from the circulation via receptor-mediated uptake in the liver (Havel and Hamilton, 1988). VLDL goes through similar conversions (Eisenberg, 1990). The remnant particle formed is called a low-density lipoprotein (LDL). In this conversion also the hepatic lipase participates (Olivecrona and Bengtsson-Olivecrona, 1993). LDL is taken up in many types of cells via the well characterized LDL-receptors which binds apolipoprotein B, the main apolipoprotein constituent of LDL (Haven and Hamilton, 1988). Both chylomicrons and VLDL contain a minor apolipoprotein called apolipoprotein CII (apo CII), which acts as an activator protein for lipoprotein lipase. Deficiency of functional apo CII results in massive hyperchylomicronemia with symptoms much resembling those seen in total deficiency of lipoprotein lipase (Santamarina-Fojo and Dugi, 1994). Nascent chylomicrons carry some apo CII, but more is obtained from HDL, which serves as a reservoir of recycled apo CII (Eisenberg, 1990).

A new interesting aspect on the role of lipoprotein lipase has emerged recently. Because of its ability to bind to cell surfaces in a way that also allows interaction with lipoproteins, lipoprotein lipase has the ability to attract and accumulate lipoproteins at cell surfaces (Chajek-Shaul et al., 1982; Beisiegel et al., 1991; Eisenberg et al., 1992; Mulder et al., 1992; Rumsey et al., 1992). In addition, this stimulated binding allows interaction between heparin-binding apolipoprotein constituents of the lipoprotein particle (apo B and/or apo E) and the proteoglycans on the cell surface, thus generating a multipoint attachment of the particle (Olivecrona and Bengtsson-Olivecrona, 1993). The duration of the attachment in the vicinity of lipoprotein lipase must be a main determinant for the rate of lipolysis of the particle. Furthermore, concentration of lipoproteins at the cell surface may facilitate contacts with receptors that mediate that ultimate uptake of the particle in the cell. Interestingly, lipoprotein lipase has been shown to directly mediate binding of lipoproteins to one of the candidate receptors, the low-density lipoprotein receptor-related protein (LRP) (Beisiegel et al., 1991; Nykjaer et al., 1993; Chappell et al., 1992; Willnow et al., 1992; Chappell et al., 1993).

It follows from this brief presentation that lipoprotein lipase is engaged in several interactions. Some of these functions may have evolved because of the complexity of the lipid transport system and because of the need to regulate energy metabolism and plasma lipid levels.

Structural properties of lipoprotein lipase as compared to pancreatic lipase

The high degree of homology between lipoprotein lipase and pancreatic lipase (Fig. 2), especially in the N-terminal domain, has allowed predictions of the three-dimensional arrangement of lipoprotein lipase based on the crystal structure of pancreatic lipase (Persson et al., 1991; Derewenda and Cambillau, 1991; van Tilbeurgh et al., 1994). Human lipoprotein lipase contains 448 amino acid residues. The regions around the active site in pancreatic lipase are among the best conserved among the members of the lipase family. Therefore, it was concluded that the lipases must have a closely similar arrangement in their active sites. This has been illustrated by mutagenesis of residues in the proposed active site triad of lipoprotein lipase, Ser 132, Asp 156 and His 241 (Emmerich et al., 1992; Faustinella et al., 1991; Wong et al., 1994) which resulted in total loss of activity.

In contrast to the strong similarity around the active site region, there is only one identical residue in the lid region (Gly 229 in human lipoprotein lipase, shown to be not essential (Henderson et al., 1993)), and the lid of the lipoprotein lipase is one residue shorter (Fig. 2). The disulphide bond which closes the lid is, conserved. Substitution of the lid in lipoprotein lipase with that from pancreatic lipase results in complete loss of activity against long-chain triglyceride substrates

(Faustinella et al., 1992), while activity against tributyrin is only reduced by about 50% (Dugi et al., 1992). Deletion of the lid leads to loss of activity against long-chain triglycerides, while the activity against tributyrin is 2.4-fold higher than in the intact lipase (Dugi et al., 1992). Prediction of the secondary structure of the lid in lipoprotein lipase indicates that it is composed of two amphipathic helices (Dugi et al., 1992). Mutations in these segments show that its structure is essential for hydrolysis of emulsified substrates, but not for hydrolysis of short-chain esters. This is in concert with results from proteolytic scission of the peptide bond at the tip of the lid, which leads to loss of activity against long-chain triglycerides, while activity against short-chain esters is not affected (Bengtsson-Olivecrona et al., 1986).

The similarities in the C-terminal domain between lipoprotein lipase and pancreatic lipase are less stringent than those in the N-terminal domain. Analysis of the hydrophobicity patterns by means of the hydrophobic cluster method suggested, however, that this domain has a β-sandwich structure similar to that in pancreatic lipase (van Tilbeurgh et al., 1994). Further support for a similar folding comes from limited proteolytic cleavage of lipoprotein lipase with chymotrypsin (Lookene and Bengtsson-Olivecrona, 1993). A particularly sensitive region corresponds to the loop between the $\beta 5$ and the $\beta 6$ strands in the C-terminal domain of pancreatic lipase (Fig. 2, at the residues Phe 388 and Trp 390 in human lipoprotein lipase). Cleavage at this site results in loss of activity against long-chain triglyceride substrates, while activity against soluble substrates is retained. This is similar to what is found after cleavage of the lid region by trypsin (Bengtsson-Olivecrona et al., 1986). Lipase truncated by chymotrypsin does not bind to lipoproteins, which explains the loss of activity (Lookene and Bengtsson-Olivecrona, 1993). Substitutions of two tryptophan residues (Trp 392 and Trp 394) close to the chymotryptic cleavage site by alanine in a recombinant fragment corresponding to the C-terminal domain of lipoprotein lipase leads to loss of lipoprotein binding ability of the fragment (Williams et al., 1994). Thus, several lines of evidence suggest that the C-terminal region of lipoprotein lipase is involved in lipid binding, possibly in a concerted action with the lid region (Santamarina-Fojo and Dugi, 1994).

The structure of lipoprotein lipase is different from that of pancreatic lipase in that lipoprotein lipase is a non-covalent dimer (Iverius and Östlund-Lindqvist, 1976; Osborne, Jr. et al., 1985). The dimeric structure appears to be absolutely necessary for function. On monomerization, the lipase goes through a minor conformational change, which is practically irreversible (Osborne, Jr. et al., 1985). Pancreatic lipase is reported to be monomeric in its functional form, but the crystal unit is head-to-tail dimer (van Tilbeurgh et al., 1993). Speculation about the quaternary structure of the lipoprotein lipase dimer has led to the conclusion that a head-to-tail dimer with the two active sites facing the same side is the most likely arrangement (Wong et al., 1991; van Tilbeurgh et al., 1994; Santamarina-Fojo and Dugi, 1994). This would allow both active sites to face the lipid/water interface simultaneously. Stoichiometry studies with the active-site inhibitor tetrahydrolipstatin have demonstrated that both active sites on the dimer are functional (Lookene et al., 1994). The site for subunit interaction in lipoprotein lipase is not known, but it does not involve the C-terminal 60 residues, since lipase truncated by chymotrypsin is still dimeric (Lookene and Bengtsson-Olivecrona, 1993). It was pointed out by Derewenda and Cambillau that most of the deleterious mutations in lipoprotein lipase are in the closely packed N-terminal domain, while mutations in exposed surface loops are not as harmful to the catalytic activity of the enzyme (Derewenda and Cambillau, 1991). Many of the natural and site-directed inactive mutants of lipoprotein lipase that have been studied are in fact monomers (Hata et al., 1992). Therefore, conclusions regarding the functional importance of individual amino acid residues cannot be drawn directly.

One clear difference between lipoprotein lipase and pancreatic lipase is that the former has high affinity for polyanions like heparin and related substances. Interaction with cell surface heparan sulfate proteoglycans is important for attachment of the lipase to the vascular endothelium (Olivecrona and Bengtsson-Olivecrona, 1985; Bensadoun, 1991). The site for heparin interaction has been predicted to be in the interconnection between the two folding units, based on the pancreatic lipase structure (Persson et al., 1991; van Tilbeurgh et al., 1994). Here, there are several regions with clusters of positive charges which are not present in pancreatic lipase. Site-directed mutagenesis of both avian and human lipoprotein lipases have shown that at least two regions are involved in heparin binding, Arg 279–Arg 282 and Lys 296–Arg 297 (Hata et al., 1993; Berryman and Bensadoun, 1993). Studies with chimeras between lipoprotein lipase and hepatic lipase have led to opposing conclusions regarding which parts of the molecule that are most important for the high heparin affinity of lipoprotein lipase (Davis et al., 1992; Dichek et al., 1993). Most likely, multiple sites are involved and the proper cooperation between these sites is probably conformation dependent. van Tilbeurgh et al. have calculated the electrostatic potential of lipoprotein lipase and showed that there is a pronounced maximum of positive charges situated in front of a loop at the junction between the two domains and opposite to the entrance to the active site (van Tilbeurgh et al., 1994). Pancreatic lipase has a less pronounced electrostatic potential surface due to a more even distribution of positive and negative charges. In lipoprotein lipase the positive charges are grouped in four clusters, two of which come into close proximity due to the folding of the protein in the junction between the two folding domains. Another cluster, situated between residues 148 and 152, should also come into close contact and might therefore contribute to the heparin affinity. A fourth cluster is situated in the C-terminal domain. This differs from the others, since it is not composed of a continuous stretch of positive charges. This region was suggested to determine the high heparin affinity of lipoprotein lipase (Davis et al., 1992). This cannot be the whole story, since lipoprotein lipase truncated by chymotrypsin was found to bind to heparin (Lookene and Bengtsson-Olivecrona, 1993). Lipoprotein lipase has, however, to be dimeric to display high affinity for heparin (Bengtsson-Olivecrona and Olivecrona, 1985; Peterson et al., 1992; Hata et al., 1992). The reason is not fully understood, but perhaps only the dimeric structure has the proper conformation to allow efficient cooperation between all the clusters of positive charges. It has previously been demonstrated that octasaccharides can interact with lipoprotein lipase (Bengtsson et al., 1980).

Interestingly, van Tilbeurgh et al. demonstrated by molecular modeling that a heparin dodecamer could satisfy the putative heparin-binding sites on both subunits of the dimer (van Tilbeurgh et al., 1994). It is possible that one reason for the strong stabilizing effect of heparin on lipoprotein lipase is that it prevents dissociation of the dimer by holding the two subunits together.

Lipoprotein lipase at the interface

Several lines of evidence indicate that lipoprotein lipase behaves differently from pancreatic lipase at the lipid/water interface. First of all, lipoprotein lipase is a more surface-active molecule able to bind to lipid monolayers at higher surface pressures than pancreatic lipase (Vainio et al., 1983; Jackson et al., 1986). With pancreatic lipase, the main function of the activator, colipase, is to anchor the enzyme to interfaces even at high surface pressures as well as in the presence of bile salts and other competing proteins. With lipoprotein lipase and apolipoprotein CII this is not the case. Under most conditions lipoprotein lipase binds well to the lipid substrate and apo CII does not add to the binding (Olivecrona and Bengtsson-Olivecrona, 1987). Interaction with apo CII at the interface increases the catalytic efficiency of lipoprotein lipase. Details about how this is accomplished are not known. The need for the activator differs between different substrates. With short-chain triglycerides, like tributyrin, there is little or no additional effect because the activity of the lipase alone is high (Rapp and Olivecrona, 1978). With other substrates, the basal activity of the enzyme alone is depressed, and apo CII is required to restore it to a high level. Based on studies with natural and synthetic fragments, the ability of apo CII to interact with lipoprotein lipase and to activate is found to reside in the C-terminal third of the molecule (Smith and Pownall, 1984). This region is well conserved in apo CII:s from all species studied so far (Andersson et al., 1991; Hoffer et al., 1993). The N-terminal two-thirds of apo CII most likely form amphipathic helices of the type that is found also in several of the other apolipoproteins (Segrest et al., 1994). These helices, which typically span about 20 residues, turn their hydrophobic side towards the lipid and thereby anchor the protein at the lipid/water interface. The lipid-binding part of apo CII is required for efficient activation with natural lipoproteins. This was the case with chylomicrons from a CII-deficient patient, demonstrating that with complex biologically packed lipids also the interfacial binding of apo CII is necessary (Olivecrona and Beisiegel, unpublished).

There is no structural relationship between apo CII and colipase and it follows from the discussion above that the activators fulfill different functions. Furthermore, available evidence from lipoprotein lipase/hepatic lipase chimeras (Davis et al., 1992; Dichek et al., 1993) indicate that apo CII most likely is bound to the N-terminal domain of lipoprotein lipase. This further accentuates differences in the activation mechanisms between the lipases. Explanations can probably be found in differences in lid structures and in their movements on binding to lipids (van Tilbeurgh et al., 1994). There are no data yet to support the view that apo CII binds to the lid of lipoprotein lipase as in the case of the open form of the colipase/pancreatic lipase complex (van Tilbeurgh et al., 1993).

Around the active site of lipoprotein lipase

Compared to pancreatic lipase, lipoprotein lipase does not reach as high specific catalytic activities *in vitro*. When compared with tributyrin as substrate, the maximal activity differs by a factor of at least 10 (about 7500 U/mg for pancreatic lipase and about 400 U/mg for lipoprotein lipase as measured at pH 8.5, 20°C). In contrast, lipoprotein lipase has relatively high activity against soluble substrates and does not show the classical interfacial activation seen with pancreatic lipase. Figure 5 shows a typical experiment with triacetin. The activity of porcine pancreatic lipase is low at the lowest substrate concentrations, but increases in the manner predicted above 0.3 M triacetin when interfaces being to be formed. In contrast, the activity of lipoprotein lipase increases in a saturable manner and approaches V_{MAX} (corresponding to about 10 U/mg) well below 0.3 M triacetin. This must imply that the oxyanion hole is performed in lipoprotein lipase and thus is not affected by lid movements on binding to lipid/water interfaces, as was discussed above for members of the PL-RP2 subfamily.

Lipoprotein lipase hydrolyzes phospholipids. The interface of lipoproteins is composed to more than 95% of phospholipids and of only a

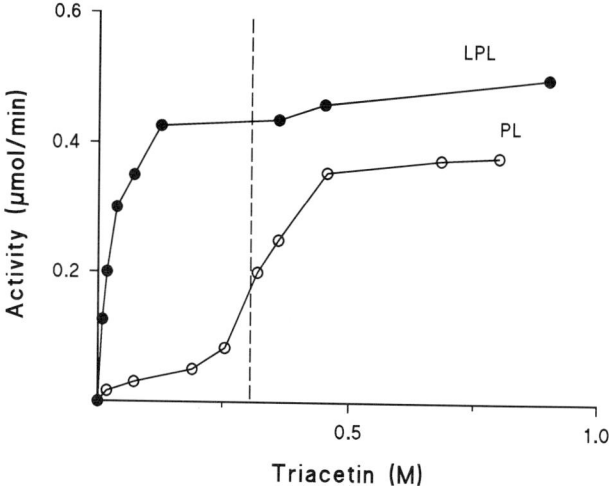

Figure 5. Lack of interfacial activation of lipoprotein lipase by triacetin. Activity versus concentration of triacetin for bovine LPL (filled symbols) and for porcine PL (open symbols). The activities were recorded using a pH-stat. The reaction medium (total volume 5 ml) contained 0.15 M NaCl, 2 mM $CaCl_2$ and 15 IU heparin/ml at pH 7.5 and 25°C. In the case of PL, the measurements were performed in the presence of 0.2 mM colipase. The enzyme concentrations were 78 nM for PL and 71 nM for LPL (based on the amount of dimer). The dashed line shows the limit for triacetin solubility.

few percent triglycerides. The interface is believed to be similar to that of bilayered phospholipid liposomes containing a few percent triglycerides. When triglycerides are presented to lipoprotein lipase in such bilayered phospholipid vesicles, the maximal catalytic rate for triglycerides is similar to that for the phospholipids (Rojas et al., 1991). In contrast, when the lipase acts on phospholipid-enveloped triglyceride core particles, it preferentially hydrolyzes triglycerides. This indicates that the lipase stays at the interface for several catalytic events and that it is easier for a triglyceride to enter the active site from the core of the particle than from the side by lateral diffusion in the phospholipid layer (Rojas et al., 1991). With triglycerides in mixed emulsions, short-chain trioctanoin is preferred in comparison with triolein, probably because of a higher surface concentration and of a higher flexibility of trioctanoin (Deckelbaum et al., 1990). This illustrates that lipoprotein lipase makes a selection at the interface. None of these pathways are influenced by apolipoprotein CII. There are no indications that the activator directly participates in the catalytic reaction, or influences the substrate selection at the interface (Olivecrona and Bengtsson-Olivecrona, 1987; Rojas et al., 1991). In contrast, available data suggest that the activator changes the catalytic rate constant to a similar degree for all lipid substrates in the particle (Rojas et al., 1991).

Lipoprotein lipase: Receptor interactions

One function of lipoprotein lipase that clearly resides in the C-terminal domain is binding to the low-density lipoprotein receptor-related protein (LRP). This binding was first seen in crosslinking experiments (Beisiegel et al., 1991) and has since then been verified in several laboratories by different techniques (Chappell et al., 1992; Willnow et al., 1992; Nykaer et al., 1993). The ability to bind to LRP was lost in lipoprotein lipase that was truncated in the C-terminal end by chymotrypsin at residue 388 (Nykjaer et al., 1993) (residue numbering according to human LPL), indicating that the binding region should be within or close to the chymotryptic cleavage site. A recombinant fragment spanning the postulated C-terminal domain of lipoprotein lipase (residues 313–448) was shown to bind to LRP (Williams et al., 1994). Binding was partly lost when Lys 407 was exchanged for Ala. Nykjaer et al. used a combination of CNBr fragments and of chymotryptic fragments to demonstrate that the binding region is between residues 378 to 423 in human lipoprotein lipase (Nykjaer et al., 1994). Interestingly, also hepatic lipase appears to bind to LRP (Nykjaer et al., 1994). The only sequence similarity between the two lipases in the suggested region is between residues 404–412, which could indicate that this region is involved in LRP binding.

Lipoprotein lipase and fatty acids

A marked difference between lipoprotein lipase and pancreatic lipase is that the former is strongly inhibited by long-chain fatty acids (Scow and Olivecrona, 1977; Bengtsson and Olivecrona, 1980; Posner and DeSanctis, 1987), while the latter shows a normal apparent slow down of the hydrolysis due to reverse reactions (Borgström and Carlsson, 1957). In some systems, pancreatic lipase is even stimulated by the presence of free fatty acids (Larsson and Erlanson-Albertsson, 1986; Gargouri et al., 1986b). For lipoprotein lipase it has been shown that fatty acids have a marked stabilizing effect on the enzyme, most likely due to direct binding of the fatty acid to the lipase (Olivecrona and Bengtsson-Olivecrona, 1987). This binding has multiple effects: it reduces blinding of lipoprotein lipase to the substrate, it reduces the effect of apolipoprotein CII and it reduces the affinity between lipoprotein lipase and heparin. It has been proposed that lipoprotein lipase is more strongly inhibited by its product fatty acids than other lipases as a regulatory mechanism (Olivecrona and Bengtsson-Olivecrona, 1987). If fatty acids are generated more rapidly than the underlaying tissue can assimilate then, lipolysis should no longer go on at that site. Since lipoprotein lipase at the cell surface cannot be regulated by kinases or other control mechanisms, an efficient way to limit its actions at the endothelial binding sites is required. It seems as if fatty acids are used as a signal for this purpose.

Conclusions

Structure/function relationships in the mammalian triglyceride lipases cannot be inferred directly from studies on lipases from simpler organisms. A main difference is that the latter lipases do not contain a C-terminal domain of the type found in the mammalian lipases. Furthermore, both of the mammalian lipases discussed above each requires a specific protein confactor for activity under physiological conditions. It is apparent that the mammalian lipases have evolved to fulfill more complex functions, and that much of the adaptation is to allow controlled lipolysis under specific conditions. The crystal structure of the human pancreatic lipase/colipase complex gives the overall architecture also for the other members of this gene family. However, we need 3D information also on the other lipases in order to understand the different behavior of lipoprotein lipase at interfaces, details in its interactions with other ligands, and similarities and differences with the behavior of hepatic lipase.

Acknowledgements
The authors are indebted to the European Communities for financial support within the framework of the BRIDGE T-Lipase programme (BIOT-CT91-0274) and the BIOTECH G-project (BIO2-CT94-3041), and the Bank of Sweden Tercentenary Foundation.

References

Andersson, Y., Thelander, L. and Bengtsson-Olivecrona, G. (1991) Rat apolipoprotein C-II lacks the conserved site for proteolytic cleavage of the pro-form. *J. Lipid Res.* 32: 1805–1809.

Beisiegel, U., Weber, W. and Bengtsson-Olivecrona, G. (1991) Lipoprotein lipase enhances the binding of chylomicrons to low density lipoprotein receptor-related protein. *Proc. Natl. Acad. Sci. USA* 88: 8342–8346.

Bengtsson, G. and Olivecrona, T. (1980) Lipoprotein lipase. Mechanism of product inhibition. *Eur. J. Biochem.* 106: 557–562.

Bengtsson, G. and Olivecrona, T., Höök, M., Riensenfeld, J. and Lindahl, U. (1980) Interaction of lipoprotein lipase with native and modified heparin-like polysaccharides. *Biochem. J.* 189: 625–633.

Bengtsson-Olivecrona, G. and Olivecrona, T. (1985) Binding of active and inactive forms of lipoprotein lipase to heparin: effects of pH. *Biochem. J.* 226: 409–413.

Bengtsson-Olivecrona, G., Olivecrona, T. and Jörnvall, H. (1986) Lipoprotein lipases from cow, guinea-pig and man. Structural characterization and identification of protease-sensitive internal regions. *Eur. J. Biochem.* 161: 281–288.

Bensadoun, A. (1991) Lipoprotein lipase. *Annu. Rev. Nutr.* 11: 217–237.

Berryman, D.E. and Bensadoun, A. (1993) Site-directed mutagenesis of a putative heparin binding domain of avian lipoprotein lipase. *J. Biol. Chem.* 268: 3272–3276.

Borgström, B. and Carlsson, L.A. (1957) Lipolytic action of the lipemia-clearing factor. *Biochim. Biophys. Acta* 24: 638–639.

Borgström, B. and Erlanson, C. (1971) Pancreatic juice colipase: physiological importance. *Biochim. Biophys. Acta* 242: 509–513.

Borgström, B. and Erlanson, C. (1973) Pancreatic lipase and colipase. Interactions and effects of bile salts and other detergents. *Eur. J. Biochem.* 37: 60–68.

Bousset-Risso, M., Bonicel, J. and Rovery, M. (1985) Limited proteolysis of porcine pancreatic lipase. Lability of the Phe 335–Ala 336 bond towards chymotrypsin. *FEBS Lett.* 182: 323–326.

Brady, L., Brzozowski, A.M., Derewenda, Z.S., Dodson, E., Dodson, G., Tolley, S., Turkenburg, J.P., Christiansen, L., Huge-Jensen, B., Norskov, L., Thim, L. and Menge, U. (1990) A serine protease triad forms the catalytic centre of a triacylglycerol lipase. *Nature* 343: 767–770.

Brzozowski, A.M., Derewenda, U., Derewenda, Z.S., Dodson, G.G., Lawson, D.M., Turkenburg, J.P., Bjorkling, F., Huge-Jensen, B., Patkar, S.A. and Thim, L. (1991) A model for interfacial activation in lipases from the structure of a fungal lipase-inhibitor complex. *Nature* 351: 491–494.

Carrière, F., Thirstrup, K., Boel, E., Verger, R. and Thim, L. (1994) Structure-function relationships in naturally occurring mutants of pancreatic lipases. *Prot. Eng.* 7: 563–569.

Chajek-Shaul, T., Friedman, G., Stein, O., Olivecrona, T. and Stein, Y. (1982) Binding of lipoprotein lipase to the cell surface is essential for the transmembrane transport of chylomicron cholesteryl ester. *Biochim. Biophys. Acta* 712: 200–210.

Chappell, D.A., Fry, G.L., Waknitz, M.A., Iverius, P.-H., Williams, S.E. and Strickland, D.K. (1992) The low density lipoprotein receptor-related protein/α2-macroglobulin receptor binds and mediates catabolism of bovine milk lipoprotein lipase. *J. Biol. Chem.* 267: 25764–25767.

Chappell, D.A., Fry, G.L., Waknitz, M.A., Muhonen, L.E., Pladet, M.W., Iverius, P.-H. and Strickland, D.K. (1993) Lipoprotein lipase induces catabolism of normal triglyceride-rich lipoprotein via the low density lipoprotein receptor-related protein/α2-macroglobulin receptor in vitro. A process facilitated by cell-surface proteoglycans. *J. Biol. Chem.* 268: 14168–14175.

Davis, R.C., Wong, H., Nikazy, J., Wang, K., Han, Q. and Schotz, M.C. (1992) Chimeras of hepatic lipase and lipoprotein lipase. Domain localization of enzyme-specific properties. *J. Biol. Chem.* 267: 21499–21504.

De Caro, J., Boudouard, M., Bonicel, J., Guidoni, A., Desnuelle, P. and Rovery, M. (1981) Porcine pancreatic lipase: completion of the primary structure. *Biochim. Biophys. Acta* 671: 129–138.

Deckelbaum, R.J., Hamilton, J.A., Moser, A., Bengtsson-Olivecrona, G., Butbul, E., Carpentier, Y.A., Gutman, A. and Olivecrona, T. (1990) Medium-chain vs long-chain triacylglycerol emulsion hydrolysis by lipoprotein lipase and hepatic lipase: Implications for the mechanism of lipase action. *Biochemistry* 29: 1136–1142.

Deckelbaum, R.J., Ramakrishnan, R., Eisenberg, S., Olivecrona, T. and Bengtsson-Olivecrona, G. (1992) Triacylglycerol and phospholipid hydrolysis in human plasma lipoproteins: Role of lipoprotein and hepatic lipase. *Biochemistry* 31: 8544–8551.

Derewenda, Z.S. and Cambillau, C. (1991) Effects of gene mutations in lipoprotein and hepatic lipases as interpreted by a molecular model of the pancreatic triglyceride lipase. *J. Biol. Chem.* 266: 23112–23119.

Derewenda, Z.S. and Sharp, A.M. (1993) News from the interface: The molecular structures of triacylglyceride lipases. *Trends Biochem. Sci.* 18: 20–25.

Dichek, H.L., Parrott, C., Ronan, R., Brunzell, J.D., Brewer, H.B., Jr. and Santamarina-Fojo, S. (1993) Functional characterization of a chimeric lipase genetically engineered from human lipoprotein lipase and human hepatic lipase. *J. Lipid Res.* 34: 1393–1401.

Dugi, K.A., Dichek, H.L., Talley, G.D., Brewer, H.B., Jr. and Santamarina-Fojo, S. (1992) Human lipoprotein lipase: The loop covering the catalytic site is essential for interaction with lipid substrates. *J. Biol. Chem.* 267: 25086–25091.

Eisenberg, S. (1990) Metabolism of apolipoproteins and lipoproteins. *Curr. Opin. Lipidol.* 1: 205–221.

Eisenberg, S., Sehayek, E., Olivecrona, T. and Vlodavsky, I. (1992) Lipoprotein lipase enhances binding of lipoproteins to heparan sulfate on cell surfaces and extracellular matrix. *J. Clin. Invest.* 90: 2013–2021.

Emmerich, J., Beg, O.U., Peterson, J., Previato, L., Brunzell, J.D., Brewer, H.B.J. and Santamarina-Fojo, S. (1992) Human lipoprotein lipase. Analysis of the catalytic triad by site-directed mutagenesis of Ser-132, Asp-156, and His-241. *J. Biol Chem.* 267: 4161–4165.

Entressangles, B. and Desnuelle, P. (1968) Action of pancreatic lipase on aggregated glyceride molecules in an isotropic system. *Biochim. Biophys. Acta* 159: 285–295.

Faustinella, F., Chang, A., Van Biervliet, J.P., Rosseneu, M., Vinaimont, N., Smith, L.C., Chen, S.-H. and Chan, L. (1991) Catalytic triad residue mutation (Asp156 $\rightarrow$ Gly) causing familial lipoprotein lipase deficiency. Co-inheritance with a nonsense mutation (Ser447 $\rightarrow$ Ter) in a Turkish family. *J. Biol. Chem.* 266: 14418–14424.

Faustinella, F., Smith, L.C. and Chan, L. (1992) Functional topology of a surface loop shielding the catalytic center in lipoprotein lipase. *Biochemistry* 31: 7219–7223.
Gargouri, Y., Piéroni, G., Lowe, P.A., Sarda, L. and Verger, R. (1986a) Human gastric lipase: the effect of amphiphils. *Eur. J. Biochem.* 156: 305–310.
Gargouri, Y., Piéroni, G., Rivère, C., Lowe, P.A., Saunière, J.F., Sarda, L. and Verger, R. (1986b) Importance of human gastric lipase for intestinal lipolysis: an *in vitro* study. *Biochim. Biophys. Acta* 879: 419–423.
Giller, T., Buchwald, P., Kaelin, D.B. and Hunziker, W. (1992) Two novel human pancreatic lipase related proteins, hPLRP1 and hPLRP2. Differences in colipase dependence and in lipase activity. *J. Biol. Chem.* 267: 16509–16516.
Grochulski, P., Li, Y., Schrag, J.D., Bouthillier, F., Smith, P., Harrison, D., Rubin, B. and Cygler, M. (1993) Insights into interfacial activation from an open structure of *Candida rugosa* lipase. *J. Biol. Chem.* 268: 12843–12847.
Grusby, M.J., Nabavi, N., Wong, H., Dick, R.F., Bluestone, J.A., Schotz, M.C. and Glimcher, L.H. (1990) Cloning of an Interleukin-4 inducible gene from cytotoxic T lymphocytes and its identification as a lipase. *Cell* 60: 451–459.
Guidoni, A., Benkouka, F., DeCaro, J. and Rovery, M. (1981) Characterization of the serine reacting with diethyl p-nitrophenyl phosphate in porcine pancreatic lipase. *Biochim. Biophys. Acta* 660: 148–150.
Hata, A., Ridinger, D.N., Sutherland, S.D., Emi, M., Kwong, L.K., Shuhua, J., Lubbers, A., Guy-Grand, B., Basdevant, A., Iverius, P.-H., Wilson, D.E. and Lalouel, J.-M. (1992) Missense mutations in exon 5 of the human lipoprotein lipase gene. Inactivation correlates with loss of dimerization. *J. Biol. Chem.* 267: 20132–20139.
Hata, A., Ridinger, D.N., Sutherland, S., Emi, M., Shuhua, Z., Myers, R.L., Ren, K., Cheng, T., Inoue, I., Wilson, D.E., Iverius, P.-H. and Lalouel, J.-M. (1993) Binding of lipoprotein lipase to heparin. Identification of five critical residues in two distinct segments of the amino-terminal domain. *J. Biol. Chem.* 268: 8447–8457.
Havel, R.J. and Hamilton, R.L. (1988) Hepatocytic lipoprotein receptors and intracellular lipoprotein catabolism. *Hepatology* 8: 1689–1704.
Henderson, H.E., Ma, Y., Liu, M.-S., Clark-Lewis, I., Maeder, D.L., Kastelein, J.J.P., Brunzell, J.D. and Hayden, M.R. (1993) Structure-function relationships of lipoprotein lipase: Mutation analysis and mutagenesis of the loop region. *J. Lipid Res.* 34: 1593–1602.
Hide, W.A., Chan, L. and Li, W.-H. (1992) Structure and evolution of the lipase superfamily. *J. Lipid Res.* 33: 167–178.
Hjorth, A., Carrière, F., Cudrey, C., Wöldike, H., Boel, E., Lawson, D.M., Ferrato, F., Cambillau, C., Dodson, G.G., Thim, L. and Verger, R. (1993) A structural domain (the lid) found in pancreatic lipases is absent in the guinea pig (phospho) lipase. *Biochemistry* 32: 4702–4707.
Hoffer, M.J.V., Van Eck, M.M., Havekes, L.M., Hofker, M.H. and Frants, R.R. (1993) Structure and expression of the mouse apolipoprotein C2 gene. *Genomics* 17: 45–51.
Iverius, P.-H. and Östlund-Lindqvist, A.M. (1976) Lipoprotein lipase from bovine milk. Isolation procedure, chemical characterization and molecular weight analysis. *J. Biol. Chem.* 251: 7791–7795.
Jackson, R.L., Ponce, E., McLean, L.R. and Demel, R.A. (1986) Comparison of the triacylglycerol hydrolase activity of human post-heparin plasma lipoprotein lipase and hepatic triacylglycerol lipase. A monolayer study. *Biochemistry* 25: 1166–1170.
Kirchgessner, T.G., Chaut, J.C., Heinzamann, C., Etienne, J., Guilhot, S., Svenson, K., Ameis, D., Pilon, C., D'Auriol, L., Andalibi, A., Schotz, M.C., Galibert, F. and Lusis, A.J. (1989) Organization of the human lipoprotein lipase gene and evolution of the lipase gene family. *Proc. Natl. Acad. Sci. USA* 86: 9647–9651.
Larsson, A. and Erlanson-Albertsson, C. (1986) Effect of phosphatidylcholine free fatty acids on the activity of pancreatic lipase-colipase. *Biochim. Biophys. Acta* 876: 543–550.
Lawson, D.M., Brzozowski, A.M. and Dodson, G.G. (1992) Protein structures: Lifting the lid off lipases. *Curr. Biol.* 2: 473–475.
Lookene, A., and Bengtsson-Olivecrona, G. (1993) Chymotryptic cleavage of lipoprotein lipase – Identification of cleavage sites and functional studies of the truncated molecule. *Eur. J. Biochem.* 213: 185–194.
Lookene, A., Skottova, N. and Olivecrona, G. (1994) Interactions of lipoprotein lipase with the active-site inhibitor tetrahydrolipstatin (Orlistat)[R]. *Eur. J. Biochem.* 222: 395–403.

Lowe, M.E., Rosenblum, J.L. and Strauss, A.W. (1989) Cloning and characterization of human pancreatic lipases cDNA. *J. Biol. Chem.* 264: 20042–20048.

Luthi-Peng, Q. and Winkler, F.K. (1992) Large spectral changes accompany the conformational transition of human pancreatic lipase induced by acylation with the inhibitor tetrahydrolipstatin. *Eur. J. Biochem.* 205: 383–390.

Martinez, C., De Geus, P., Lauwereys, M., Matthyssens, G. and Cambillau, C. (1992) *Fusarium solani* cutinase is a lipolytic enzyme with catalytic serine accessible to solvent. *Nature* 356: 615–618.

Maylié, M.F., Charles, M., Gache, C. and Desnuelle, P. (1971) Isolation and partial identification of a pancreatic colipase. *Biochim. Biophys. Acta* 229: 286–289.

Mulder, M., Lombardi, P., Jansen, H., Van Berkel, T.J.C., Frants, R.R. and Havekes, L.M. (1992) Heparan sulphate proteoglycans are involved in the lipoprotein lipase-mediated enhancement of the cellular binding of very low density and low density lipoproteins. *Biochem. Biophys. Res. Commun.* 185: 582–587.

Noble, M.E.M., Cleasby, A., Johnson, L.N., Egmond, M.R. and Frenken, L.G.J. (1993) The crystal structure of triacylglycerol lipase from *Pseudomonas glumae* reveals a partially redundant catalytic aspartate. *FEBS Lett.* 331: 123–128.

Nykjaer, A., Bengtsson-Olivecrona, G., Lookene, A., Moestrup, S.K., Petersen, C.M., Weber, W., Beisiegel, U. and Gliemann, J. (1993) The α2-macroglobulin receptor/low density lipoprotein receptor-related protein binds lipoprotein lipase and – migrating very low density lipoprotein associated with the lipase. *J. Biol. Chem.* 268: 15048–15055.

Nykjaer, A., Nielsen, M., Lookene, A., Meyer, N., Roigaard, H., Etzerodt, M., Beisiegel, U., Olivecrona, G. and Gliemann, J. (1994) A C-terminal fragment of lipoprotein lipase binds to LDL receptor-related protein and inhibits lipase mediated uptake of lipoprotein in cells. *J. Biol. Chem.* 269: 31747–31755.

Olivecrona, T. and Bengtsson-Olivecrona, G. (1985) Lipoprotein lipase and heparin. *In*: V.V. Kakkar (ed.): *Antheroma and Thrombosis*, Pitman, London, pp 250–256.

Olivecrona, T. and Bengtsson-Olivecrona, G. (1987) Lipoprotein lipase from milk – The model enzyme in lipoprotein lipase research. *In*: J. Borensztajn (ed.): *Lipoprotein Lipase*, Evener Publishers, Chicago, pp 15–58.

Olivecrona, T. and Bengtsson-Olivecrona, G. (1993) Lipoprotein lipase and hepatic lipase. *Curr. Opin. Lipidol.* 4: 187–196.

Ollis, D.L., Cheah, E., Cygler, M., Dijkstra, B., Frolow, F., Franken, S.M., Harel, M., Remington, S.J., Silman, I., Schrag, J., Sussman, J.L., Verschueren, K.H.G. and Goldman, A. (1992) The α/β hydrolase fold. *Prot. Eng.* 5: 197–211.

Osborne, J.C., Jr., Bengtsson-Olivecrona, G., Lee, N.S. and Olivecrona, T. (1985) Studies on inactivation of lipoprotein lipase: role of the dimer to monomer dissociation. *Biochemistry* 24: 5606–5611.

Persson, B., Bengtsson-Olivecrona, G., Enerbäck, S., Olivecrona, T. and Jörnvall, H. (1989) Structural features of lipoprotein lipase. Lipase family relationships, binding interactions, non-equivalence of lipase cofactors, vitellogenin similarities, and functional subdivisions of lipoprotein lipase. *Eur. J. Biochem.* 179: 39–45.

Persson, B., Jörnvall, H., Olivecrona, T. and Bengtsson-Olivecrona, G. (1991) Lipoprotein lipases and vitellogenins in relation to the known three-dimensional structure of pancreatic lipase. *FEBS Lett.* 288: 33–36.

Peterson, J., Fujimoto, W.Y. and Brunzell, J.D. (1992) Human lipoprotein lipase: Relationship of activity, heparin affinity, and conformation as studied with monoclonal antibodies. *J. Lipid Res.* 33: 1165–1170.

Posner, I. and DeSanctis, J. (1987) Kinetics of product inhibition and mechanisms of lipoprotein lipase activation by apolipoprotein C-II. *Biochemistry* 26: 3711–3717.

Rapp, D. and Olivecrona, T. (1978) Kinetics of milk lipoprotein lipase. Studies with tributyrin. *Eur. J. Biochem.* 91: 379–385.

Rojas, C., Olivecrona, T. and Bengtsson-Olivecrona, G. (1991) Comparison of the action of lipoprotein lipase on triacyglycerols and phospholipids when presented in mixed liposomes or in emulsion particles. *Eur. J. Biochem.* 197: 315–321.

Rumsey, S.C., Obunike, J.C., Arad, Y., Deckelbaum, R.J. and Goldberg, I.J. (1992) Lipoprotein lipase-mediated uptake and degradation of low density lipoproteins by fibroblasts and macrophages. *J. Clin. Invest.* 90: 1504–1512.

Santamarina-Fojo, S. and Dugi, K.A. (1994) Structure, function and role of lipoprotein lipase in lipoprotein metabolism. *Curr. Opin. Lipidol.* 5: 117–125.

Sarda, L. and Desnuelle, P. (1958) Action de la lipase pancréatique sur les esters en émulsion. *Biochim. Biophys. Acta* 30: 513–521.
Schaefer, E.J., Wetzel, M.G., Bengtsson, G., Scow, R.O., Brewer, H.B., Jr. and Olivecrona, T. (1982) Transfer of human lymph chylomicron constituents to other lipoprotein density fractions during in vitro lipolysis. *J. Lipid Res.* 23: 1259–1273.
Schrag, J.D., Li, Y., Wu, S. and Cygler, M. (1991) Ser-His-Glu triad forms the catalytic site of the lipase from *Geotrichum candidum*. *Nature* 351: 761–764.
Scott, D.L., White, S.P., Otwinowski, Z., Yaun, W., Gelb, M.H. and Sigler, P.B. (1990) Interfacial catalysis: the mechanism of phospholipase A2. *Science* 250: 1541–1546.
Scow, R.O. and Olivecrona, T. (1977) Effect of albumin on products formed from chylomicron triacylglycerol by lipoprotein lipase in vitro. *Biochim. Biophys. Acta* 487: 472–486.
Segrest, J.P., Garber, D.W., Brouillette, C.G., Harvey, S.C. and Anantharamaiah, G.M. (1984) The amphipathic α helix: A multifunctional structural motif in plasma apolipoproteins. *Adv. Protein Chem.* 45: 303–369.
Smith, L.C. and Pownall, H.J. (1984) Lipoprotein lipase. *In*: B. Borgström and H. Brockman (eds): *Lipases*. Elsevier, Amsterdam, pp 263–305.
Soldatova, L., Kochoumian, L. and King, T.P. (1993) Sequence similarity of a hornet (D. maculata) venom allergen phospholipase A1 with mammalian lipases. *FEBS Lett.* 320: 145–149.
Thirstrup, K., Carrière, F., Hjorth, S., Rasmussen, P.B., Wöldike, H., Nielsen, P.F. and Thim, L. (1993) One-step purification of human pancreatic lipase expressed in insect cells. *FEBS Lett.* 327: 79–84.
Thirstrup, K., Verger, R. and Carrière, F. (1994) Evidence for a pancreatic lipase subfamily with new kinetic properties. *Biochemistry* 33: 2748–2756.
Uppenberg, J., Hansen, M.T., Patkar, S. and Jones, T.A. (1994) The sequence, crystal structure determination and refinement of two crystal forms of lipase B from *Candida antarctica*. *Structure* 2: 293–308.
Vainio, P. Virtanen, J.A., Kinnunen, P.K.J., Voyta, J.C., Smith, L.C., Gotto, A.M., Jr, Sparrow, J.T., Pattus, F. and Verger, R. (1983) Action of lipoprotein lipase on phospholipid monolayers. Activation by apolipoprotein C-II. *Biochemistry* 22: 2270–2275.
van Tilbeurgh, H., Sarda, L., Verger, R. and Cambillau, C. (1992) Structure of the pancreatic lipase-colipase complex. *Nature* 359: 159–162.
van Tilbeurgh, H., Egloff, M.-P., Martinez, C., Rugani, N., Verger, R. and Cambillau, C. (1993) Interfacial activation of the lipase-procolipase complex by mixed micelles revealed by X-ray crystallography. *Nature* 362: 814–820.
van Tilbeurgh, H., Roussel, A., Lalouel, J.-M. and Cambillau, C. (1994) Lipoprotein lipase. Molecular model based on the pancreatic lipase X-ray structure: consequences for heparin binding and catalysis. *J. Biol. Chem.* 269: 4626–4633.
Verger, R. and deHaas, G.H. (1976) Interfacial enzyme kinetics of lipolysis. *Annu. Rev. Bioeng.* 5: 77–117.
Verger, R., Rietsch, J. and Desnuelle, P. (1977) Effects of colipase on hydrolysis of monomolecular films by lipase. *J. Biol. Chem.* 252: 4319–4325.
Williams, S.E., Inoue, I., Tran, H., Fry, G.L., Pladet, M.W., Iverius, P.-H., Lalouel, J.-M., Chappell, D.A. and Strickland D.K. (1994) The carboxyl-terminal domain of lipoprotein lipase binds to the low density lipoprotein receptor-related protein/α_2-macroglobulin receptor (LRP) and mediates binding of normal very low density lipoproteins to LRP. *J. Biol. Chem.* 269: 8653–8658.
Willnow, T.E., Goldstein, J.L., Orth, K., Brown, M.S. and Herz, J. (1992) Low density lipoprotein receptor-related protein and gp330 bind similar ligands, including plasminogen activator-inhibitor complexes and lactoferrin, an inhibitor of chylomicron remnant clearance. *J. Biol. Chem.* 267: 26172–26180.
Winkler, F.K., D'Arcy, A. and Humziker, W. (1990) Structure of human pancreatic lipase. *Nature* 343: 771–774.
Wishart, M.J., Andrews, P.C., Nichols, R., Blevins, G.T., Logsdon, C.D. and Williams, J.A. (1993) Identification and cloning of GP-3 from rat pancreatic acinar zymogen granules as a glycosylated membrane-associated lipase. *J. Biol. Chem.* 268: 10303–10311.
Wong, H., Davis, R.C., Nikazy, J., Seebart, K.E. and Schotz, M.C. (1991) Domain exchange: Characterization of a chimeric lipase of hepatic lipase and lipoprotein lipase. *Proc. Natl. Acad. Sci. USA* 88: 11290–11294.
Wong, H., Davis, R.C., Thuren, T., Goers, J.W., Nikazy, J., Waite, M. and Schotz, M.C. (1994) Lipoprotein lipase domain function. *J. Biol. Chem.* 269: 10319–10323.

Structure and orientation of hydrophobic surfactant-associated proteins in a lipid environment

G. Vandenbussche[1], J. Johansson[2], A. Clercx[3], T. Curstedt[4] and J.-M. Ruysschaert[1]

[1]*Laboratoire de Chimie-Physique des Macromolécules aux Interfaces, Université Libre de Bruxelles, Boulevard du Triomphe CP 206/2, B-1050 Bruxelles, Belgium*
[2]*Department of Medical Biochemistry and Biophysics, Karolinska Institutet, S-17177 Stockholm, Sweden*
[3]*Service de Néonatologie, Hôpital Universitaire des Enfants Reine Fabiola, Université Libre de Bruxelles, 15 Avenue J.J. Crocq, B-1020 Bruxelles, Belgium*
[4]*Department of Clinical Chemistry, Karolinska Institutet at Danderyd Hospital, S-18288 Danderyd, Sweden*

Summary. The structures of hydrophobic surfactant-associated proteins SP-B and SP-C have been estimated by attenuated total reflection Fourier transform infrared spectroscopy, nuclear magnetic resonance and/or circular dichroism spectroscopy. The conformation of the lipopeptide SP-C in a mixed organic solvent consists of a 26-residue continuous α-helix composed mainly of valyl residues. The α-helical content apparently increases upon insertion in a lipid bilayer but decreases upon removal of the two thioester linked palmitoyl side chains. Polarized infrared spectroscopy data provide evidence that the long axis of this α-helical segment is oriented parallel to the lipid acyl chains and this orientation is strongly supported by comparison of the length of the SP-C helix with the thickness of a fluid DPPC bilayer, showing that SP-C is a transmembrane-spanning polypeptide. In SP-B, nearly half of the polypeptide chain adopts an α-helical conformation but no significant change of structure was observed upon interaction with the lipid vesicles. SP-B does not modify the orientation and the conformation of the lipid molecules and 60–80% of the α-helices are parallel to the lipid acyl chains.

Introduction

During the respiratory cycle, the collapsing tendency of the alveoli depends largely on the surface tension at the air/liquid interface. The main role of pulmonary surfactant, a surface-active material lining the alveoli, is to reduce this interfacial force and prevent atelectasis (Goerke, 1974; van Golde et al., 1988). The pulmonary surfactant is synthesized in type II pneumocyte cells and secreted into the alveolar space in the form of multilamellar structures called lamellar bodies. The material is transformed into a lattice-like structure, the tubular myelin (Williams, 1977), which is the precursor of the surface-active lipid monolayer at the alveolar air/liquid interface (Goerke, 1974).

Extracellular surfactant consists of 85–90% lipids, mainly phospholipids, about 10% proteins, and a small amount of carbohydrates. The major component of surfactant is dipalmitoylphosphatidylcholine (DPPC), which is the principal constituent of the interfacial monolayer and also responsible for the maintenance of low surface tension (Hildebran et al., 1979). The lipid composition is also characterized by an unusually high percentage of phosphatidylglycerol (PG) (Shelley et al., 1984; Rooney, 1985). Four proteins, SP-A, SP-B, SP-C and SP-D appear to be closely related to the structure and properties of the pulmonary surfactant (Hawgood and Shiffer, 1991; Weaver and Whitsett, 1991; Johansson et al., 1994a).

Respiratory distress syndrome (RDS) results from deficiency of pulmonary surfactant and is the principal cause of morbidity and mortality in premature infants. This disease can be effectively treated by intratracheal instillation of exogenous surfactant (Fujiwara et al., 1980; Enhorning et al., 1985; Hallman et al., 1985; Jobe and Ikegami, 1987; Collaborative European Multicenter Group, 1988, 1991, 1992; Speer et al., 1992; Horbar et al., 1993). Several studies have demonstrated the importance of the proteins SP-B and SP-C in exogenous surfactants (Cummings et al., 1992; Hall et al., 1992).

The water-insoluble proteins SP-B and SP-C are isolated with the surfactant lipids during extraction with organic solvents (Whitsett et al., 1986; Yu and Possmayer, 1986; Curstedt et al., 1987; Hawgood et al., 1987). In vitro, they induce the rapid spreading of lipids at the air/liquid interface (Suzuki et al., 1986; Takahashi and Fujiwara, 1986; Curstedt et al., 1987; Hawgood et al., 1987; Oosterlaken-Dijksterhuis et al., 1991a,b).

Mature SP-C (Fig. 1) is a 33–35 residue polypeptide, rich in valine and leucine and highly conserved (Warr et al., 1987; Glasser et al., 1988; Johansson et al., 1988a,b, 1991b; Curstedt et al., 1990). The N-terminal part of the protein contains one or two Cys residues, stoichiometrically palmitoylated (Curstedt et al., 1990; Johansson et al., 1991b; Stults et al., 1991). Palmitoylation of SP-C occurs in the precursor (Vorbroker et al., 1992) but its functional role is not understood (see Beers and Fisher, 1992). SP-B (Fig. 1) contains more polar and positively charged residues throughout its 79-residue structure (Glasser et al., 1987; Hawgood et al., 1987; Olafson et al., 1987; Curstedt et al., 1988). SP-B has three intrachain disulfide bridges and is found as a disulfide-linked homodimer in natural surfactant (Curstedt et al., 1990; Johansson et al., 1991a, 1992).

SP-B and SP-C clearly interact with the surfactant phospholipids (for reviews see Hawgood and Shiffer, 1991; Weaver and Whitsett, 1991; Keough, 1991; Waring et al., 1993), but only for SP-C is a molecular description of this interaction available (Johansson et al., 1995b). Such detailed descriptions of the lipid/protein interactions should contribute to the understanding of its surface-active properties and will very likely aid in developing synthetic peptide analogs for use in surfactant preparations for treatments of RDS.

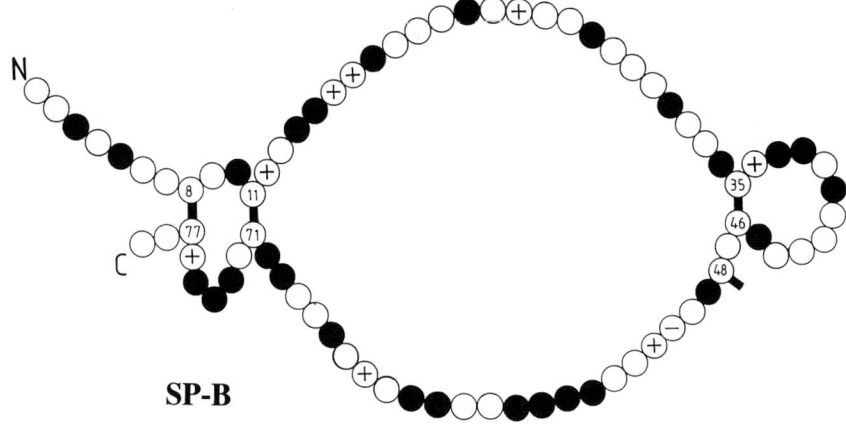

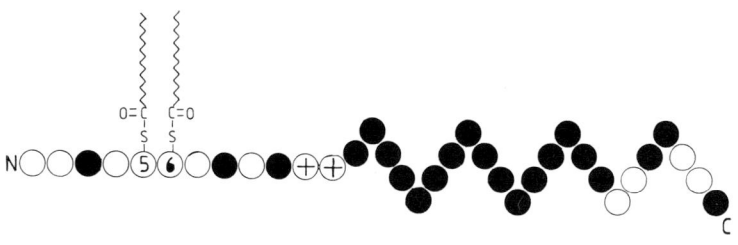

Figure 1. Schematic presentation of the SP-B monomer (upper panel) and SP-C (lower panel) covalent structures. Each circle represents one amino acid residue and black circles are either valine, isoleucine or leucine residues. Positively or negatively charged side-chains are denoted with plus and minus signs, respectively. In SP-B, disulfide bridges are indicated with black bars and in the native homodimer Cys-48 is disulfide-linked to a second polypeptide chain. In SP-C, the cysteine residues at positions 5 and 6 contain thioester-linked palmitoyl groups.

In addition to X-ray analyses of membrane proteins structural information on proteins in lipid environments can be obtained from circular dichroism (CD), Fourier transform infrared (FTIR) and nuclear magnetic resonance (NMR) spectroscopy. These techniques have somewhat different advantages, making a combined use particularly attractive. CD spectroscopy affords a rapid evaluation of secondary structure content and can be used with proteins in different lipid environments, preferentially micelles to avoid extensive light scattering, but data on proteins in oriented lipid films have been presented (see Gierasch, 1989). The recent availability of Fourier transform infrared instrumentation has allowed the structure and orientation of peptides with respect to the bilayer plane to be determined (Goormaghtigh and Ruysschaert, 1990; Goormaghtigh et al., 1990; Goormaghtigh et al., 1994).

Concerning the protein conformation, CD and FTIR spectroscopy provide only the overall secondary structure content but no sequence-specific information. Such data can however be obtained for lipid-associated polypeptides in solution by NMR spectroscopy (Wüthrich, 1986).

The purpose of this chapter is to review the present knowledge about the conformation of the hydrophobic proteins SP-B and SP-C in organic solvents and in different phospholipid environments determined by CD (Oosterlaken-Dijksterhuis et al., 1991b; Pérez-Gil et al., 1993; Morrow et al., 1993a; Shiffer et al., 1993; Johansson et al., 1995a,b), FTIR (Pastrana et al., 1991; Baatz et al., 1992; Vandenbussche et al., 1992a,b), and NMR (Johansson et al., 1994b) spectroscopy, and to illustrate how the orientation of hydrophobic surfactant proteins can be determined in a lipid environment on thin hydrated films by attenuated total reflection (ATR) spectroscopy (Goormaghtigh and Ruysschaert, 1990; Goormaghtigh et al., 1990; Pastrana et al., 1991; Vandenbussche et al., 1992a,b; Goormaghtigh et al., 1994).

Structure of SP-C

To evaluate the structure of SP-C by FTIR spectroscopy, the isolated protein was spread at the surface of a germanium ATR plate from a chloroform/methanol solution. To reconstitute SP-C, a surfactant protein-lipid mixture was dissolved in organic solvents. After evaporation to dryness, proteoliposomes were formed by resuspension in aqueous phase. The unbound protein constituents were separated by sucrose gradient centrifugation and undesirable sucrose was removed by dialysis (Vandenbussche et al., 1992a).

The spectra of these samples were recorded after hydrogen/deuterium exchange. A mathematical analysis (Fourier self-deconvolution and least-square iterative curve-fitting) allows to evaluate the percentage of each secondary structure. Polarized spectra enable the calculation of the orientation of these structures with respect to the lipid bilayer plane (Goormaghtigh and Ruysschaert, 1990; Goormaghtigh et al., 1990; Goormaghtigh et al., 1994).

In the FTIR spectra of native and depalmitoylated porcine SP-C (Fig. 2), the peak between $1600\ cm^{-1}$ and $1700\ cm^{-1}$ represents the amide I band ($v(C=O)$ of the peptide bond) whose shape is closely related to the secondary structure of the protein. The maximal adsorption at $1657\ cm^{-1}$ is typical for the α-helix domain. The Fourier self-deconvolution curve-fitting procedure applied to the spectra of both deuterated forms of SP-C clearly confirms the high α-helical content (Fig. 2).

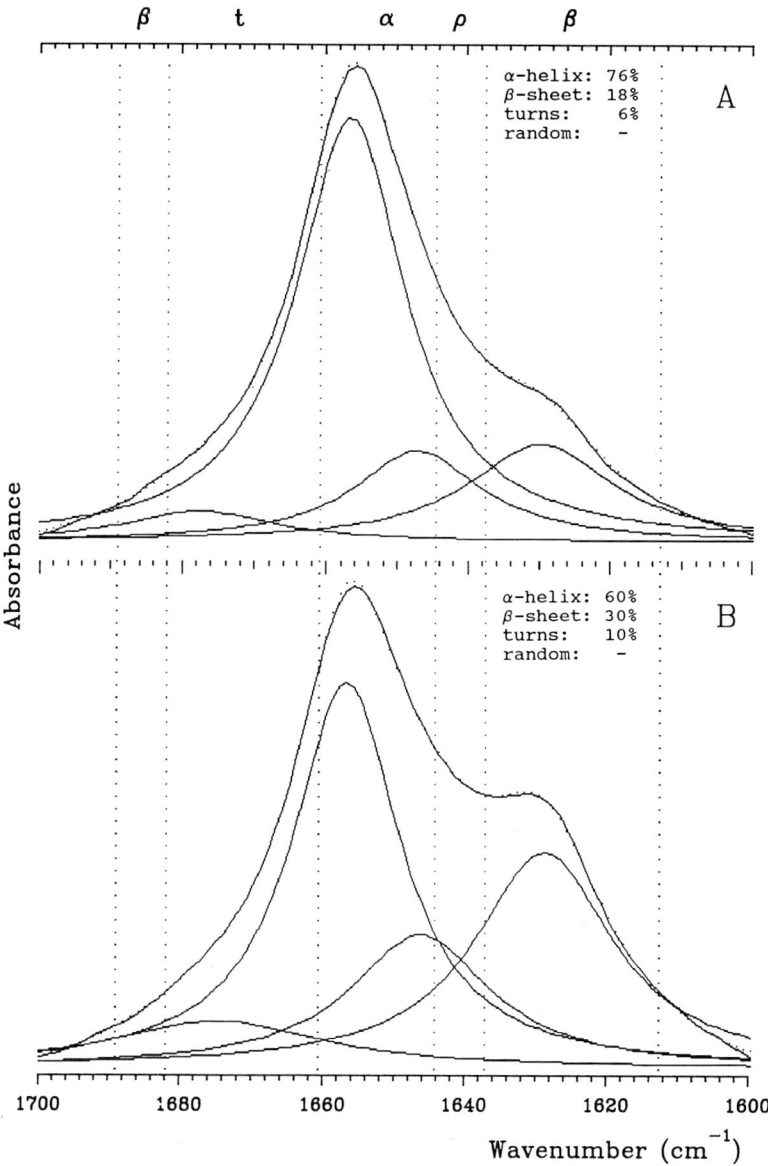

Figure 2. Secondary structure determination after curve-fitting of amide I' of deuterated (A) native and (B) depalmitoylated porcine SP-C. The spectrum resulting from the sum of the Lorentzian line shapes is represented by the dotted line. The vertical dotted lines limit the regions assigned to the different secondary structures: α, α-helix; β, β-sheet; t, turns and ρ, random coil.

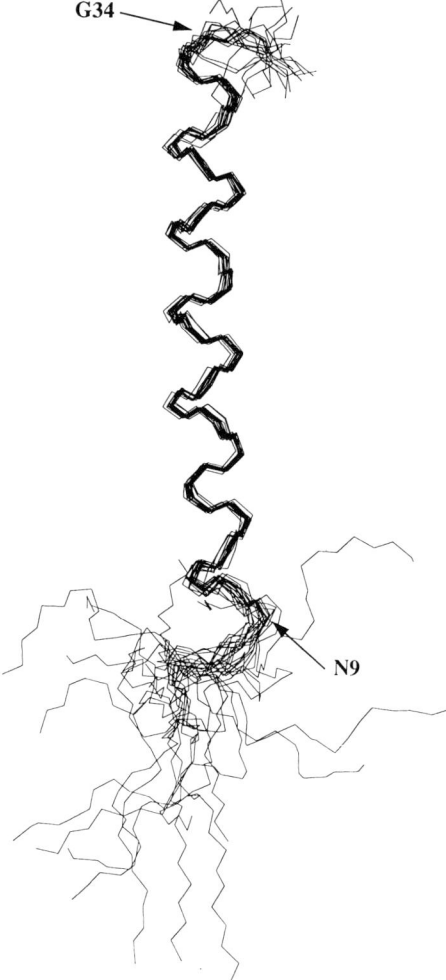

Figure 3. Polypeptide backbone fold of the 20 NMR-derived conformers of native SP-C with lowest target functions. The conformers 2–19 have been superimposed with conformer 1 for pairwise minimum root mean square deviations of the polypeptide backbone heavy atoms of residues 9–34. The start and end of the helix is identified with one-letter amino acid code and sequence location.

The calculated three-dimensional NMR structure of native porcine SP-C in chloroform/methanol/0.1 M HCl (32:64:5 by vol.) reveals a highly regular and continuous α-helix between positions 9 and 34, while the N-terminal eight residues, including the palmitoylcysteines, are flexibly disordered (Fig. 3) (Johansson et al., 1994b). This helical part constitutes 74% of the polypeptide chain and is thus in excellent agreement with the FTIR results of SP-C in chloroform/methanol (Pastrana et al., 1991; Vandenbussche et al., 1992a). The α-helical contents of this hydro-

Table 1. α-Helical contents reported for SP-C

Sample	α (%)	Environment	Method	Reference
Native SP-C	52	butanol	CD	Shiffer et al., 1993
	51	80% acetonitrile	CD	Pérez-Gil et al., 1993
	80	trifluoroethanol	CD	Johansson et al., 1995a
	22	70% trifluoroethanol	CD	Pérez-Gil et al., 1993
	74	chloroform/methanol/ 0.1 M HCl	NMR	Johansson et al., 1994b
	52	chloroform/methanol (1:1 v/v)	FTIR	Pastrana et al., 1991
	76	chloroform/methanol (1:1 v/v)	FTIR	Vandenbussche et al., 1992a
	80	DPC	CD	Johansson et al., 1995a,b
	60	lyso-PC	CD	Pérez-Gil et al., 1993
	71	DPPC	CD	Shiffer et al., 1993
	46	DPPC	CD	Shiffer et al., 1993
	46	DPPC	FTIR	Baatz et al., 1992
	67	DPPC/DPPG (7:1 w/w)	FTIR	Horowitz et al., 1993
	60	DPPC/DPPG (7:3 w/w)	FTIR	Pastrana et al., 1991
	92	DPPC/PG (8:2 w/w)	FTIR	Vandenbussche et al., 1992a
Non-palmitoylated SP-C	60	chloroform/methanol (1:1 v/v)	FTIR	Vandenbussche et al., 1992a
	40–70[1]	DPC	CD	Johansson et al., 1995a
	86	DPPC	CD	Shiffer et al., 1993
	77	DPPC/PG (8:2 w/w)	FTIR	Vandenbussche et al., 1992a
	60	monolayer	CD	Creuwels et al., 1993

[1]Depending on whether synthetic SP-C or chemically depalmitoylated SP-C was analyzed

phobic lipopeptide SP-C and its non-palmitoylated form as determined by different experimental approaches are summarized in Table 1.

In the FTIR spectra, the band around 1543 cm^{-1} was used to monitor the H/D exchange for different times. We observed that after 22 h, 60% of the polypepide chain of native SP-C or depalmitoylated SP-C escaped deuteration which remains constant for further time of H/D exchange. NMR spectroscopy of SP-C in all-deuterated solvents substantiated this finding by showing that all amide protons from position 10 to 32 exchange slowly. These data indicate a strongly hydrogen-bonded secondary structure of SP-C.

The FTIR approach revealed a higher helical content when the two acyl chains are present. To our knowledge, such an increase of the helical content has not been reported for palmitoylated proteins or peptides. It may be speculated that the acyl chains of mature SP-C can stabilize the α-helix structure by interaction between hydrophobic amino acids and the lipid hydrocarbon backbone. Such interactions are however not confirmed by the NMR analysis since no NOEs between palmitoyl methy-lene or methyl groups and peptide side-chains protons could be identified.

The modification of the secondary structure of SP-C and depalmitoylated SP-C in a lipid environment was studied by FTIR spectroscopy after association of proteins to multilamellar vesicles of DPPC/PG (8:2 w/w). The main difference between the spectra of SP-C obtained in the presence and absence of lipid is that the shoulder observed in the β-sheet region (between 1615 cm^{-1} and 1637.5 cm^{-1}) vanishes upon interaction of the native protein with a lipid bilayer. The disappearance of the β-sheet component is observed to a lesser extent with the depalmitoylated SP-C. The results of the Fourier self-deconvolution curve-fitting procedure indicate that both forms of SP-C seem to be affected by the presence of lipids (Vandenbussche et al., 1992a). The secondary structures in lipid bilayers are characterized by an increase of the α-helical content (Tab. 1) which has also been reported for viral peptides (Lear and Degrado, 1987; Martin et al., 1991), signal peptides (Goormaghtigh et al., 1989), and synthetic amphipathic α-helical peptides (McLean et al., 1991). However, in the case of SP-C the increase in α-helical content should be interpreted cautiously since it may be caused by factors other than direct effects on the peptide conformation. Thus, aggregated SP-C peptides in β-sheet conformations may associate less well with the lipid vesicles than α-helical native SP-C and, hence, be selectively removed during the sucrose density gradient centrifugation procedure. Such effects could, at least partly, cause the observed higher α-helical content of lipid-associated SP-C compared to SP-C in chloroform/methanol solution. The reasons that such an alternative explanation can not be ruled out are that the conformation of SP-C is apparently labile and forms aggregates in solution (Johansson et al., 1994b) and that it has been claimed that SP-C preparations contain about 15% of dimeric peptides that lack palmitoyl groups and are in β-sheet conformation (Baatz et al., 1992).

The ATR-FTIR spectroscopy technique provides information about the orientation of the protein in a lipid bilayer. The spectra of SP-C inserted into DPPC/PG liposomes were recorded with two orthogonal polarizations of the incident light, 0° and 90° (Fig. 4) (Goormaghtigh and Ruysschaert, 1990). 0° is polarized in the plane of the germanium plate and 90° is perpendicular to 0° polarization. In lipid bilayers, the transition dipole moment of the $\gamma_w(CH_2)$ peak to 1200 cm^{-1} lies along the hydrocarbon chains; this band was therefore chosen to characterize the lipid acyl chain orientation. The dichroic spectra 90°–0° show a positive peak at 1200 cm^{-1}, indicating that the lipid chains are organized preferentially perpendicular to the germanium surface. Confirmation of this orientation comes from the $\delta(CH_2)$ peak near 1468 cm^{-1}. The dipole moment of this vibration is perpendicular to the lipid chains. In the difference spectrum, a negative deviation in this region is the image of a larger absorbance at 0° than at 90° and therefore indicates an

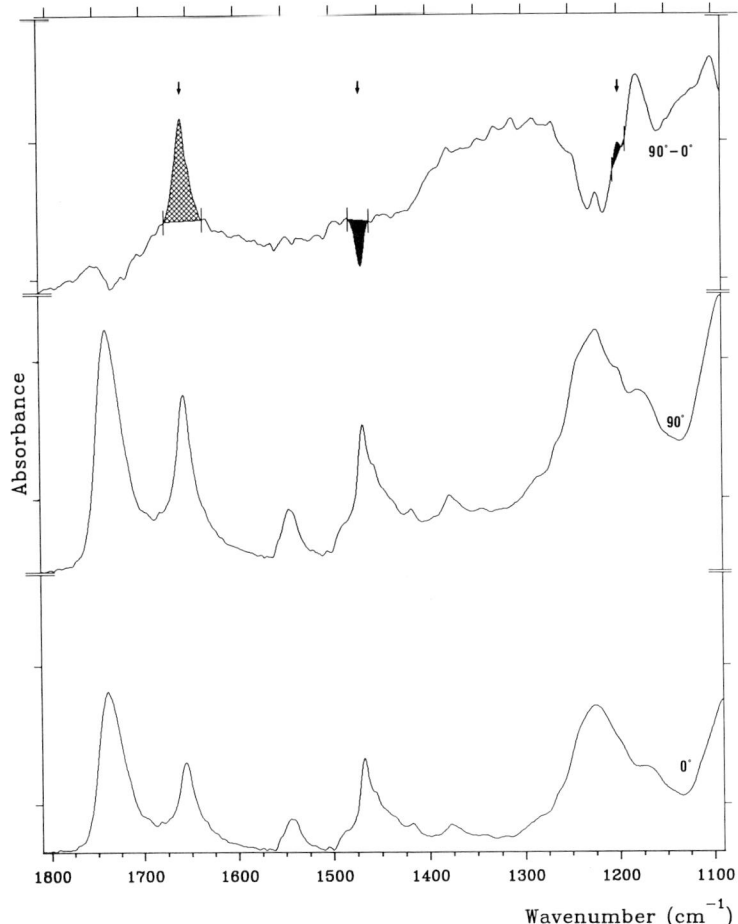

Figure 4. IR dichroic spectra of native porcine SP-C inserted in DPPC/PG (8:2 w/w) (protein/lipid molar ratio 1:27). The upper spectrum is obtained by the subtraction (90°–0°). The arrows indicate the protein amide I, the phospholipid $\delta(CH_2)$ at 1468 cm^{-1} and the $\gamma_W(CH_2)$ at 1200 cm^{-1}.

orientation of the dipole nearly parallel to the crystal plane and of the acyl chains nearly normal to the cystal plane (Fig. 4).

The difference spectra 90°–0° show an important positive deviation around 1657 cm^{-1} assigned to α-helical structure. The curve-fitting data allowed to calculate the dichroic ratio of the α-helix component ($R_{ATR} = 1.9$ for native SP-C and $R_{ATR} = 2.1$ for depalmitoylated SP-C). From these dichroic ratios, we have calculated a maximum tilt of 20° for the long helix axis with respect to a normal to the crystal surface (Vandenbussche et al., 1992a) (for more information about the determination of the angle see Goormaghtigh and Ruysschaert, 1990). Since in oriented multilayers, the average tilt between the lipid hydrocarbon chains and a normal to the germanium surface is smaller than 25°, the angle found for the α-helix corresponds to a predominantly parallel orientation of its long axis with respect to the lipid acyl chains. A similar orientation of the α-helix was described for calf SP-C inserted in DPPC/DPPG lipid bilayers (Pastrana et al., 1991).

The size parameters of the NMR-derived SP-C α-helix determined in organic solution strongly support a transmembranous orientation of this polypeptide in lipid bilayers. The length of the entire helix is 37 Å and the all-aliphatic part between positions 13 and 28, which is composed of aliphatic residues with branched side chains only, measures 23 Å. This is remarkably close to the thickness (Lewis and Engelman, 1983) of an entire fluid DPPC bilayer (37 Å) and its hydrocarbon part (26 Å) and very strongly supports a transmembranous orientation of the SP-C helix in the DPPC-rich pulmonary surfactant bilayers. This conformation is probably already formed in pro SP-C and a transmembrane domain was shown to be important for membrane insertion of the precursor of SP-C in the endoplasmic reticulum, processing and intracellular transport of the mature protein (Keller et al., 1991).

Porcine SP-C contains exclusively Val residues between positions 15 and 28, except for residues 22 and 27 which are Leu and Ile. This poly-valyl-α-helix constitutes a previously unknown structural motif for transmembrane helices and is somewhat unexpected considering that Val is generally underrepresented in α-helices (Levitt, 1978) and that Leu is the most common residue in transmembrane α-helices (von Heijne and Gavel, 1988).

Initial NMR studies of native SP-C in $^2H_{38}$-dodecylphosphocholine (DPC) micelles were unsuccessful due to pronounced line-broadening. This is somewhat unexpected since CD spectra of native SP-C in this environment reveal a high content of α-helical structure (Johansson et al., 1995b) (Tab. 1) and several other lipid-associated polypeptides of size similar to SP-C have been structurally characterized in DPC micelles (see Wüthrich, 1986). The reasons behind the observed line-broadening are unknown, but disruption of the micelle structure caused by the all-hydrophobic helix of SP-C and/or possible micelle cross-linking by the palmitoylcysteines are plausible explanations.

Furthermore, NMR spectroscopy was used to investigate the secondary structure of a synthetic polypeptide corresponding to SP-C positions 1–17 (Johansson et al., 1995b). In $^2H_{38}$-DPC micelles the onset of α-helical structure (Lys11) in this shorter non-palmitoylated peptide is very close to the N-terminal end of the α-helix (Asn9) of native SP-C in organic solvents. This together with the perfect match of SP-C in DPPC bilayers and the FTIR data described support that reliable conclusions about the functional properties of SP-C can be drawn from the MNR structure in organic solvents (Fig. 3). In a transmembranous orientation, the N-terminal end of SP-C contains both the palmitoylcysteines at positions 5 and 6, and two evolutionarily conserved basic residues at positions 11 and 12. No information about the orientation in the membrane of the two palmitic chains linked to the adjacent cysteines is available, but they are either inserted into the same bilayer as the polypeptide or inserted into another juxtaposed bilayer. Both the palmitoylcysteines and the basic residues thus probably will restrict lateral movements of the lipopolypeptide in the bilayer phase. In contrast, the C-terminal end contains only residues with small and unpolar side-chains, making this part potentially more mobile. It is tempting to speculate that the combination of the probably rigid poly-valyl-α-helix of SP-C and this mobility gradient are important for the functional features of SP-C in pulmonary surfactant phospholipids (Johansson et al., 1995b). This is consistent with fluorescence measurements indicating that SP-C increases rigidity in the head group region but disrupts the interior acyl-chain packing of the bilayer (Horowitz et al., 1992) and that the N-terminus is located in the vicinity of the membrane surface (Horowitz et al., 1993). ^{2}H NMR confirms also that the positively charged amino acids in the N-terminal domain of porcine SP-C are located close to the lipid headgroup (Morrow et al., 1993b).

Simatos et al. (1990) reported by ^{2}H NMR that SP-C has no effect on lipid chain order in the liquid-crystal phase but reduced orientational order in the gel phase. This is nicely explained by the size parameters of the SP-C helix in relation to the thickness of a DPPC bilayer in the gel phase and fluid phase, respectively. In gel phase lipids, the SP-C helix is significantly shorter than the thickness of the lipids and such mismatches are known to cause phase separations of peptides and lipids (Zhang et al., 1992). Accordingly, Horowitz et al. (1993) recently found that SP-C aggregates in gel phase bilayers composed of DPPC/PG, but is monomeric in the same lipids in the fluid phase. In conclusion, it was proposed that aggregated SP-C peptides in gel phase lipids disrupt the lipid packing and that the near perfect fit of SP-C in fluid phase bilayers promotes a monomeric state (Johansson et al., 1995b). Perturbations of the acyl chain packing have also been described by monolayer epifluorescence microscopy (Pérez-Gil et al., 1992), fluorescence measurements of phospholipid probes (Horowitz et al., 1992), and by differential scanning calorimetry (Shiffer et al., 1993). On the other hand, acyl chain C-H and C-D stretching frequencies indicate that SP-C has a minor effect on the thermotropic properties of a binary lipid mixture (Pastrana et al., 1991). These discrepancies may be explained in terms of the variety of lipid compositions, lipid/protein ratios, and technical approaches used.

Structure of SP-B

The secondary structure of SP-B has been analyzed mainly by FTIR and CD spectroscopy (Tab. 2). The strategy used for the evaluation of

Table 2. Survey of reported overall secondary structure contents of SP-B

α (%)	β (%)	Others (%)	Environment	Method	Reference
45	NR[1]	NR	70% trifluoroethanol	CD	Pérez-Gil et al., 1993
47	1	52	methanol	CD	Morrow et al., 1993a
44	23	33	chloroform/methanol (1:1 v/v)	FTIR	Vandenbussche et al., 1992b
47	3	50	DPPC	CD	Morrow et al., 1993b
48	NR	NR	DPPC	CD	Pérez-Gil et al., 1993
27	42	31	DPPC	FTIR	Baatz et al., 1992
42	22	36	DPPG	FTIR	Vandenbussche et al., 1992b
52	17	31	DPPC/PG (7:3 w/w)	FTIR	Vandenbussche et al., 1992b

[1] Not reported

the secondary structure of a SP-C by FTIR spectroscopy was applied to SP-B. After hydrogen/deuterium exchange, the FTIR spectrum is characterized by a maximum at 1654 cm^{-1} and a shoulder near 1629 cm^{-1} associated to α-helical and β-sheet structures, respectively (Vandenbussche et al., 1992b). Secondary structures percentages were quantified from the amide I area as described above. The data confirm that nearly half of porcine SP-B adopts an α-helical structure, which is also supported by CD spectroscopy (Fig. 5, Tab. 2).

SP-B was reconstituted in multilamellar vesicles of DPPC/PG (7:3 w/w) or DPPG. The shape of the amide I band in the ATR spectra of the reconstituted samples was similar to that of the isolated protein, as confirmed by curve-fitting data (Vandenbussche et al., 1992b). Recent CD analyses of synthetic human or native porcine SP-B confirm that the folding of the protein is independent of the presence of phospholipids; values of the α-helix content are in good agreement with the data presented here (Morrow et al., 1993a; Pérez-Gil et al., 1993).

Like SP-C, an important fraction of peptide bond protons of SP-B escaped hydrogen/deuterium exchange. After 6 h in a D_2O-saturated N_2 stream, about 60% of the polypeptide chain still remain unexchanged. This slow exchange is closely related to the hydrophobic nature of the protein, and/or to strong hydrogen bonding in ordered secondary structure elements.

The curve-fitting applied to the polarized spectra in the amide I' (the prime indicated that a H/D exchange has been performed) region allows the calculation of the dichroic ratios for the α-helical structure of SP-B associated to the liposomes and of the angle between the long axis of the α-helix and a normal to the germanium plate (Vandenbussche et al., 1992b). Although potential amphipathic helical domains of SP-B have been proposed, their number and exact position vary as reported by various authors (Glasser et al., 1987; Takahashi et al., 1990; Bruni et

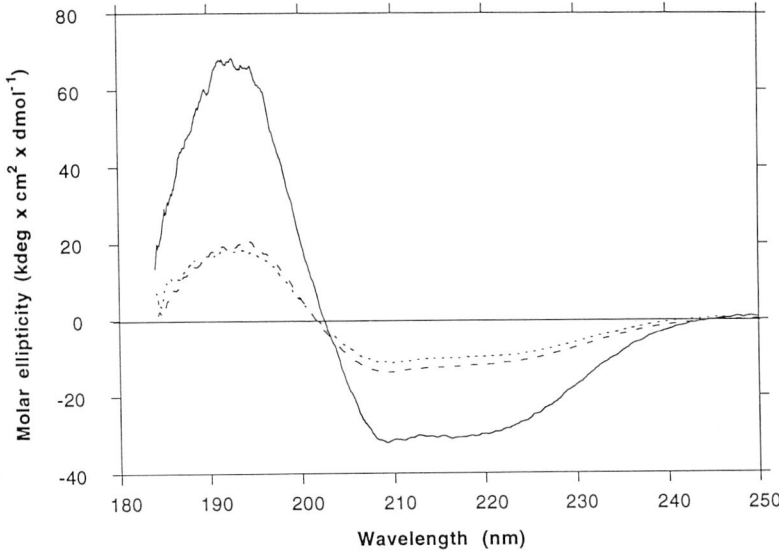

Figure 5. CD spectra of native SP-B and SP-C. Spectra were recorded between 184 and 250 nm for native porcine SP-B in the absence (dashed line) and in the presence of reducing agents (dotted line), and for native SP-C (solid line). The spectra emphasize the overall lower helical content of SP-B compared to SP-C. Both polypeptides were solubilized in 10 mM DPC/50 mM sodium phosphate, pH 6.0.

al., 1991; Whitsett and Baatz, 1992; Waring et al., 1993). It seems likely that several regions of the polypeptidic chain adopt a helical conformation. On the basis of a mathematical approach (Goormaghtigh and Ruysschaert, 1990) and assuming that the α-helices adopt either a parallel or a perpendicular orientation with respect to the membrane plane, we have proposed a model where 60–80% of the α-helices are parallel to the acyl chains (Vandenbussche et al., 1992b). This proportion could represent several stretches of hydrophobic residues slightly anchored near the membrane interface. The existence of these hydrophobic stretches was postualted from studies describing the position in a lipid matrix of simplified peptides containing hydrophilic and hydrophobic residues mimicking the SP-B polypeptide chain (Cochrane and Revak, 1991; Vincent et al., 1991). The other part of the α-helix content would be associated to charged residues and oriented parallel to the lipid/water interface.

Lipid organization upon interaction with SP-B can be detected by polarized infrared spectroscopy. The dichroic ratios corresponding to the $\gamma_w(CH_2)$ vibration at 1200 cm^{-1} are 4.0 ± 0.2 for a film of pure DPPC/PG and 4.2 ± 0.2 and 3.5 ± 0.5 for the SP-B/DPPC/PG complex with a protein/lipid molar ratio of 1:291 and 1:111, respectively. The dichroic ratios are 2.3 ± 0.1 for a film of pure DPPG and 2.9 ± 0.2 for the SP-B/DPPG complex with a protein/lipid molar ratio of 1:250. From these values, the corresponding mean angles between the axis of the hydrocarbon chains and a normal to the ATR plate surface were calculated for an order parameter equal to 1. The DPPC/PG acyl chains have a maximum tilt of 20° with respect to the normal to the germanium surface, whereas angles of 19° and 22° are obtained for the SP-B/DPPC/PG complex with protein/lipid molar ratios of 1:291 and 1:111, respectively. Angles of 29° and 25° are found for the acyl chains' orientation in the film of DPPG and of the SP-B/DPPG complex, respectively.

Additional information can be gained from the dichroic spectra. In the gel state, the hydrocarbon chain in the α-position of DPPC or DPPG is all-trans from the ester group down to the methyl group. This conformation allows a resonance to occur between the ester group and the methylene groups of the chain, giving rise to the so-called $\gamma_w(CH_2)$ progression between 1180 cm^{-1} and 1345 cm^{-1}. The number of peaks in the progression is equal to n/2 in all-trans acyl chain where n is the number of methylene in the chain. Deviation of the hydrocarbon chains conformation from an all-trans orientation induces modifications in the wagging progression (Fringeli and Günthard, 1981). The number of bands in the progression are conserved for the SP-B/DPPC/PG vesicles (peaks at 1200, 1221, 1246, 1266, 1286, 1309, and 1330 cm^{-1}) and no modifications of the shape or intensity of these bands are observed (Fig. 6). Likewise, no modification of the $\gamma_w(CH_2)$ progression was observed for the SP-B/DPPG complex. As a conformational modification of the acyl chains would be associated to a perturbation of this progression, our data indicate the absence of conformational change in the gel-phase DPPC or DPPG molecules upon interaction with the SP-B protein. These results confirm that the protein does not affect the organization and orientation of the lipid acyl chains in the gel-phase membrane. Fluorescence anisotropy measurements indicate that native bovine or synthetic human SP-B in a 7:1 DPPC/DPPG model membrane induces ordering at the bilayer surface with little effect on the organization of the bilayer interior (Baatz et al., 1990, 1991). Only slight perturbations of the packing of the acyl chains of DPPC in the presence of synthetic human SP-B was also reported by ^{2}H NMR (Morrow et al., 1993a).

Affinity of SP-B for PG molecules (Baatz et al., 1990; Yu and Possmayer, 1990) is likely associated to the charge complementarity between phospholipid head group and positively charged residues. Such interactions of SP-B with the PG phospholipid head group is suggested by the analysis of the bands associated to (PO_2^-) stretching between 1050 cm^{-1} and 1250 cm^{-1} (for background see Fringeli and Günthard, 1981; Dluhy et al., 1983; Casal and Mantsch, 1984; Casal et al., 1987). Upon association of SP-B with DPPG, the symmetric (PO_2^-) stretching peak shifts from 1094 cm^{-1} to 1110 cm^{-1} and is broadened (Vanden-

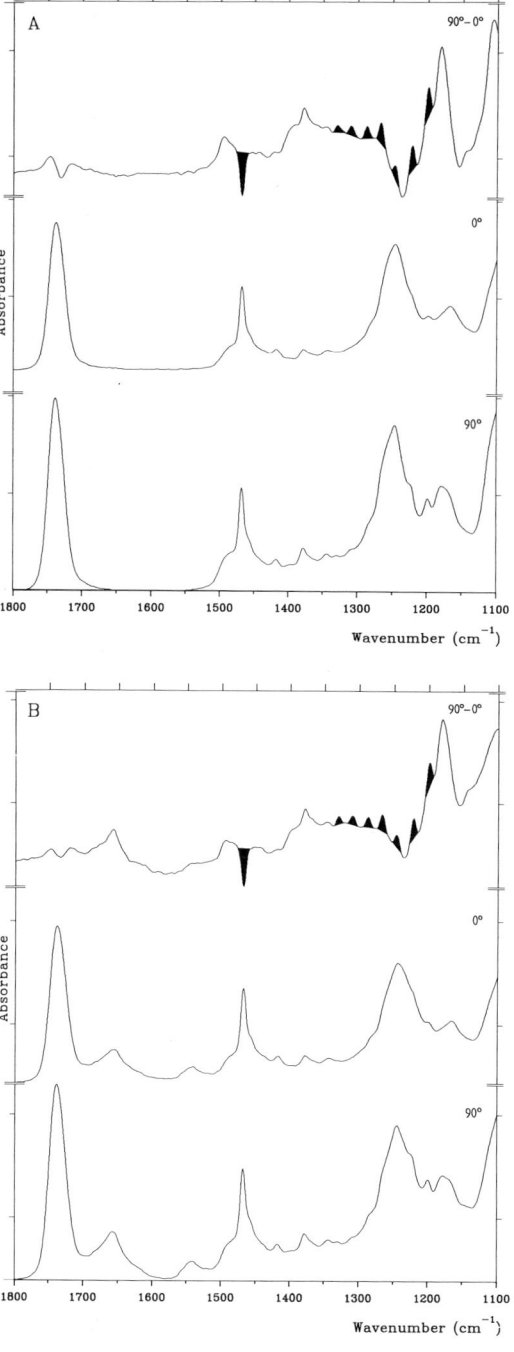

Figure 6. IR dichroic spectra of (A) DPPC/PG (7:3 w/w) and (B) native porcine SP-B associated to DPPC/PG (7:3 w/w) (protein/lipid molar ratio 1:291). In each figure the upper spectrum was obtained by (90°− 0°) subtraction.

bussche et al., 1992b). The modification of $\nu_S(PO_2^-)$ very likely denotes an interaction of the surfactant protein SP-B with lipid polar head groups. Although these data support the concept that SP-B is not a membrane-spanning protein, it is not an easy task to elaborate a model of association of the protein within the lipid matrix, especially due to the fact that the protein exists in an homodimeric form (Curstedt et al., 1990; Johansson et al., 1991a, 1992).

It has been suggested that SP-B lies on the membrane surface (Keough, 1992; Whitsett and Baatz, 1992) where the lipid/protein complex is stabilized by both electrostatic interactions between lipid head groups and basic amino acids, and hydrophobic interactions between apolar faces of amphipathic helices and the domains of the acyl chains located in the vicinity of the interface.

The models proposed should be considered as preliminary and more information is required to describe the association of dimers of SP-B with the surfactant phospholipids.

Conclusion

Extensive effort has been made during recent years to relate the structures of SP-B and SP-C with their surfactant activity. Several studies report the use of synthetic peptides, mostly based on SP-B, to determine the constraints of primary structure favorable to protein/lipid interaction and required to induce a rapid adsorption of phospholipids at the air/water interface. Information about the topology of SP-B and SP-C in their lipid environment is now emerging and will help to elucidate the mechanism of lipid spreading at the alveolar interface.

The unusual hydrophobic nature of SP-B and SP-C prevents their study by X-ray diffraction, but combination of ATR-FTIR, CD and NMR spectroscopy has provided detailed structural information as well as information on the orientation of segments of the protein in a lipid membrane.

The strategy described here could be extended to domains of SP-B and SP-C produced by solid-phase synthesis and should provide information helpful to better define requirements for optimal surfactant function. In particular, the available atomic resolution NMR structure of SP-C will facilitate rational design of analogs. Such analogs will be valuable for further characterizing the biological functions of SP-C and for attempts to find synthetic SP-C analogs to be used in surfactant preparations for treatment of RDS.

Acknowledgements
We are indebted to Erik Goormaghtigh for his constructive remarks. This work was supported by Oscar II:s Jubileumsfond, the Swedish Medical Research Council, the Swedish Society for Medical Research, and a grant from ARC (Action de Recherches Concertées).

References

Baatz, J.E., Elledge, B. and Whitsett, J.A. (1990) Surfactant protein SP-B induces ordering at the surface of model membrane bilayers. *Biochemistry* 29: 6714–6720.

Baatz, J.E., Sarin, V., Absolom, D.R., Baxter, C. and Whitsett, J.A. (1991) Effects of surfactant-associated protein SP-B synthetic analogs on the structure and surface activity of model membrane bilayers. *Chem. Phys. Lipids* 60: 163–178.

Baatz, J.E., Smyth, K.L., Whitsett, J.A., Baxter, C. and Absalom, D.R. (1992) Structure and functions of a dimeric form of surfactant protein SP-C: a Fourier transform infrared and surfactometry study. *Chem. Phys. Lipids* 63: 91–104.

Beers, M.F. and Fisher, A.B. (1992) Surfactant protein C: a review of its unique properties and metabolism. *Am. J. Physiol.* 263: L151–L160.

Bruni, R., Taeusch, H.W. and Waring, A.J. (1991) Surfactant protein B: lipid interactions of synthetic peptides representing the amino-terminal amphipathic domain. *Proc. Natl. Acad. Sci. USA* 88: 7451–7455.

Casal, H.L. and Mantsch, H.H. (1984) Polymorphic phase behaviour of phospholipid membranes studied by infrared spectroscopy. *Biochim. Biophys. Acta* 779: 381–401.

Casal, H.L., Mantsch, H.H. and Hauser, H. (1987) Infrared studies of fully hydrated phosphatidylserine bilayers. Effect of Li^+ and Ca^{2+}. *Biochemistry* 26: 4408–4416.

Cochrane, C.G. and Revak, S.D. (1991) Pulmonary surfactant protein B (SP-B): structure-function relationships. *Science* 254: 566–568.

Collaborative European Multicenter Study Group (1988) Surfactant replacement therapy for severe neonatal respiratory distress syndrome: an international randomized clinical trial. *Pediatrics* 82: 683–691.

Collaborative European Multicenter Study Group (1991) Factors influencing the clinical response to surfactant replacement therapy in babies with severe respiratory distress syndrome. *Eur. J. Pediatr.* 150: 433–439.

Collaborative European Multicenter Study Group (1992) A 2-year follow up of babies enrolled in a European multicenter trial of porcine surfactant replacement for severe neonatal respiratory distress syndrome. *Eur. J. Pediatr.* 151: 372–376.

Creuwels, L.A.J.M., Demel, R.A., van Golde, L.M.G., Benson, B. and Haagsman, H.P. (1993) Effect of acylation on structure and function of surfactant protein C at the air–liquid interface. *J. Biol. Chem.* 268: 26752–26758.

Cummings, J.J., Holm, B.A., Hudak, M.L., Hudak, B.B., Ferguson, W.H. and Egan, E.A. (1992) A controlled clinical comparison of four different surfactant preparations in surfactant-deficient preterm lambs. *Am. Rev. Respir. Dis.* 145: 999–1004.

Curstedt, T., Jörnvall, H., Robertson, B., Bergman, T. and Berggren, P. (1987) Two hydrophobic low-molecular-mass protein fractions of pulmonary surfactant. Characterization and biophysical activity. *Eur. J. Biochem.* 168: 255–262.

Curstedt, T., Johansson, J., Barros-Söderling, J., Robertson, B., Nilsson, G., Westberg, M. and Jörnvall, H. (1988) Low-molecular-mass surfactant protein type 1. The primary structure of a hydrophobic 8-kDa polypeptide with eight half-cystine residues. *Eur. J. Biochem.* 172: 521–525.

Curstedt, T., Johansson, J., Persson, P., Eklund, A., Robertson, B., Löwendaler, B. and Jörnvall, H. (1990) Hydrophobic surfactant-associated polypeptides: SP-C is a lipopeptide with two palmitoylated cysteine residues whereas SP-B lacks covalently linked fatty acyl groups. *Proc. Natl. Acad. Sci. USA* 87: 2985–2989.

Dluhy, R.A., Cameron, D.G., Mantsch, H.H. and Mendelsohn, R. (1983) Fourier transform infrared spectroscopic studies of the effect of calcium ions on phosphatidylserine. *Biochemistry* 22: 6318–6325.

Enhorning, G., Shennan, A., Possmayer, F., Dunn, M., Chen, C.P. and Milligan, J. (1985) Prevention of neonatal respiratory distress syndrome by tracheal instillation of surfactant: a randomized clinical trial. *Pediatrics* 76: 145–153.

Fringeli, U.P. and Günthard, H.H. (1981) Infrared membrane spectroscopy. *In:* E. Grell (ed.): *Membrane Spectroscopy*, Springer-Verlag, Berlin, pp 270–332.

Fujiwara, T., Maeta, H., Chida, S., Morita, T., Watabe, Y. and Abe, T. (1980) Artifical surfactant therapy in hyaline-membrane disease. *Lancet* 1: 55–59.

Gierasch, L.M. (1989) Signal sequences. *Biochemistry* 28: 923–930.

Glasser, S.W., Korfhagen, T.R., Weaver, T., Pilot-Matias, T., Fox, J.L. and Whitsett, J.A. (1987) cDNA and deduced amino acid sequence of human pulmonary surfactant-associated proteolipid SPL(Phe). *Proc. Natl. Acad. Sci. USA* 84: 4007–4011.

Glasser, S.W., Korfhagen, T.R., Weaver, T.E., Clark, J.C., Pilot-Matias, T.R., Meuth, J., Fox, J.L. and Whitsett, J.A. (1988) cDNA, deduced polypeptide structure and chromosomal assignment of human pulmonary surfactant proteolipid, SPL (pVal). *J. Biol. Chem.* 263: 9–12.

Goerke, J. (1974) Lung surfactant. *Biochim. Biophys. Acta* 344: 241–261.

Goormaghtigh, E., Martin, I., Vandenbranden, M., Brasseur, R. and Ruysschaert, J.-M. (1989) Secondary structure and orientation of a chemically synthesized mitochondrial signal sequence in phospholipids bilayers. *Biochem. Biophys. Res. Commun.* 158: 610–615.

Goormaghtigh, E. and Ruysschaert, J.-M. (1990) Polarized attenuated total reflection as a tool to investigate the conformation and orientation of membrane components. *In:* R. Brasseur (ed.): *Molecular Description of Biological Components*, Vol. 1, CRC Press, Boca Raton, Florida, pp 285–329.

Goormaghtigh, E., Cabiaux, V. and Ruysschaert. J.-M. (1990) Secondary structure and dosage of soluble and membrane proteins by attenuated total reflection Fourier-transform infrared spectroscopy on hydrated films. *Eur. J. Bioch.* 193: 409–420.

Goormaghtigh, E., Cabiaux, V. and Ruysschaert, J.-M. (1994) Determination of soluble and membrane protein structure by Fourier transform infrared spectroscopy. *In:* H.J. Hilderson and G.B. Ralston (eds): Subcellular biochemistry: Physicochemical methods in the study of biomembranes, Vol. 23, Plenum Press, New York, pp 329–450.

Hall, S.B., Venkitaraman, A.R., Whitsett, J.A., Holm, B.A. and Notter, R.H. (1992) Importance of hydrophobic apoproteins as constituents of clinical exogenous surfactants. *Am. Rev. Respir. Dis.* 145: 24–30.

Hallman, M., Meritt, A., Jarvenpaa, A.-L., Boynton, B., Mannino, F., Gluck, L., Moore, T. and Edwards, D. (1985) Exogenous human surfactant for treatment of severe respiratory distress syndrome: a randomized prospective clinical trial. *J. Pediatr.* 106: 963–969.

Hawgood, S., Benson, B.J., Schilling, J., Damm, D., Clements, J.A. and White, R.T. (1987) Nucleotide and amino acid sequences of pulmonary surfactant protein SP 18 and evidence for cooperation between SP 18 and SP 28–36 in surfactant lipid adsorption. *Proc. Natl. Acad. Sci. USA* 84: 66–70.

Hawgood, S. and Shiffer, K. (1991) Structures and properties of the surfactant-associated proteins. *Annu. Rev. Physiol.* 53: 375–394.

Hildebran, J.N., Goerke, J. and Clements, J.A. (1979) Pulmonary surface film stability and composition. *J. Appl. Physiol.* 47: 604–611.

Horbar, J.D., Wright, L.L., Soll, R.F., Wright, E.C., Fanaroff, A.A., Korones, S.B., Shankaran, S., Oh, W., Fletcher, B.D., Bauer, C.R., Tyson, J.E., Lemons, J.A., Donovan, E.F., Stoll, B.J., Stevenson, D.D., Papile, L.-A. and Philips, J. (1993) A multicenter randomized trial comparing two surfactants for the treatment of neonatal respiratory distress syndrome. *J. Pediatr.* 123: 757–766.

Horowitz, A.D., Elledge, B., Whitsett, J.A. and Baatz, J.E. (1992) Effects of lund surfactant proteolipid SP-C on the organization of model membrane lipids: a fluorescence study. *Biochim. Biophys. Acta* 1107: 44–54.

Horowitz, A.D., Baatz, J.E. and Whitsett, J.A. (1993) Lipid effects on aggregation of pulmonary surfactant protein SP-C studied by fluorescence energy transfer. *Biochemistry* 32: 9513–9523.

Jobe, A. and Ikegami, M. (1987) Surfactant for the treatment of respiratory distress syndrome. *Am. Rev. Respir. Dis.* 136: 1256–1275.

Johansson, J., Curstedt, T., Robertson, B. and Jörnvall, H. (1988a) Size and structure of the hydrophobic low molecular weight surfactant-associated polypeptide. *Biochemistry* 27: 3544–3547.

Johansson, J., Jörnvall, H., Eklund, A., Christensen, N., Robertson, B. and Curstedt, T. (1988b) Hydrophobic 3.7 kDa surfactant polypeptide: structural characterization of the human and bovine forms. *FEBS Lett.* 232: 61–64.

Johansson, J., Curstedt, T. and Jörnvall, H. (1991a) Surfactant protein B: disulfide bridges, structural properties, and kringle similarities. *Biochemistry* 30: 6917–6921.

Johansson, J., Persson, P., Löwenadler, B., Robertson, B., Jörnvall, H. and Curstedt, T. (1991b) Canine hydrophobic surfactant polypeptide SP-C. A lipopeptide with one thioester-linked palmitoyl group. *FEBS Lett.* 281: 119–122.

Johansson, J., Jörnvall, H. and Curstedt, T. (1992) Human surfactant polypeptide SP-B. Disulfide bridges, C-terminal end, and peptide analysis of the airway form. *FEBS Lett.* 301: 165–167.

Johansson, J., Curstedt, T. and Robertson, B. (1994a) The proteins of the surfactant system. *Eur. Respir. J.* 7: 372–391.

Johansson, J., Szyperski, T., Curstedt, T. and Wüthrich, K. (1994b) The NMR solution structure of the pulmonary surfactant-associated polypeptide SP-C in an apolar solvent contains a valyl-rich-α-helix. *Biochemistry* 33: 6015–6023.

Johansson, J., Nilsson, G., Strömberg, R., Robertson, B., Jörnvall, H. and Curstedt, T. (1995a) Secondary structure and biophysical activity of synthetic analogs to the pulmonary surfactant-associated polypeptide SP-C. *Biochem. J.* 307: 535–541.

Johansson, J., Szyperski, T. and Wüthrich, K. (1995b) Pulmonary surfactant-associated polypeptide SP-C in lipid micelles: CD studies of intact SP-C and NMR secondary structure determination of depalmitoyl-SP-C (1–17). *FEBS Lett.* 362: 261–265.

Keller, A., Eistetter, H.R., Voss, T. and Schäfer, K.-P. (1991) The pulmonary surfactant protein C (SP-C) precursor is a type II transmembrane protein. *Biochem. J.* 277; 493–499.

Keough, K.M.W. (1992) Physical chemistry of pulmonary surfactant in the terminal air spaces. *In*: B. Robertson, L.M.G. van Golde and J.J. Batenburg (eds): *Pulmonary Surfactant: From Molecular Biology to Clinical Practice*, Elsevier Science Publishers B.V., Amsterdam, The Netherlands, pp 109–164.

Lear, J.D. and Degrado, W.F. (1987) Membrane binding and conformational properties of peptides representing the NH_2 terminus of Influenza HA-2. *J. Biol. Chem.* 262: 6500–6505.

Levitt, M. (1978) Conformational preferences of amino acids in globular proteins. *Biochemistry* 17: 4277–4285.

Lewis, B.A. and Engelman, D.M. (1983) Lipid bilayer thickness varies linearly with acyl chain length in fluid phosphatidylcholine vesicles. *J. Mol. Biol.* 166; 211–217.

Martin, I., Defrise-Quertain, F., Mandieau, V. Nielsen, N.M., Saermark, T., Burny, A., Brasseur, R., Ruysschaert, J.-M. and Vandenbranden, M. (1991) Fusogenic activity of SIV (Simian Immunodeficiency Virus) peptides located in the gp32 NH_2 terminal domain. *Biochem. Biophys. Res.Commun.* 175: 872–879.

McLean, L.R., Hagaman, K.A., Owen, T.J. and Krstenansky, J.L. (1991) Minimal peptide length for interaction of amphipathic α-helical peptides with phosphatidylcholine liposomes. *Biochemistry* 30: 31–37.

Morrow, M.R., Pérez-Gil, J., Simatos, G., Boland, C., Stewart, J., Absolom, D., Sarin, V. and Keough, K.M.W. (1993a) Pulmonary surfactant-associated protein SP-B has little effect on acyl chains in dipalmitoylphosphatidylcholine dispersions. *Biochemistry* 32: 4397–4402.

Morrow, M.R., Taneva, S., Simatos, G.A., Allwood, L.A. and Keough, K.M.W. (1993b) ^{2}H NMR studies of the effect of pulmonary surfactant SP-C on the 1,2-dipalmitoyl-sn-glycero-3-phosphocholine headgroup: a model for transbilayer peptides in surfactant and biological membranes. *Biochemistry* 32: 11338–11344.

Olafson, R.W., Rink, U., Kielland, S., Yu, S.-H., Chung, J., Harding, P.G.R. and Possmayer, F. (1987) Protein sequence analysis studies on the low molecular weight hydrophobic proteins associated with bovine pulmonary surfactant. *Biochem. Biophys. Res. Commun.* 148: 1406–1411.

Oosterlaken-Dijksterhuis, M.A., Haagsman, H.P., van Golde, L.M.G. and Demel, R.A. (1991a) Interaction of lipid vesicles with monomolecular layers containing lung surfactant proteins SP-B or SP-C. *Biochemistry* 30: 8276–8281.

Oosterlaken-Dijksterhuis, M.A., Haagsman, H.P., van Golde, L.M.G. and Demel, R.A. (1991b) Characterization of lipid insertion into monomolecular layers mediated by lung surfactant proteins SP-B and SP-C. *Biochemistry* 30: 10965–10971.

Pastrana, B., Mautone, A.J. and Mendelsohn, R. (1991) Fourier transform infrared studies of secondary structure and orientation of pulmonary surfactant SP-C and its effect on the dynamic surface properties of phospholipids. *Biochemistry* 30: 10058–10064.

Pérez-Gil, J., Nag, K., Taneva, S. and Keough, M.W. (1992) Pulmonary surfactant protein SP-C causes packing rearrangements of dipalmitoylphosphatidylcholine in spread monolayers. *Biophys. J.* 63: 197–204.

Pérez-Gil, J., Cruz, A. and Casals, C. (1993) Solubility of hydrophobic surfactant proteins in organic solvent/water mixtures. Structural studies on SP-B and SP-C in aqueous organic solvents and lipids. *Biochim. Biophys. Acta* 1168: 261–270.

Rooney, S.A. (1985) The surfactant system and lung phospholipid biochemistry. *Am. Rev. Respir. Dis.* 131: 439–460.

Shelley, S.A. Paciga, J.E. and Balis, J.U. (1984) Lung surfactant phospholipids in different animal species. *Lipids* 19: 857–862.

Shiffer, K., Hawgood, S., Haagsman, H.P., Benson, B., Clements, J.A. and Goerke, J. (1993) Lung surfactant proteins, SP-B and SP-C, alter the thermodynamic properties of phospholipid membranes: a differential calorimetry study. *Biochemistry* 32: 590–597.

Simatos, G.A., Forward, K.B., Morrow, M.R., and Keough, K.M.W. (1990) Interaction between perdeuterated dimyristoylphosphatidylcholine and low molecular weight pulmonary surfactant protein SP-C. *Biochemistry* 29: 5807–5814.

Speer, C.P., Robertson, B., Curstedt, T., Halliday, H.L., Compagnone, D., Gefeller, O., Harms, K., Herting, E., McClure, G., Reid, M., Tubman, R., Herin, P., Noack, G., Kok, J., Koppe, J., van Sonderen, L., Laufkötter, E., Köhler, W., Boenisch, H., Albrecht, K., Hanssler, L., Haim, M., Oetomo, S.B., Okken, A., Altfeld, P.C., Groneck, P., Kachel, W., Relier, J.-P. and Walti, H. (1992) Randomized European multicenter trial of surfactant replacement therapy for severe neonatal respiratory distress syndrome: Single versus multiple doses of Curosurf. *Pediatrics* 89: 13–20.

Stults, J.T., Griffin, P.R., Lesikar, D.D., Naidu, A., Moffat, B. and Benson, B.J. (1991) Lung surfactant protein SP-C from human, bovine, and canine sources contains palmityl cysteine thioester linkages. *Am. J. Physiol.* 261: L118–L125.

Suzuki, Y., Curstedt, T., Grossmann, G., Kobayashi, T., Nilsson, R., Nohara, K. and Robertson, B. (1986) The role of the low-molecular weight ($\leq 15\,000$ daltons) apoproteins of pulmonary surfactant. *Eur. J. Respir. Dis.* 69: 336–345.

Takahashi, A. and Fujiwara, T. (1986) Proteolipid in bovine lung surfactant: its role in surfactant function. *Biochem. Biophys. Res. Commun.* 135: 527–532.

Takahashi, A., Waring, A.J., Amirkhanian, J., Fan, B. and Taeusch, H.W. (1990) Structure-function relationships of bovine pulmonary surfactant proteins: SP-B and SPC. *Biochim. Biophys. Acta* 1044: 43–49.

Vandenbussche, G., Clercx, A., Curstedt, T., Johansson, J., Jörnvall, H. and Ruysschaert, J.-M. (1992a) Structure and orientation of the surfactant-associated protein C in a lipid bilayer. *Eur. J. Biochem.* 203: 201–209.

Vandenbussche, G., Clercx, A., Clercx, M., Curstedt, T., Johansson, J., Jörnvall, H. and Ruysschaert, J.-M. (1992b) Secondary structure and orientation of the surfactant protein SP-B in a lipid environment. A Fourier transform infrared spectroscopy study. *Biochemistry* 31: 9169–9176.

van Golde, L.M.G., Batenburg, J.J. and Robertson, B. (1988) The pulmonary surfactant system: biochemical aspects and functional significance. *Physiol. Rev.* 68: 374–455.

Vincent, J.S., Revak, S.D., Cochrane, C.G. and Levin, I.W. (1991) Raman spectroscopic studies of model human pulmonary surfactant systems: phospholipid interactions with peptide paradigms for the surfactant protein SP-B. *Biochemistry* 30: 8395–8401.

von Heijne, G. and Gavel, Y. (1988) Topogenic signals in integral membrane proteins. *Eur. J. Biochem.* 174: 671–678.

Vorbroker, D.K., Dey, C., Weaver, T.E. and Whitsett, J.A. (1992) Surfactant protein C precursor is palmitoylated and associates with subcellular membranes. *Biochim. Biophys. Acta* 1105: 161–169.

Waring, A.J., Gordon, L.M., Taeusch, H.W. and Bruni, R. (1993) Amphipatic helical segments in lung surfactant proteins. *In:* R.M. Epand (ed.): *The Amphipatic Helix*, CRC Press, Boca Raton, Florida, pp 143–171.

Warr, R.G., Hawgood, S., Buckley, D.I., Crisp, T.M., Schilling, J., Benson, B.J., Ballard, P.L., Clements, J.A. and White, R.T. (1987) Low molecular weight human pulmonary surfactant protein (SP5): isolation, characterization, and cDNA and amino acid sequences. *Proc. Natl. Acad. Sci. USA* 84: 7915–7919.

Weaver, T.E. and Whitsett, J.A. (1991) Function and regulation of expression of pulmonary surfactant-associated proteins. *Biochem. J.* 273: 249–264.

Whitsett, J.A., Ohning, B.L., Ross, G., Meuth, J., Weaver, T., Holm, B.A., Shapiro, D.L. and Notter, R.H. (1986) Hydrophobic surfactant-associated protein in whole lung surfactant and its importance for biophysical activity in lung surfactant extracts used for replacement therapy. *Pediatr. Res.* 20: 460–467.

Whitsett, J.A. and Baatz, J.E. (1992) Hydrophobic surfactant proteins SP-B and SP-C: molecular biology, structure and function. *In:* B. Robertson, L.M.G. van Golde and J.J. Batenburg (eds): *Pulmonary Surfactant: From Molecular Biology to Clinical Practice*, Elsevier Science Publisher B.V., Amsterdam, The Netherlands, pp 55–75.

Williams, M.C. (1977) Conversion of lamellar body membranes into tubular myelin in alveoli of fetal rat lungs. *J. Cell. Biol.* 72: 260–277.

Wüthrich, K. (1986) *NMR of Proteins and Nucleic Acids.* Wiley, New York.
Yu, S.-H. and Possmayer, F. (1986) Reconstitution of surfactant activity by using the 6 kDa apoprotein associated with pulmonary surfactant. *Biochem. J.* 236: 85–89.
Yu, S.-H. and Possmayer, F. (1990) Role of bovine pulmonary surfactant-associated proteins in the surface-active property of phospholipid mixtures. *Biochim. Biophys. Acta* 1046: 233–241.
Zhang, Y.-P., Lewis, R.N.A.H., Hodges, R.S. and McElhaney, R.N. (1992) Interaction of a peptide model of a hydrophobic transmembrane α-helical segment of a membrane protein with phosphatidylcholine bilayers: a differential scanning calorimetric and FTIR spectroscopic studies. *Biochemistry* 31: 11579–11588.

Enzyme function in organic solvents

M.N. Gupta

Chemistry Department, IIT Delhi, India 110 016

Summary. Enzyme catalysis in nearly anhydrous organic solvents has several advantages. While biocatalyst engineering is aimed at designing enzyme "form" with optimum function, medium engineering involves altering the polarity, water content, etc. for the same purpose. Enhanced thermostability, pH memory, and possibility of protein imprinting are some interesting features associated with enzyme function in these media. These and several other applications in the area of bioanalysis and organic synthesis have made this an important topic in the area of applied biocatalysis.

Introduction

Khmelnitsky et al. (1988) have pointed out that solid enzyme suspension in organic solvent was first used in 1913 by Bourquelot and Bridel. These workers carried out stereospecific synthesis of various alkyl glucosides from glucose and alcohols in 80–85% yield using dry emulsin in alcohol. About 50 years later, Price and coworkers again used similar systems. For example, chymotrypsin (Dastoli et al., 1966) and Xanthine oxidase (Dastoli and Price, 1967a, 1967b) powders were found to be catalytically active in various dry organic solvents. A few years later, Yokozeki et al. (1982) showed that immobilized lipase could carry out regiospecific interesterification of triglycerides in n-hexane. Soon after, there was considerable activity in this area (Zaks and Klibanov, 1984, 1985, 1988a, 1988b; Reslow et al., 1987, 1988a, 1988b) and since then, enzyme function in organic solvents has become an extremely important topic in the area of applied biocatalysis. This article focusses primarily on enzyme function in nearly anhydrous organic systems.

There are several reasons why enzymes are being increasingly used in anhydrous organic solvents. Table 1 lists several advantages associated with this approach. A detailed discussion on these is already available from several works (Klibanov, 1986, Khmelnitsky et al., 1988; Dordick, 1992).

Broadly speaking, in the case of immiscible organic solvents, water should form a separate phase in order to be termed a low water system. In the case of miscible organic solvents, it is slightly more difficult to differentiate but systems with water up to 10% have been described as low water systems. On the other hand, Halling is of the opinion that such terms should be reserved for systems where "the

Table 1. Advantages associated with the use of enzyme in organic solvents

— When substrates have greater solubility in organic solvents. In specific instances, it is possible to also use substrates as medium so that solvents are totally eliminated.
— Shift of reaction equilibria in desirable directions. Many hydrolases can be used for synthetic purposes, and they are readily available.
— Covalent immobilization is not required as enzymes are practically insoluble in most of the organic solvents. In many cases, no immobilization is necessary as solid powders give adequate rate.
— Reduced risk of microbial growth.
— Side reactions involving water are reduced.
— Convenient to use "moisture-sensitive" reagents like acid anhydrides.
— More energy efficient downstream processing in the case of volatile solvents.
— Enzyme catalyzed synthetic step can be easily integrated with other steps involving conventional organic chemistry.
— Enhanced storage stability and thermostability.
— Greater control of substrate specificity, regiospecificity and enantioselectivity.
— Higher solubility of oxygen in organic solvents should facilitate oxidase reactions involving molecular oxygen.

thermodynamic water activity (a_w) is significantly less than 1" (Halling, 1987). Some of the methods which have been used for measurement of water in such "anhydrous" systems are: Karl Fischer method (Zaks and Klibanov, 1988a), gas chromatography (Reslow et al., 1988a) and membrane inlet mass spectrometry (Degan et al., 1992).

The various forms in which enzymes have been used in the low water systems are described in Table 2.

Hydration induced conformational flexibility and protein dynamics

Rupley et al. (1983) have used heat capacity data to examine the hydration of lysozyme. The gradual hydration event can be viewed in

Table 2. Forms of enzymes used in low water systems

— Solid enzyme powder suspended in Organic solvents	Cambou and Klibanov, 1984 Skrika-Alexopoulos and Freedman, 1993.
— Solid enzyme adsorbed on support particles	Reslow et al., 1988a Ncube et al., 1993 Burton et al., 1993
— Poly (ethylene glycol)-modified enzymes soluble in aromatic hydrocarbons	Takahashi et al., 1984a,b,c, 1985. Matsushima et al., 1984 Inada et al., 1986.
— Complex between enzyme and polymer soluble in organic solvent	Otamiri et al., 1992
— Enzyme modified by neutral or changed amphiphiles	Cabaret et al., 1992
— Enzyme modified by hydrophobic groups	Ampon et al., 1992
— Immobilized enzymes suspended in organic solvents	Tanaka and Kawamoto, 1991. Guisan et al., 1992, 1993.
— Crosslinked enzyme crystals	Navia et al., 1993.

three stages: (i) Hydration of charged groups, between 0–0.7 g water per g of protein; (ii) clustering of water around these polar patches, between 0.07–0.25 g water per g of protein; (iii) water interacting with the rest of the protein surface, between 0.025–0.38 g water per g of protein. Thus the fully hydrated protein molecule has about 300 water molecules. This barely forms a monolayer on the protein surface. The biological activity of lysozyme becomes detectable at 0.2 g/g. The change in protein mobility is complete around 0.25 g/g (nearly 220 water molecules/lysozyme molecule). Thus onset of biological activity correlates well with approach of total mobility. This is in agreement with general notions about flexibility being necessary for enzymic catalysis (Karplus and McCammon, 1983; Poole and Finney, 1983). It has been pointed out (Zaks and Klibanov, 1988a) that the main role of water is to form bonds with functional groups which otherwise interact with each other to form "locked" conformation. Perhaps, the most critical event is the hydration of polar groups (Gupta, 1993) as this would explain why even at the level of 50 molecules of water/protein molecules, chymotrypsin and subtilisin show biological activity in the organic solvents (Zaks and Klibanov, 1988a).

In this context, it may be mentioned that Valivety et al. (1992) have shown that Rhizomucor miehei lipase adsorbed on polymer beads remains highly active even at water activity below 0.0001. This observation probably records the most anhydrous organic system to show enzymatic activity.

Affleck et al. (1992) have used EPR spectroscopy of nitroxide labelled enzymes to show that molecular flexibility and enzyme activity correlates well in low water organic solvent systems. Burke et al. (1993), using solid-state NMR, have examined the tryosine ring motions in alpha-lytic protease suspended in organic solvents. Earlier (Burke et al., 1992), the same group had shown that the active centre of the enzyme remains intact under these conditons. The latter work demonstrates that "librational motions are fastest in aqueous crystals, intermediate in the hydrated solvents, and slowest in the anhydrous solvents". This agrees well with the existing view that dynamic freedom is restricted in low water systems and hydration relieves rigidity. However, more significant is their finding that "the results presented here are not consistent with the notion that water stripping by hydrophilic solvents causes further enzyme rigidification". The reduced catalytic efficiencies in hydrophilic solvents compared to hydrophobic ones may not be due to rigidification and may be due to a partial active center disruption on hydrophilic solvent addition as suggested by Burke et al. (1992). They add "the above results provide no support to a direct link between enzyme conformational flexibility and either solvent dielectric constant or catalytic activity. This is in contrast to results obtained with spin labeled alpha chymotrypsin" (Affleck et al., 1992). Perhaps, this work illustrates that, in view of the near absence of any precise structural information so far, the picture of enzyme catalysis in water-poor media has been rather oversimplified. Undoubtedly, we will see more structural work in this area in the near future, which will ultimately remove the apparent contradictions.

Kinetics of enzymatic reactions in anhydrous solvents

Zaks and Klibanov (1988a) mentioned that the kinetics of subtilisin catalyzed transesterification in octane followed Michalis-Menten kinetics. Similar agreement with Michalis-Menten kinetics has been found by many other workers. For example, Burton et al. (1993) with polyphenol oxidase in choloroform; Skrika-Alexopoulous and Freedman (1993) with bilirubin oxidase in chloroform-heptane mixture and Yamamoto and Kise (1993) with chymotrypsin and subtilisin in ethanol. An interesting observation reported by Yamamoto and Kise (1993) is that catalytic activities of enzymes markedly increased by decreasing the amount of enzymes due to formation of smaller enzyme aggregates. Kanerva and Klibanov (1989) have carried out Hammett analysis and shown that rho values were independent of the solvent and the same as in water. Similar results were obtained with peroxidase (Ryu and Dordick, 1992) and indicate that native active site structures are not affected by the solvent. ^{15}N magic angle spinning NMR also shows that alpha-lytic protease active site is not affected by octane or acetone (Burke et al., 1989).

The work of Ryu and Dordick (1992) has focussed much needed attention on the interdependence of substrate structure and solvent hydrophobicity in achieving catalytic efficiency. Working with peroxidase, they showed that partitioning of hydrophobic phenols to the active site diminishes as both substrate and solvent hydrophobicities increase. This, of course, is reflected in higher Km.

Enhanced thermostability

The highly rigid structures of proteins at low water content show enhanced stability. Also, absence of sufficient water also inhibits inactivation process such as protease action and many thermoinactivation mechanisms. The latter includes deamidation, peptide hydrolysis and cystine decomposition (Gupta, 1991; Gupta and Gupta, 1993). It is for this reason that the thermostabilization observed is quite dramatic. The pancreatic lipase remained stable for many hours at 100°C (Zaks and Klibanov, 1984). A more comprehensive list of various enzymes examined so far has been provided elsewhere recently (Gupta, 1992).

Reslow et al. (1988a) have shown that the thermostability of celite adsorbed chymotrypsin at 50°C was better in solvents having log P values greater than 0.7. Garza-Ramos et al. (1990) reported that the optimal temperature of ATP hydrolysis by mitochondrial ATPase is the same (i.e., around 58°C) in aqueous medium as in low water medium. As has been mentioned earlier (Gupta, 1992), this issue needs to be examined with a larger set of enzymes. It is possible that high thermal stability may not always translate into proportionate increase in catalytic rates.

Inactivation mechanisms

Volkin et al. (1991) reported that protein aggregation was the main reason of loss of activity beyond 110°C in the case of ribonuclease A, chymotrypsin and lysozyme in low water situations. The aggregation was caused by formation of scrambled structures due to disulfide interchange reactions and transamidation reaction leading to crosslinking of asparagine with lysine residues. The differential scanning calorimetry data with ribonuclease A show that the enzyme is more thermostable in hydrophobic solvents as compared to hydrophilic solvents. More recently, Estrada et al. (1992) have studied the thermoinactivation of polyphenol oxidase in toluene (0.8%, v/v) at 60–75°C. The first order kinetics observed for thermoinactivation suggests the monomolecular nature of the inactivation process. The high values of 111.5 Kcal/ml for activation energy for denaturation process points towards an unfolding process followed by formation of a scrambled structure. This view is further supported by the observation that presence of p-nitrophenol and N-acetyl-L-tyrosine-ethyl ester protected the enzyme.

Medium engineering

Correlation between enzyme activity in non-aqueous surroundings and the nature of the medium is an important issue. Zaks and Klibanov (1988a) had stated that for subtilisin catalyzed transesterification "the best solvents are water immiscible hydrophobic solvents and the worst are highly hydrophilic water-miscible ones". More recently however, the same group (Narayan and Klibanov, 1993), working with four different hydrolases, concluded that "neither solvent apolarity nor water-immiscibility by themselves are essential for optimal enzymatic activity". The latter work was carried out with commercial enzyme preparations. While probing such issues, it is advisable to work with pure enzymes. Gorman and Dordick (1992) measured the stripping of water from enzymes by organic solvents by placing tritiated enzyme in organic solvents. Their data showed "Polar solvents resulted in the highest degree of T_2O desorption (e.g., methanol desorbed from 56%–62% of the bound T_2O), while nonpolar solvents resulted in the lowest degree of desorption (e.g., hexane desorbed from 0.4%–2% of the bound T_2O). Desorption is nearly immediate with most of the desorbable T_2O being released from the enzymes within the first 5 minutes". Dordick (1992), in a well written review, has very rationally stated that solvent engineering "involves altering the polarity, hydrophobicity, water content etc. of the organic milieu".

Laane (1987) and Laane et al. (1987) stated that the best correlation between enzyme activity and medium is obtained when log P is chosen as the medium parameter. P is the partition coefficient for the standard octanol/water two-phase systems. While the solvents having log $P < 2$ are poor choices, the ones with log $P < 4$ are most suitable. Solvents in the intermediate range are unpredictable. One should also remember that if the microenvironments of the enzyme promotes solubility of the substrate and has low product solubility, the reaction rates would be higher. Over the years, log P has indeed emerged as the most acceptable parameter for the purpose (Reslow et al., 1987; Adlercreutz and Mattiasson, 1987; Mattiasson and Adlercruetz, 1991; Valivety et al., 1991). Gorman and Dordick (1992) have observed that the capacity to strip off water from the enzymes is independent of log P for water miscible or polar water immiscible solvents having log P below 1.5. According to their data, solvent dielectric correlates better with water desorption power. Schneider (1991) has suggested a three-dimensional solubility parameter. Recently, Yang et al. (1992) have reported that no obvious relationship was found between enzyme activity and log P at constant water activity in the case of tyrosinase. Their work reiterates the guidelines by Laane (1987) and Laane et al. (1987) that log P for the substrate and product should be taken into account.

In another important paper, Valivety et al. (1991) have examined fairly comprehensively the issue of the best parameter. These authors rightly point out that a severe limitation of some studies is "that they have simply measured variations in product yield after a set of incubation time. Differences of this type may reflect either or both equilibrium position and reaction rates while the latter may be influenced by catalyst activity and/or stability". They mention that two kinds of solvent polarity parameters are found to correlate with both equilibrium and rates. The first kind, water solubility in the solvent and the sum of normalized Gutmann donor number and normalized electron pair acceptance index reflect (i) the bulk behaviour of the solvent and (ii) interactions based upon specific functional groups of the solvent. The second kind consists of log P and reverse solubility of the solvent in water. Such parameters measure (i) the behaviour of molecules of the solvent dissolved in the water (around the biocatalyst) and the overall balance between polar functional groups and hydrocarbon groups. To sum up, few guidelines have emerged over the years but one should take into account totality of the individual system. Another exciting development in solvent engineering has been that not only may it be possible to replace essential water by polar solvents, moderate replacements may even lead to increase in activity (Mattiasson and Adlercruetz, 1991). More recently (Triantafyllou et al., 1993), it has been found that enhanced transesterification yields and increased selectivity of the lipase were obtained by adding small amounts of various solvents having log P values between 1.08–1.93 to the medium isooctane.

Biocatalyst engineering

Yamane et al. (1990) have examined the effect of adding various additives to lipase before lyophilizing for subsequent use in benzene. Sugar alcohols such as erythitol, arabiotol and sorbitol led to significant increase in activity. Addition of lactose, albumin, casein, dextran, polyvinyl alcohol, phosphate or sodium chloride caused a decrease in the enzyme activity. Yang and Robb (1993) have described the use of salt hydrates for controlling activity of Tyrosinase in chloroform and toluene. Three salts viz. sodium sulphate, sodium phosphate and sodium carbonate, were effective activator whereas 11 other salts proved ineffective. All these results by various workers underlines the importance of working with pure enzymes.

Tanaka and Kawamoto (1991) have reviewed the behaviour of a variety of immobilized enzymes in organic solvents. It is generally agreed that immobilization leads to faster rates (Halling, 1987) as this reduces the mass transfer limitations. Norin (1988) evaluated the reactions rates of the lipase immobilized by adsorption to various hydrophilic and hydrophobic supports. The esterification rates with the secondary alcohol were highest in the case of hydrophobic supports; with primary alcohol heptanol, "free lipase" gave the highest reaction rate! In organic solvents, immobilization by simple adsorption is good enough unless one is using hydrophilic solvents with high water contents. In the latter situation, covalent attachment (Tanaka and Kawamoto, 1991) or entrapment into polymer gels (Yokozeki et al., 1982) may be carried out. Reslow et al. (1988a) found that in the case of both chymotrypsin and alcohol dehydrogenase, hydrophobic supports gave better results than hydrophilic supports. Greater partitioning of water to enzymes as compared to the supports in such cases was the obvious explanation. The considerable experience gained by now indicates that no universal generalizations are possible, and each system needs to be tested individually. For example, Bloomer et al. (1990)

point out that of the four commercial lipases immobilized on five supports, some preparations were more affected by the choice of support and no support gave best results with all lipases. The factors which may be considered while choosing a support are: (a) hydrophobicity of the substrate, (b) nature of the solvent, (c) hydrophobicity of the active site. Apart from giving higher reaction rates, immobilization is expected to increase the operational stability as well. Tanaka and Kawamoto (1991) give several examples where this has indeed been observed.

Protein engineering shows great promise for designing efficient enzymes for functioning in hydrophilic solvents. A few years back, Arnold (1988; 1990) had stated some design principles for this purpose. More recent work by this group (Arnold et al., 1993) and others (Zhong et al., 1991) validates those rules. With subtilisin E, substitutions of Asp 248 → Asn (to diminish the surface charge) and Asn 218 → Ser (improved H-bonding) individually and additively showed improved stability in 40% dimethylformamide (Chen et al., 1991). In subsequent work (Martinez et al., 1992), it has been shown that increase in surface hydrophobicity and improvement of internal H-bonding gave stabilization in 80% dimethylformamide as well. With alpha-lytic protease (Martinez and Arnold, 1991) replacements of Arg 45 → Glu, or Ser or Leu or Ile and that of Arg 78 → Leu, Tyr or Phe increased stability in 84% dimethylformamide. Zhong et al. (1991) have examined the stability of engineered subtilisins in anhydrous dimethylformamide. The best was subtilisin mutant 8397 which had a half-life of 350 h in dimethylformamide. In general, mutagenesis had little effect on catalysis as indicated by K_{cat}/K_m values. Yet another exciting approach for enzyme stabilization had been recently described by Navia et al. (1993). Thermolysin in microcrystalline form was crosslinked by glutaraldehyde. The resulting preparation could be lyophilized and used in organic solvents as "monodispersed, nonaggregated suspensions". The enzyme preparation remained active in harsh environments, "including prolonged exposure to high temperatures, extremes of pH, near anhydrous organic solvents and aqueous-organic solvent mixtures".

Finally, one should mention that enzymes from thermophilic microorganisms show not only inherently greater thermostability but also resistance to organic solvent denaturation and hence can be used with advantage in anhydrous organic solvents (Owusu and Cowan, 1990).

pH Memory

This refers to the phenomenon that exposing the enzyme to optimum pH in aqueous environment before removing water and transferring to nearly anhydrous media enhances the rate of the reaction. It is as if "the enzyme remembers the pH of the latest aqueous solution to which it has been exposed" (Klibanov, 1986). This pH tuning increased the rate of subtilisin catalyzed transesterification by 300 fold (Klibanov, 1986). According to Zaks and Klibanov (1988a), the phenomena of "pH memory" points to the existence of kinetically trapped enzyme structures in organic solvents". A more likely and simple explanation seems to be that pH of the environment from which enzyme is taken before being placed in the water-poor environment of an organic solvent determines the ionization state of the enzyme (Mattiasson and Adlercreutz, 1991). Mention should be made of results with bilirubin oxidase (Skrika-Alexopoulos and Freedman, 1993) which showed that the activity of the enzyme suspension following lyophilization of enzyme solutions buffered to the same pH varied significantly with the amount of buffer and nature of the buffer species. A more striking result was that lyophilization from a volatile buffer gave an enzyme which showed no activity in the organic solvent. Curiously, in a recent publication (Narayan and Klibanov, 1993), the group which originally described "pH memory" has not used any pH tuning.

Another phenomena which is worth mentioning is the observation that "the half-life of subtilisin BPN in dimethyl formamide dramatically depends on the pH of the aqueous solution from which the enzyme was lyophilized, increasing from 48 min to 20 h when the pH is raised from 6.0 to 7.9" (Schulze and Klibanov, 1991). This supplements the earlier observation of the group that "the 'pH memory' of chymotrypsin applies not only to its catalytic activity but also to thermal stability" (Zaks and Klibanov, 1988a).

Protein imprinting

The high rigidity of protein molecules in anhydrous organic solvents makes it possible to imprint them with stereospecific ligands. Stahl et al. (1990) showed that after precipitating chymotrypsin in n-propanol in the presence of N-acetyl-D-tryptophan, chymotrypsin accepted N-Acetyl-D-tryptophan as substrate for the ethyl ester synthesis in cyclohexane. These results indicated that protein/enzymes can also be imprinted similar to polymers. Subsequently, the same group (Stahl et al., 1991) described the effect of water on this induced stereoselectivity of the imprinted alpha-chymotrypsin and found that the D-specific enzyme activity was totally lost at 4 mM water.

Around the same time, Russel and Klibanov (1988), had found that lyophilization of subtilisin from aqueous solution containing the competitive inhibitor gave a preparation which was up to 100 times more active than the enzyme unexposed to the "imprinting" by the inhibitor. Again, the presence of even 0.4% water was found to erase the imprinted "enzyme memory". Braco et al. (1990) extended this approach

Table 3. Some important applications of use of enzymes in organic solvents

Application	Examples	References
Analysis	Cholesterol estimation by flow injection analysis.	Braco et al., 1992
	Determination of p-cresol	Hall et al., 1988
Oligomerization and polymerization	Sucrose-containing polymers	Patil et al., 1991a,b
	Large scale synthesis of oligosaccharides	Wong et al., 1993
Organic synthesis (Wong and Whitesides, 1989)	*Peptide Synthesis* Use of alpha-chymotryspin and thermolysin	Reslow et al., 1988a
	Application of PEG-modified enzymes	Inada et al., 1986
	Comparative study of water-miscible and water immiscible solvents using chymotrypsin	Clapes et al., 1986
	Molecular modelling of stereoselectivity in peptide bond formation by chymotrypsin	Crout et al., 1993
	Isopeptide bond formation	Kitaguchi et al., 1988
	Esterification/Interesterification Intramolecular esterification	Yamane et al., 1990
	Triglyceride interesterification	Bloomer et al., 1990, 1991
	Transesterification reactions Synthesis of monobutyryl glycerol by transeserification with lipase	Otero et al., 1990
	Achieving high equilibrium conversion in lipase catalysed transesterication by use of reduced pressure	Hult and Norin, 1993
	Transesterification reactions of alcohols catalyzed by lipase below zero degree	Hogan et al., 1992
Regioselective/ enantioselective synthesis/ resolution of enantiomers (Review: Klibanov, 1990)	Preparative enatio- and regioselective synthesis of alcohols, glycerol derivatives, sugars and organometallics.	Wang et al., 1988

Table 3. (Continued)

Application	Examples	References
	Complete reversal of enantioselectivity in transesterification of A. oryzae protease upon change in the solvent.	Tawaki and Klibanov, 1992
	Effect of solvent on the enantioselectivity of subtilisin catalyzed transesterification.	Fitzpatrick and Klibanov, 1991
	Control and prediction of enantioselectivity of lipases.	Hult and Norin, 1993
	Control of enzyme enantioselectivity by medium	Sakurai et al., 1988
	Regioselectivity of glycosyl transfer	Crout et al., 1993
	Synthesis of D-mandelonitrile	Wehtje et al., 1990

to "imprint" bovine serum albumin with L-tartaric acid and p-hydroxybenzoic acid. The imprinted albumin was found to bind upto 30-fold more of the ligand. Addition of as little as 2% water decreased the extent of binding by half. Similar imprinting could be carried out on bovine hemoglobin and lysozme. More recently, Dabulis and Klibanov (1992) have examined the imprinting of a wider range of materials. Many H-bonding polymers like dextrans, partially hydrolyzed starch and poly (methacrylic acid) could be successfully imprinted.

Applications

In the last two decades, enzymes have found a large number of applications in analysis, synthesis and monitoring processes. This aspect has been well covered by many others in recent years (Klibanov, 1990; Wong and Whitesides, 1989). Table 3 briefly illustrates some recent applications.

Conclusions

Despite the explosive growth of this area, few basic questions remain unanswered. Is there any three-way correlation among water stripping by solvents, protein rigidification and enzyme activity? What is the absolute minimum water that enzymes need for catalysis? (Is Rhizomucor miehei lipase best designed for functioning in organic solvents as it remains highly active at water activity below 0.0001?) What is the universally applicable medium parameter (if any) which can help in medium engineering?

This is undoubtedly an area where "applications" have preceded "basic knowledge". However further applications and full exploitation will be dependent upon further understanding of some of the issues raised above. Continued use of solid-state NMR and FT-IR will enhance our knowledge of the structural aspects. Some recent work on protein structure is also quite relevant. Steinback and Brooks (1993) have looked at protein hydration by molecular dynamics simulation. Distribution of solvent molecules around apolar side chains in protein cyrstals has been examined by Walshaw and Goddellow (1993). Gerstein (1993) has raised the fundamental issue of defining natural boundary of protein in solution.

Finally, we would like to mention an early work which is not as widely known in this area as it should be, not only because it provided some of the information which was subsequently mentioned in more well known and often cited publications, but also because it discusses the in vivo significance of looking at enzymatic reactions under low water conditions. Stevens and Stevens (1979) mentioned "For glucose-6-phosphate dehydrogenase, hexokinase and fumarase, enzyme activity became detectable (about 0.05% of the fully hydrated rate) when the water content was about 0.2 g/g of reaction mixture... It is concluded that to detect enzyme activity a certain minimal amount of water is required and that above this minimum the rate is still restricted by diffusion limitation... Air-dry seeds and fungal spores have a very low rate of metabolism. This is undoubtedly in part due to low water content". Stevens and Stevens (1979), of course, were looking at enzymes in dry states and were not working with organic solvents. Their work was carried out before the renaissance of enzyme action in organic solvents started.

Acknowledgements
I would like to thank Mr. G.K. Pandita in diligently typing this manuscript, and other members of my research group, Ms. Renu Batra and Dr. Renu Tyagi, for assisting in the preparation of this review. Valuable discussions with Prof. Bo Mattiasson, Chemical Center, Lund, Sweden, were of great help in the writing of this review. I also acknowledge the support of the Department of Biotechnology (Govt. of India) and the Swedish Agency for Research Cooperation with Developing Countries.

References

Adlercruetz, P. and Mattiasson, B. (1987) Aspects of biocatalyst stability in organic solvents. *Biocatalysis* 1: 99–108.
Affleck, R., Xu, Z.F., Suzawa, V., Focht, K., Clark, D.S. and Dordick, J.S. (1992) Enzymatic catalysis and dynamics in low water environments. *Proc. Natl. Acad. Sci. USA* 89: 1100–1104.

Ampon, K., Basri, M., Razak, C.N.A., Yunus, W.M.Z. and Salleh, A.B. (1992) The effect of attachment of hydrophobic modifiers on the catalytic activities of lipase and trypsin. In: J. Tramper, M.H. Vermue, H.H. Beeftink and V. von Stockar (eds): *Biocatalysis in Non-conventional Media*, Elsevier, Amsterdam, p 331.
Arnold, F.H. (1988) Protein design for non-aqueous solvents. *Protein Eng.* 2: 21–25.
Arnold, F.H. (1990) Enzyme engineering for non-aqueous solvents. *Trends Biotechnol.* 8: 244–249.
Arnold, F.H. Chen, K., Economou, C., Chen, W., Martinez, P., Yoon, K.P. and Dam, M.V. (1993) Engineering nonaqueous solvent-compatible enzymes. In: M.E. Himmel and G. Georgiou (eds): *Biocatalyst Design for Stability and Specificity*, Am. Chem. Soc., Washington, D.C. pp 109–113.
Bloomer, S., Adlercruetz, P. and Mattiasson, B. (1990) Triglyceride intersterification by lipases 1. Cocoa butter equivalents from a fraction of palm oil. *J. Am. Oil. Chem. Soc.* 67: 519–524.
Bloomer, S., Adlercruetz, P. and Mattiasson, B. (1991) Triglyceride interesterification by lipases 2. Reaction parameters for the reduction of trisaturated impurities and diglycerides in batch reactions. *Biocatalysis* 5: 145–162.
Braco, L., Dabulis, K. and Klibanov, A.M. (1990) Production of abiotic receptors by molecular imprinting of proteins. *Proc. Natl. Acad. Sci. USA* 87: 274–277.
Braco, L., Daros, J.A. and de la Guardia, M. (1992) Enzymatic flow injection analysis in non aqueous media. *Anal. Chem.* 64: 129–133.
Burke, P.A., Smith, S.O., Bachovchin, W.W. and Klibanov, A.M. (1989) Demonstration of structural integrity of an enzyme in organic solvents by solid state NMR. *J. Am. Chem. Soc.* 111: 8290–8291.
Burke, P.A., Griffin, R.G. and Klibanov, A.M. (1992) Solid state NMR assessment of enzyme active center structure under non-aqueous conditions. *J. Biol. Chem.* 267: 20057–20064.
Burke, P.A., Griffin, R.G. and Klibanov, A.M. (1993) Solid state nuclear magnetic resonance investigation of solvent dependence of tyrosyl ring motion in an enzyme. *Biotechnol. Bioeng.* 42: 87–94.
Burton, S.G., Duncan, J.R., Kaye, P.T. and Rose, P.D. (1993) Activity of mushroom polyphenol oxidase in organic media. *Biotechnol. Bioeng.* 42: 938–944.
Cambou, B. and Klibanov, A.M. (1984) Preparative production of optically active esters and alcohols using esterase-catalyzed stereospecific transesterification in organic media. *J. Am. Chem. Soc.* 106: 2687–2692.
Chen, K., Robinson, A.C., vanDam, M.E., Martinez, P. Economou, C. and Arnold, F.H. (1991) Enzyme engineering for nonaqueous solvents II. Additive effects of mutations on the stability and activity of subtilisin E in polar organic media. *Biotechnol. Progress* 7: 125–129.
Cabaret, D., Boucier, S., Maillot, S. and Wakselman, M. (1992) Chymotrypsins modified by neutral or charged amphiphiles as catalyst for ester synthesis in polar organic solvents. *Biocatalysis* 6: 191–199.
Clapes, P., Adlercruetz, P. and Mattiasson, B. (1990) Enzymatic peptide synthesis in organic media: a comparative study of water-miscible and water immiscible solvent systems. *J. Biotechnol.* 15: 323–338.
Crout, D.H.G., MacManus, D.A., Ricca, J.M., Singh, S., Critchlay, P. and Gibson, W.T. (1993) Biotransformation in the peptide and carbohydrate fields. *Ind. J. Chem.* 32B: 195–201.
Dabulis, K. and Klibanov, A.M. (1992) Molecular imprinting of proteins and other macromolecules resulting in new adsorbents. *Biotechnol. Bioeng.* 39: 176–185.
Dastoli, F.R. and Musto, N.A. and Price, S. (1966) Reactivity of active sites of chymotrypsin suspended in an organic medium. *Arch. Biochem. Biophys.* 115: 44–47.
Dastoli, F.R. and Price, S. (1967a) Catalysis by xanthine oxidase suspended in organic media. *Arch. Biochem. Biophys.* 118: 163–165.
Dastoli, F.R. and Price, S. (1967b) Further studies on xanthine oxidase in non polar media. *Arch. Biochem. Biophys.* 122: 289–299.
Degn, H., Bohatka, S. and Loyd, D. (1992) Enzyme activity in organic solvent as a function of water activity determined by membrane mass spectrometry. *Biotechnol. Tech.* 6: 161–164.
Dordick, J.S. (1992) Designing enzymes for use in organic solvents. *Biotechnol. Prog.* 8; 259–267.

Estrada, P., Baroto, W., Sanchez-Muniz, R., Acebal, C., Castillon, M.P. and Arche, R. (1992) Thermoinactivation of polyphenol oxidase in organic solvents with low water content. *In*: J. Tramper, M.H. Vermue, H.H. Beeftink and V. von. Stockar (eds): *Biocatálysis in Non-conventional Media*, Elsevier, Amsterdam, pp 483–489.

Fitzpatrick, P.A. and Klibanov, A.M. (1991) How can the solvent affect enzyme enantioselectivity? *J. Am. Chem. Soc.* 113: 3166–3171.

Garza-Ramos, G., Darszon, A., Gomez-Puyou, M.T.D. and Gomez-Puyou, A. (1990) Enzyme catalysis in organic solvents with low water content at high temperatures. The adenotriphosphatase of submitochondrial particles. *Biochemistry* 29: 751–757.

Gerstein, M. (1993) What is natural boundary of protein in solution? *J. Mol. Biol.* 230: 641–650.

Gorman, L.A.S. and Dordick, J.S. (1992) Organic solvents strip water off enzyme function in organic solvents. *Eur. J. Biochem.* 203: 25–32.

Guisan, J.M., Bastida, A., Blanco, R.M., Cuesta, C., Rodriguez, V. and Fernandez-Lafuente, R. (1992) Insolubilized enzyme derivatives in organic solvents: mechanisms of inactivation and strategies for reactivation. *In*: J. Tramper, M.H. Vermue, H.H. Beeftink and V. von Stockar (eds): *Biocatalysis in Non-conventional Media*, Elsevier, Amsterdam, pp 221–228.

Guisan, J.M., Fernandez, Lafuente, R., Rodriguez, V., Bastida, A., Blanco, R.M. and Alvaro, G. (1993) Enzyme stabilization by multipoint covalent attachment to activated pre-existing supports. *In*: W.J.J. van den Tweel, A. Harder and R.M. Buitelaar (eds): *Stability and Stabilization of Enzymes*. Elsevier, Amsterdam, pp 55–62.

Gupta, M.N. (1991) Thermostabilization of proteins. *Biotechnol. Appl. Biochem.* 14: 1–11.

Gupta, M.N. (1992) Enzyme function in organic solvents. *Eur. J. Biochem.* 203: 25–32.

Gupta, S. and Gupta, M.N. (1993) Mechanisms of irreversible thermoinactivation and medium engineering. *In*: M.N. Gupta (ed.): *Thermostability of Enzymes*, Springer-Verlag, Heidelberg, pp 114–122.

Hall, G.F., Best, D.J. and Turner, A.P.F. (1988) The determination of p-cresol in chloroform with an enzyme electrode used in the organic phase. *Anal. Chim. Acta.* 213: 113–119.

Halling, P.J. (1987) Rates of enzymatic reactions in predominantly organic, low water systems. *Biocatalysis* 1: 109–115.

Hogan, V.F., O'Hagan, D. and Sanoisin, J. (1992) Rate enhancement of the candida cylindracea lipase catlayzed transferifications in organic solvents: Enzymatic reaction below zero. *Ind. J. Chem.* 31B: 883–885.

Hult, K. and Norin, T. (1993) Enantioselectively of some lipases, control and prediction. *Ind. J. Chem.* 32B: 123–126.

Inada, Y., Takahashi, K., Yoshimoto, T., Ajima, A., Matsushima, A. and Saito, Y. (1986) Application of polyethylene glycol modified enzymes in biotechnological processes: Organic solvent soluble enzymes. *Trends Biotechnol.* 4: 190–194.

Kanerva, L.T. and Klibanov, A.M. (1989) Hammett analysis of enzyme action in organic solvents. *J. Am. Chem. Soc.* 111: 6864–6865.

Karplus, M. and McCammon, J.A. (1983) Dynamics of proteins: elements and function. *Ann. Rev. Biochem.* 53: 263–300.

Khmelnitsky, Y.L., Levashov, A.V., Klyachko, N.L. and Martinek, K. (1988) Engineering biocatalytic systems in organic media with low water content. *Enzyme Microb. Technol.* 10: 710–724.

Kitaguchi, H., Tai, D.F. and Klibanov, A.M. (1988) Enzymatic formation of an isopetide bond involving the epsilon-amino group lysine. *Tetrahedron Lett.* 29: 5487–5488.

Klibanov, A.M. (1986) Enzymes that work in organic solvents. *CHEMTECH* 16: 354–359.

Klibanov, A.M. (1987) Enzymatic conversions in organic solvents. *Pharm. Technol.* March issue: 32–35.

Klibanov, A.M. (1990) Asymmetric transformations catalyzed by enzymes in organic solvents. *Acc. Chem. Res.* 23: 114–120.

Laane, C. (1987) Medium engineering for bioorganic synthesis. *Biocatalysis* 1: 17–32.

Laane, C., Boeren, S., Vos, K. and Veeger, C. (1987) Rules for optimization of biocatalysis in organic solvents. *Biotechnol. Bioeng.* 30: 81–87.

Martinez, P. and Arnold, F.H. (1991) Surface charges substitutions increase the stability of alpha protease in organic solvents. *J. Am. Chem. Soc.* 113: 6336–37.

Martinez, P., Dam, M.E.V., Robinson, A.C., Chen, K. and Arnold, F.H. (1992) Stabilization of subtilisin E in organic solvents by site directed mutagenesis. *Biotechnol. Bioeng.* 39: 141–147.

Mattiasson, B. and Adlercreutz, P. (1991) Tailoring the microenvironment of enzymes in water-poor systems. *Trends Biotechnol.* 9: 394–398.

Matsushima, A., Okada, M. and Inada, Y. (1984) Chymotrypsin modified with polyethylene glycol catalyzes peptide synthesis reaction in benzene. *FEBS Lett.* 178: 275–277.

Narayan, V.S. and Klibanov, A.M. (1993) Are water-immiscibility and apolarity of the solvent relevant to enzyme efficiency? *Biotechnol. Bioeng.* 41: 390–393.

Navia, M.A., St. Clair, N.L. and Griffith, J.P. (1993) Crosslinked enzyme crystals (CLEC-STM) as immobilized enzyme particles. *In*: W.J.J. van-den Tweel, A. Harder and R.M. Buitelaar (eds): *Stability and Stabilization of Enzymes*. Elsevier, Amsterdam, pp 63–73.

Ncube, I. Adlercreutz, P., Read, J. and Mattiasson, B. (1993) Purification of rape (Brassica napus) seedling lipase and its use in organic media. *Biotechnol. Appl. Biochem.* 17: 327–336.

Norin, M., Boutelje, J., Holmberg, E. and Hult, K. (1988) Lipase immobilized by adsorption. Effect of support hydrophobicity on the reaction rate of ester synthesis in cyclohexane. *Appl. Microbiol. Biotechnol.* 28: 527–530.

Otamiri, M., Adlercruetz, P. and Mattiasson, B. (1992) Complex formation between chymotrypsin and ethyl cellulose as a means to solubilize the enzyme in active form in toluene. *Biocatlaysis* 4: 291–305.

Otero, C., Pastor, E. and Ballesteros, A. (1990) Synthesis of monobutyryl glycerol by transesterification with soluble and immobilized lipase. *Appl. Biochem. Biotechnol.* 26: 35–44.

Owusu, R.K. and Cowan, D.A. (1990) Biocatalysis in organic solvent systems using thermostable enzymes: esterase catalyzed transesterification of Z-L-tyrosine-p-nitrophenyl ester. *Enzyme Microb. Technol.* 12: 374.

Patil, D.R., Dordick, J.S. and Rethwisch, D.G. (1991a) Enzymatic synthesis of a sucrose containing linear polyester in organic media. *Biotechnol. Bioeng.* 37: 639–646.

Patil, D.R., Rethwisch, D.G. and Dordick, J.S. (1991b) Chemoenzymatic synthesis of novel sucrose containing polymers. *Macromolecules* 24: 3462–3463.

Poole, P.L. and Finney, J.L. (1983) Hydration induced conformational flexibility changes in lysozme at low water content. *Int. J. Biol. Macromol.* 5: 308–310.

Reslow, M., Adlercreutz, P. and Mattiasson, B. (1987) Organic solvents for bioorganic synthesis. *Appli. Microb. Biotechnol.* 26: 1–8.

Reslow, M., Adlercreutz, P. and Mattiasson, B. (1988a) On the importance of the support material for bioorganic synthesis. *Eur. J. Biochem.* 172: 573–578.

Reslow, M., Adlercreutz, P. and Mattiasson, B. (1988b) The influence of water on protease-catalyzed peptide synthesis in acetonitrile/water mixtures. *Eur. J. Biochem.* 177: 313–318.

Rupley, J.A., Gratton, E. and Careri, G. (1983) Water and globular proteins. *Trends Biochem. Sci.* 8: 18–22.

Russel, A.J. and Klibanov, A.M. (1988) Inhibitor-induced enzyme activation in organic solvents. *J. Biol. Chem.* 263: 11624–11626.

Ryu, K. and Dordick, J.S. (1992) How do organic solvents affect peroxidase structure and function? *Biochemistry* 31: 2588–2598.

Schneider, L.V. (1991) A three dimensional solubility parameter approach to non-aqueous enzymology. *Biotechnol. Bioeng.* 37; 627–638.

Sakurai, T., Margolin, A.L., Russel, A.J. and Klibanov, A.M. (1988) Control of enzyme enantioselectivity by the reaction medium. *J. Am. Chem. Soc.* 110: 7236–7237.

Schulze, B. and Klibanov, A.M. (1991) Inactivation and stabilization of subtilisins in neat organic solvents. *Biotechnol. Bioeng.* 38: 1001–1006.

Skrika-Alexopoulos, E. and Freedman, R.B. (1993) Factors affecting enzyme characteristics of bilirubin oxidase suspensions in organic solvents. *Biotechnol. Bioeng.* 41: 887–893.

Stahl, M., Mansson, M.O. and Mosbach, K. (1990) The synthesis of a D-amino acid ester in an organic media with alpha-chymotrypsin modified by a bio-imprinting procedure. *Biotechnol. Letts.* 12: 161–166.

Stahl, M., Jeppsson-Wistrand, V., Mansson, M.-O. and Mosbach, K. (1991) Induced stereoselectivity and substrate selectivity of bio-imprinted alpha-chymotrypsin in anhydrous organic media. *J. Am. Chem. Soc.* 113: 9366–9368.

Steinbach, P.J. and Brooks, B.R. (1993) Protein hydration clucidated by molecular dynamics simulation. *Proc. Natl. Acad. Sci.* USA 90: 9135–9139.

Stevens, E. and Stevens, L. (1979) The effect of restricted hydration on the rate of reaction of glucose-6-phosphate dehydrogenase, phosphoglucose isomerase, hexokinase and fumarase. *Biochem. J.* 179: 161–167.

Takahashi, K., Ajima, A., Yoshimoto, T. and Inada, Y. (1984a) Polyethylene glycol modified catalase oxhibits unexpectedly high activity in benzene. *Biochem. Biophys. Res. Commum.* 125: 761–766.

Takahashi, K., Nishimura, H., Yoshimoto, T. Saito, Y. and Inada, Y. (1984b) A chemical modification to make horseradish peroxidase soluble and active in benzene. *Biochem. Biophys. Res. Commun.* 121: 261–265.

Takahashi, K., Nishimura, H., Yoshimoto, T., Okada, M., Ajima, A., Matsushima, A., Tamaura, Y., Saito, Y. and Inada, Y. (1984c) Polyethylene glycol modified enzymes trap water on their surface and exert enzymatic activity in organic solvents. *Biotechnol. Lett.* 6: 765–770.

Takahashi, K., Kodera, Y., Yoshimoto, T., Ajima, A., Mattiasson, A. and Inada, Y. (1985) Ester exchange catalyzed by lipase modified with polyethylene glycol. *Biochem. Biophys. Res. Commun.* 131: 532–536.

Tanaka, A. and Kawamoto, T. (1991) Immobilized enzymes in organic solvents. *In*: R.F. Taylor (ed.): *Protein Immobilization*, Marcel Dekker, Inc., New York, pp 183–208.

Tawaki, S. and Klibanov, A.M. (1992) Inversion of enzyme enantioselectivity mediated by the solvent. *J. Am. Chem. Soc.* 114: 1882–1884.

Triantafyllou, A.O., Aldercreutz, P. and Mattiasson, B. (1993) Influence of the reaction medium on enzyme activity in bioorganic synthesis: behaviour of lipase from Candida rugosa in the presence of polar additives – effect of organic solvent log P value on transesterification activity. *Biotechnol. Appl. Biochem.* 17: 167–179.

Valivety, R.H., Johnston, G.A., Suekling, C.J. and Halling, P.J. (1991) Solvent effects on biocatalysis in organic systems: Equilibrium position and rates of lipase catalyzed esterification. *Biotechnol. Bioeng.* 38: 1137–1143.

Valivety, R.H., Halling, P.J. and Macrae, A.R. (1992) Rhizomucer miehi lipase remains highly active at water activity below 0.0001. *FEBS Lett.* 3: 258–260.

Volkin, D.B., Staubli, A., Langer, R. and Klibanov, A.M. (1991) Enzyme thermoinactivation in anhydrous organic solvents. *Biotechnol. Bioeng.* 37: 843–853.

Walshaw, J. and Gooddellow, J.M. (1993) Distribution of solvent molecules around apolar side chains in protein crystals. *J. Mol. Biol.* 231: 392.

Wang, Y.F., Lalonde, J.J., Momongan, M., Bergreiter, D.E. and Wong, C.H. (1988) Lipase-catalyzed inversible tranesterifications using enol esters as acylating reagents: Preparative enantio- and regioselective syntheses of alcohols, glycerol derivatives, sugars and organometallics. *J. Am. Chem. Soc.* 110: 7200–7205.

Wehtje, E., Adlercreutz, P. and Mattiasson, B. (1990) Formation of C–C bonds by mandelonitrile lyase in organic solvents. *Biotechnol. Bioeng.* 36: 39–46.

Wong, C.H. and Whitesides, G.M. (1989) *Enzymes in Synthetic Organic Chemistry*. Pergamon Press, New York.

Wong, C.H., Liu, K.K.-C, Kajimoto, T., Chen, L., Zhong, Z., Dumas, D.P., Liu, C., Ichikawa, Y. and Shen, G.-J. (1993) Enzymes for carbohydrate and peptide synthesis. *Ind. J. Chem.* 32B: 135–139.

Yamamoto, Y. and Kise, H. (1993) Catalysis of enzyme aggregates in organic solvents. An attempt at evaluation of intrinsic activity of proteases in ethanol. *Biotechnol. Letts.* 15: 647–652.

Yamane, T., Ichiryu, T., Nagata, M., Veno, A. and Shimizu, S. (1990) Intramolecular esterification by lipase powder in microaqueous benzene: factors affecting activity of pure enzyme. *Biotechnol. Bioeng.* 36: 1063–1069.

Yang, Z., Robb, D.A. and Halling, P.J. (1992) Variation of tyrosinase activity with solvents at a constant water activity. *In*: J. Tramper, M.H. Vermüe and H.H. Beeftink (eds): *Biocatalysis in Non-conventional Media*, Elsevier, Amsterdam, pp 585–592.

Yang, Z. and Robb, D.A. (1993) Use of salt hydrates for controlling activity of tyrosinase in organic solvents. *Biotechnol. Tech.* 7: 37–42.

Yokezeki, K., Yamanaka, S., Takinomi, K., Tanaka, A., Sonomoto, K. and Fukui, S. (1982) Application of immobilized lipase to regio-specific ilnteresterification of triglyceride in organic solvent. *Eur. J. Appl. Microb. Biotechnol.* 14: 1–4.

Zaks, A. and Klibanov, A.M. (1984) Enzymatic catalysis in organic media at 100°C. *Science* 224: 1249–1251.

Zaks, A. and Klibanov, A.M. (1985) Enzyme catalyzed processes in organic solvents. *Proc. Natl. Acad. Sci. USA* 82: 3192–3196.

Zaks, A. and Klibanov, A.M. (1988a) Enzymatic catalysis in nonaqueous solvents. *J. Biol. Chem.* 363: 3194–3201.

Zaks, A. and Klibanov, A.M. (1988b) The effect of water on enzyme action in organic media. *J. Biol. Chem.* 263: 8017–8021.

Zhong, Z., Liu, J.L.C., Dinterman, L.M., Finkelman, M.A.J., Muller, W.T. and Wong, C.H. (1991) Engineering subtilisin for reaction in dimethylformamide. *J. Am. Chem. Soc.* 113: 683–684.

Protein sorting signals: Simple peptides with complex functions

G. von Heijne

Department of Biochemistry, Arrhenius Laboratories for Natural Sciences, Stockholm University, S-106 91 Stockholm, Sweden

Summary. Protein sorting signals provide good examples of peptides that can be studied both from a chemical and a biochemical perspective. Their simple designs and low degree of sequence conservation suggest that they are involved in rather non-specific peptide-lipid interactions, yet their ability to discriminate efficiently between the import machineries of different subcellular compartments rather points to the importance of peptide-receptor interactions. The study of protein sorting signals thus invites a cross-disciplinary approach.

Introduction

A priori, intracellular protein sorting processes would seem to require sophisticated sorting signals and complicated biochemical systems for recognition and membrane translocation. Certainly, this would not seem to be a promising area to explore interfaces between chemistry and biochemistry; rather, one would suspect that such a complex biological problem would still mainly be studied by cell biologists and biochemists and that very little would be known about the molecular details.

This, however, is only partly true. To be sure, the biochemistry of protein secretion, mitochondrial protein import, or protein uptake into the nucleus is formidable and only partly deciphered. But – and this is why the present chapter may not be completely out of character in this volume – the sorting signals turn out to be remarkably simple peptides and their action almost begs to be explained in rather simple chemical terms.

What will follow, then, is a discussion of the chemical characteristics of some of the more important and well-characterized classes of sorting signals against a sketchy background of the biochemical transport systems they provide access to. In particular, signals for protein secretion, for import into mitochondria and chloroplasts, and for membrane protein assembly will be treated from this perspective. And the bottom line, if there is one, will be as follows: chemically simple peptides can sometimes perform very important biological functions.

Protein sorting signals: Hydrophobicity, amphiphilicity and charge

In a gram negative bacterium such as *E. coli*, protein sorting is a fairly uncomplicated business: should a given polypeptide remain in the cytoplasm, insert into the inner membrane, be translocated to the periplasm, assemble into the outer membrane, or be secreted to the medium? The problems faced by the typical eukaryotic cell are much more complicated: already in the secretory pathway, there are a number of compartments where specific proteins are retained (the endoplasmic reticulum, the different Golgi stacks, the trans-Golgi network, the plasma membrane), and other organelles not only need to import their complement of proteins but further need to sort these proteins between a range of sub-organellar compartments (the outer and inner membranes, the intermembrane space, and the matrix in mitochondria; the outer and inner envelope membranes, the inter-envelope space, the stroma, the thylakoid membrane, and the thylakoid lumen in chloroplasts, to name but a few).

In all cases, proper sorting depends on signals in the polypeptide, often short, clearly delimited stretches of chain that may or may not be removed once they have served their targeting function. The different signals are thought to be recognized by receptor proteins, but may in addition interact with lipids in the target membrane; the role of lipids in protein sorting processes is a hotly debated issue.

Protein secretion

In *E. coli*, translocation of proteins through the inner membrane is catalyzed by the so-called *sec*-machinery (Schatz and Beckwith, 1990; Wickner et al., 1991). A nascent secretory protein carrying an N-terminal signal peptide first encounters a cytoplasmic chaperone such as the Ffh, SecB, or GroEL/ES proteins, is then delivered to a translocase complex in the membrane including the SecA, SecY, SecE, SecD, and SecF proteins, and is finally transported to the periplasmic side of the membrane in a process that requires ATP hydrolysis and the presence of a transmembrane electrochemical protential. Once in the periplasm, the protein folds; whether some general chaperones are involved in this process is not know, except that the DsbA-DsbB system is needed for efficient disulfide formation (Bardwell and Beckwith, 1993). At some point during the translocation process, the signal peptide is removed from the N-terminus of the chain by the leader peptidase enzyme.

The primary membrane translocation event in eukaryotic cells takes place in the endoplasmic reticulum (ER), where a machinery similar to the *sec*-machinery is present. The initial recognition of the nascent secretory protein is carried out by the signal recognition particle (SRP), one subunit of which has sequence similarity to the *E. coli* Ffh protein. The translocation complex in the ER membrane is made up from a number of subunits, including the SRP receptor, the α- and β-subunits of the SEC61 protein (with sequence similarity to the SecY and SecE

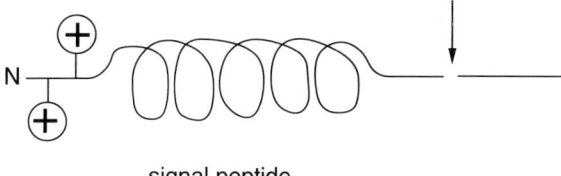

Figure 1. A collection of protein sorting signals.

proteins, respectively), the TRAM protein, the signal peptidase, and the oligosaccharyl transferase enzyme (Gilmore, 1993; Görlich and Rapoport, 1993; Hartmann et al., 1994). Translocation requires GTP hydrolysis (Miller et al., 1993).

Translocation in both prokaryotic and eukaryotic cells is triggered by the presence of an N-terminal signal peptide on the nascent chain. One may have guessed that such a complex series of events would require a correspondingly complex and highly evolved design of the signal peptide. Nature, however, has chosen to do precisely the opposite: signal peptides have no primary sequence homology but are only conserved in terms of their overall tri-partite design (von Heijne, 1990), Figure 1.

If biochemistry is characterized by precise structure-function relationships, signal peptides would hardly qualify as biochemical entities. Their design would rather suggest that they might interact in more non-spcific ways with, e.g., water-lipid interfaces, and, as shown by many biophysical studies, they in fact do (Gierasch, 1989). At the same time, however, they have also been found to interact specifically with Ffh and SRP (High and Dobberstein, 1991; Luirink et al., 1992; Lutcke et al., 1992;

Luirink and Dobberstein, 1994), and they can also be recognized and cleaved by signal peptidases (Dalbey and von Heijne, 1992). They are thus interfacial peptides, both in the literal sense and in the sense of this book.

The three salient features of a singal peptide are the positively charged N-terminus, the central hydrophobic region, and the C-terminal, slightly polar cleavage region. The C-terminal region is not necessary for the targeting function but serves only as a recognition site for the signal peptidase, and is the most higly conserved in terms of primary amino acid sequence (von Heijne, 1990). For the N-terminal and central regions, the wide latitude allowed in primary sequence is perhaps most dramatically demonstrated by studies where a signal peptide was replaced by essentially random sequences; amazingly, some 20% of all the sequences tested proved to promote a measurable degree of protein export though none was cleaved (Kaiser et al., 1987; Kaiser and Botstein, 1990).

In buffer, signal peptides have a random conformation, while in more non-polar solvents, in micelles, and in lipid vesicles the central hydrophobic region tends to form an α-helix (Rizo et al., 1993; Wang et al., 1993). The helix content in non-polar environments and the strength of interaction with lipid monolayers both correlate with the *in vivo* activity (Bruch and Gierasch, 1990), supporting the "lipidology" view of signal peptide function. On the other hand, similar correlations between signal peptide characteristics and the strength of interaction with Ffh and SRP have also been found, strengthening the "proteinology" camp. It is thus far from clear precisely what interactions signal peptides take part in, and a Solomonic compromise where both proteins and lipids play a decisive role may well be reached in the end.

Mitochondrial targeting peptides

Nuclearly encoded proteins destined for the mitochondrial matrix have to pass through two membranes on their way into the organelle, yet they only have a single targeting peptide. The current model suggests that there are distinct import machineries in the outer and inner mitochondrial membranes, but that these machineries form a transient supercomplex that guides the nascent polypeptide directly through both membranes (Glick and Schatz, 1991; Pfanner et al., 1992). Immediately after synthesis, the nascent protein is though to bind to cytoplasmic chaperones of the Hsp70 family that prevent aggregation and premature folding. The targeting peptide is then recognized by a receptor in the outer membrane (and may possibly interact also with outer membrane lipids (Endo et al., 1989)), the nascent protein is transferred to the so-called general insertion protein (GIP) that allows translocation through the outer membrane, and is then imported through both membranes in a process that requires ATP and an electrochemical potential across the inner membrane. The membrane potential drives the translocation of the targeting peptide, and translocation of the rest of the chain is thought to involve ATP-driven release of Hsp70 molecules from the part still remaining in the cytoplasm and binding of mitochondrial Hsp70 molecules to the incoming parts (Neupert et al., 1990).

The basic conserved design in the family of mitochondrial targeting peptides again evokes thoughts of peptide-lipid interactions, since these peptides have little primary sequence homology but are remarkably similar in their overall properties to classical "membrane-seeking" amphiphilic peptides such as melittin. In fact, isolated mitochondrial targeting peptides form positively charged amphiphilic helices and interact strongly with negatively charged lipid surfaces (Roise et al., 1986; Roise and Schatz, 1988), and the ability to form amphiphilic helices seem to be correlated with *in vivo* activity (Allison and Schatz, 1986; Lemire et al., 1989). Whether this means that peptide-lipid interactions in fact take place *in vivo* is of course far from certain.

Targeting peptides are in most cases cleaved from the mature chain by matrix-localized proteases. The MAS1-MAS2 protease usually cleaves after Arg-X or Arg-X-Tyr sequences (Hendrick et al., 1989; Gavel and von Heijne, 1990), and appears to recognize some part of the amphiphilic helix (Yang et al., 1991). A second protease – the mitchondrial intermediate peptidase – removes a further 8–9 amino acids from the N-terminus of some once-cleaved targeting peptides (Gavel and von Heijne, 1990; Isaya et al., 1992a; Isaya et al., 1992b; Kalousek et al., 1992); the biological significance of such two-step cleavage is not known.

The sorting of inter-membrane space localized proteins is currently the focus of an intense debate: do such proteins follow a "conservative sorting" pathway (Hartl and Neupert, 1990; Schwarz et al., 1993) where they are first completely imported into the matrix and then re-exported through the inner mitochondrial membrane, or do they use a "stop-transfer" mechanism (van Loon et al., 1987; Glick et al., 1992; Beasley et al., 1993; Glick et al., 1993) where translocation is interrrupted by a stop-transfer sequence that prevents the C-terminal part of the chain from ever entering the matrix? The design of the sorting signals responsible for inter-membrane space targeting is compatible with both models: an N-terminal matrix-targeting amphiphilic segment followed by a stretch of mostly hydrophobic residues. According to the conservative sorting idea, the hydrophobic segment would be analogous to a signal peptide and would trigger re-export across the inner membrane, whereas this same segment is postulated to serve a stop-transfer function in the competing model. Experimental data in favor of both models have been published, and this controversy serves as a nice illustration of how difficult it is to finally disprove a hypothesis: almost every result can be rationalized within both models if additional assumptions about the underlying mechanisms are allowed.

Chloroplast transit peptides

Our final example of a sorting signal concerns protein import into chloroplasts. Here, not much is known about the import machinery, although the assumption is that it is in some way analogous to the mitochondrial one (Theg and Scott, 1993). ATP is required for import across the two envelope membranes, and ATP and/or ΔpH gradient across the thylakoid membrane are necessary for import into the thylakoid lumen (Robinson et al., 1993).

There seems to exist two distinct classes of transit peptides: those of *Chlamydomonas* and similar green algae, and those of higher plants. The *Chlamydomonas* transit peptides have a design very similar to the mitochondrial targeting peptides: they are positively charged and have a strong potential to form amphiphilic helices (Franzén et al., 1990). In fact, they have been shown to be able to promote import into yeast mitochondria, at least *in vitro* (Hurt et al., 1986). Amazingly, the few known *Chlamydomonas* mitochondrial targeting peptides look very similar (Franzén and Falk, 1992), raising an interesting problem about the basis for sorting specificity in this organism.

Higher plant transit peptides have a very different appearance: they are often quite long, contain few acidic but very many hydroxylated residues, and have no clear potential for forming any regular secondary structure (von Heijne et al., 1989; von Heijne and Nishikawa, 1991). Nevertheless, in the presence of lipid monolayers of suitable composition, they become partly helical (Pilon et al., 1992; Horniak et al., 1993) and it has been suggested that they might interact specifically with galactolipids that are found only in the envelope membranes of chloroplasts (van't Hof et al., 1993). As with the previously discussed sorting signals, transit peptides are thus "chemically simple" in the sense that they have very little primary sequence conservation, yet perform a very precise targeting function.

Topological signals in integral membrane proteins

Cellular membranes not only need to be able to translocate proteins from one side to the other, they must also be able to integrate certain proteins into the lipid matrix. Such integral membrane proteins have parts protruding from both sides of the membrane, and the membrane spanning segments are either long hydrophobic helices that form bundles or amphiphilic β-strands that together form a closed β-barrel (von Heijne, 1994).

In the case of the helix bundle proteins, two distinct modes of insertion are possible: a "sequential" mode, where the protein is inserted in an N-to-C-terminal direction (Blobel, 1980) and a "local" mode where pairs of neighbouring transmembrane helices insert together as a "helical hairpin" (Engleman and Steitz, 1981; Gafvelin and von Heijne, 1994). In the sequential mode, the most N-terminal hydrophobic segment serves as a signal peptide, initiating the translocation of downstream regions. The next hydrophobic segment then interrupts the translocation process, leaving the remaining parts of the chain in the cytoplasm. Any number of signal peptides and stop-transfer sequences can be strung together in this way, resulting in a multi-spanning topology (Wessels and Spiess, 1988; Lipp et al., 1989).

In the local mode, two neighbouring hydrophobic segments and the connecting loop insert into the membrane as a unit, resulting in translocation of the loop across the bilayers. This mode is believed to operate when the connecting loop is short (< 60 residues, Andersson and von Heijne, 1993), and is evidenced by the fact that such short, translocated loops contain very few positively charged residues (von Heijne, 1986). The information required to determine the topology of a multi-spanning helix bundle protein with short connecting loops is thus encoded within a very simple pattern of hydrophobic segments connected by polar loops of alternating high and low content of positively charged residues, Figure 2. Again, this makes it tempting to envision the insertion process in terms of "chemical" protein-lipid interactions rather than "biochemical", enzyme-catalyzed mechanisms, although this is still a completely open question.

Conclusions

As shown above, many classes of protein sorting signals can be characterized as "simple peptides with complex functions". A common theme

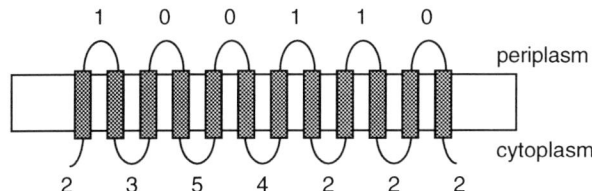

Figure 2. Membrane topology of the LacY protein from *E. coli*. The distribution of positively charged Arg + Lys residues in the connecting loops is shown.

in the examples given here is that of possible peptide-lipid interactions; in contrast, in cases where sorting depends on a much more highly conserved primary amino acid sequence (e.g., endocytosis, lysosomal targeting, peroxisomal import), it is clear that the peptide-receptor interactions provide the basic recognition function. Possibly, the deceptive simplicity of signal peptides, mitochondrial targeting peptides, and chloroplast transit peptides has an evolutionary explanation: these peptides mediate sorting functions that must have been present very early on during the evolution of cells. Processes based on simple peptide-lipid interactions must have been evolutionary more "accessible" than a highly refined receptor-dependent mechanism, and what we see today may be more efficient translocation systems built around these primitive modes of molecular recogniton: an evolving chemistry/biochemistry interface.

References

Allison, D.S. and Schatz, G. (1986) Artificial mitochondrial presequences. *Proc. Natl. Acad. Sci. USA* 83: 9011–9015.

Andersson, H. and von Heijne, G. (1993) *Sec*-dependent and *sec*-independent assembly of *E. coli* inner membrane proteins – the topological rules depend of chain length. *EMBO J.* 12: 683–691.

Bardwell, J.C.A. and Beckwith, J. (1993) The bonds that tie: catalyzed disulfide bond formation. *Cell* 74: 769–771.

Beasley, E.M., Muller, S. and Schatz, G. (1993) The signal that sorts yeast cytochrome-b2 to the mitochondrial intermembrane space contains 3 distinct functional regions. *EMBO J.* 12: 2303–2311.

Blobel, G. (1980) Intracellular protein topogenesis. *Proc. Natl. Acad. Sci. USA* 77: 1496–1500.

Bruch, M.D. and Gierasch, L.M. (1990) Comparison of helix stability in wild-type and mutant lamb signal sequence. *J. Biol. Chem.* 265: 3851–3858.

Dalbey, R.E. and von Heijne, G. (1992) Signal peptidases in prokaryotes and eukaryotes – a new protease family. *Trends Biochem. Sci.* 17: 474–478.

Endo, T., Eilers, M. and Schatz, G. (1989) Binding of a tightly folded artifical mitochondrial precursor protein to the mitochondrial outer membrane involves a lipid-mediated conformational change. *J. Biol. Chem.* 264: 2951–2956.

Engleman, D.M. and Steitz, T.A. (1981) The spontaneous insertion of proteins into and across membranes: The helical hairpin hypothesis. *Cell* 23: 411–422.

Franzén, L.G., Rochaix, J.D. and von Heijne, G. (1990) Chloroplast transit peptides from the green alga *Chlamydomonas reinhardtii* share features with both mitochondrial and higher plant chloroplast presequences. *FEBS Lett.* 260: 165–168.

Franzén, L.G. and Falk, G. (1992) Nucleotide sequence of cDNA clones encoding the beta-subunit of mitochondrial ATP synthase from the green alga *Chlamydomonas reinhardtti* – The precursor protein encoded by the cDNA contains both an N-terminal presequence and a C-terminal extension. *Plant Mol. Biol.* 19: 771–780.

Gafvelin, G. and von Heijne, G. (1994) Topological "frustration" in multi-spanning *E. coli* inner membrane proteins. *Cell* 77: 401–412.

Gavel, Y and von Heijne, G. (1990) Cleavage-site motifs in mitochondrial targeting peptides. *Protein Eng.* 4: 33–37.

Gierasch, L.M. (1989) Signal sequences. *Biochemistry* 28: 923–930.

Gilmore, R. (1993) Protein translocation across the endoplasmic reticulum – a tunnel with toll booths at entry and exit. *Cell* 75: 589–592.

Glick, B. and Schatz, G. (1991) Protein import into mitochondria: two systems acting in tandem? *Trends Cell. Biol.* 1: 99–103.

Glick, B.S., Brandt, A., Cunningham, K., Muller, S., Hallberg, R.L. and Schatz, G. (1992) Cytochromes-c1 and cytochromes-b2 are sorted to the intermembrane space of yeast mitchondria by a stop transfer mechanism. *Cell* 69: 809–822.

Glick, B.S., Wachter, C., Reid, G.A. and Schatz, G. (1993) Import of cytochrome-b(2) to the mitchondrial intermembrane space – The tightly folded heme-binding domain makes import dependent upon matrix ATP. *Protein Sci.* 2: 1901–1917.

Görlich, D. and Rapoport, T.A. (1993) Protein translocation into proteoliposomes reconstituted from purified components of the endoplasmic reticulum membrane. *Cell* 75: 615–630.

Hartl, F.U. and Neupert, W. (1990) Protein sorting to mitochondria – Evolutionary conservations of folding and assembly. *Science* 247: 930–938.

Hartmann, E., Sommer, T., Prehn, S., Görlich, D., Jentsch, S. and Rapoport, T.A. (1994) Evolutionary conservation of components of the protein translocation complex. *Nature* 367: 654–657.

Hendrick, J.P., Hodges, P.E. and Rosenberg, L.E. (1989) Survey of amino-terminal proteolytic cleavage sites in mitochondrial precursor proteins: Leader peptides cleaved by two matrix proteases share a three-amino acid motif. *Proc. Natl. Acad. Sci. USA* 86: 4056–4060.

High, S. and Dobberstein, B. (1991) The signal sequence interacts with the methionine-rich domain of the 54-kD protein of signal recognition particle. *J. Cell Biol.* 113: 229–233.

Horniak, L., Pilon, M., van't Hof, R. and de Kruijff, B. (1993) The secondary structure of the ferredoxin transit sequence is modulated by its interaction with negatively charged lipids. *FEBS Lett.* 334: 241–246.

Hurt, E.C., Soltanifar, N., Goldschmidt-Clermont, M. and Schatz, G. (1986) The cleavable pre-sequence of an imported chloroplast protein directs attached polypeptides into yeast mitochondria. *EMBO J.* 5: 1343–1350.

Isaya, G., Kalousek, F. and Rosenberg, L. (1992a) Sequence analysis of rat mitochondrial intermediate peptidase: Similarity to zinc metallopeptidase and to a putative yeast homolgue.*Proc. Natl. Acad. Sci. USA* 89: 8317–8321.

Isaya, G., Kalousek, F. and Rosenberg, L.E. (1992b) Amino-terminal octapeptides function as recognition signals for the mitochondrial intermediate peptidase. *J. Biol. Chem.* 267: 7904–7910.

Kaiser, C.A., Preuss, D., Grisafi, P. and Botstein, D. (1987) Many random sequences functionally replace the secretion signal sequence of yeast invertase. *Science* 235: 312–317.

Kaiser, C.A. and Botstein, D. (1990) Efficiency and diversity of protein localization by random signal sequences. *Mol. Cell. Biol.* 10: 3163–3173.

Kalousek, F., Isaya, G. and Rosenberg, L. (1992) Rat liver mitochondrial intermediate peptidase (MIP): purification and initial characterization. *EMBO J.* 11: 2803–2809.

Lemire, B.D., Fankhauser, C., Baker, A. and Schatz, G. (1989) The mitochondrial targeting function of randomly generated peptide sequences correlates with predicted helical amphiphilicity. *J. Biol. Chem.* 264: 20206–20215.

Lipp, J., Flint, N., Haeuptle, M.-T. and Dobberstein, B. (1989) Structural requirements for membrane assembly of proteins spanning the membrane several times. *J. Cell. Biol.* 109: 2013–2022.

Luirink, J., High, S., Wood, H., Giner, A., Tollervey, D. and Dobberstein, B. (1992) Signal-sequence recognition by an *Escherichia coli* ribonucleoprotein complex. *Nature* 359: 741–743.

Luirink, J. and Dobberstein, B. (1994) Mammalian and *Escherichia coli* signal recognition particles. *Mol. Microbiol.* 11: 9–13.

Lutcke, H., High, S., Romisch, K., Ashford, A.J. and Dobberstein, B. (1992) The methionine-rich domain of the 54 kDa subunit of signal recognition particle is sufficient for the interaction with signal sequences. *EMBO J.* 11: 1543–1551.

Miller, J.D., Wilhelm, H., Gierasch, L., Gilmore, R. and Walter, P. (1993) GTP binding and hydrolysis by the singal recognition particle during initiation of protein translocation. *Nature* 366: 351–354.

Neupert, W., Hartl, F.U., Craig, E.A. and Pfanner, N. (1990) How do polypeptides cross the mitochondrial membranes? *Cell* 63: 447–450.

Pfanner, N., Rassow, J., Vanderklei, I.J. and Neupert, W. (1992) A dynamic model of the mitochondrial protein import machinery. *Cell* 68: 999–1002.

Pilon, M., Rietveld, A.G., Weisbeek, P.J. and de Kruijff, B. (1992) Secondary structure and folding of a functional chloroplast precursor protein. *J. Biol. Chem.* 267: 19907–19913.

Rizo, J., Blanco, F.J., Kobe, B., Bruch, M.D. and Gierasch, L.M. (1993) Conformational behavior of *Escherichia coli* OmpA signal peptides in membrane mimetic environments. *Biochemistry* 32: 4881–4894.

Robinson, C., Klosgen, R.B., Herrmann, R.G. and Shackleton, J.B. (1993) Protein translocation across the thylakoid membrane – A tale of 2 mechanisms. *FEBS Lett.* 325: 67–69.

Roise, D., Horvath, S.J., Tomich, J.M., Richards, J.H. and Schatz, G. (1986) A chemically synthesized pre-sequence of an imported mitochondrial protein can form an amphiphilic helix and perturb natural and artifical phospholipid bilayers. *EMBO J.* 5: 1327–1334.

Roise, D. and Schatz, G. (1988) Mitochondrial presequences. *J. Biol. Chem.* 263: 4509–4511.

Schatz, P.J. and Beckwith, J. (1990) Genetic analysis of protein export in *Escherichia coli*. *Annu. Rev. Genet.* 24: 215–248.

Schwarz, E., Seytter, T., Guiard, B. and Neupert, W. (1993) Targeting of Cytochrome-b2 into the mitochondrial intermembrane space – specific recognition of the sorting signal. *EMBO J.* 12: 2295–2302.

Theg, S.M. and Scott, S.V. (1993) Protein import into chloroplasts. *Trends Cell. Biol.* 3: 186–190.

van Loon, A., Brändli, A., Pesold-Hurt, B., Blank, D. and Schatz, G. (1987) Transport of proteins to the mitochondrial intermembrane space: The 'matrix-targeting' and the 'sorting' domains in the cytochrome c1 presequence. *EMBO J.* 6: 2433–2439.

van't Hof, R., van Klompenburg, W., Pilon, M., Kozubek, A., de Kortekool, G., Demel, R.A., Weisbeek, P.J. and de Kruijff, B. (1993) The transit sequence mediates the specific interaction on the precursor of ferredoxin with chloroplast envelope membrane lipids. *J. Biol. Chem.* 268: 4037–4042.

von Heijne, G. (1986) The distribution of positively charged residues in bacterial inner membrane proteins correlates with the trans-membrane topology. *EMBO J.* 5: 3021–3027.

von Heijne, G., Steppuhn, J. and Herrmann, R.G. (1989) Domain structure of mitochondrial and chloroplast targeting peptides. *Eur. J. Biochem.* 180: 535–545.

von Heijne, G. (1990) The signal peptide. *J. Membr. Bio.* 115: 195–201.

von Heijne, G. and Nishikawa, K. (1991) Chloroplast transit peptides – The perfect random coil? *FEBS Lett.* 278: 1–3.

von Heijne, G. (1994) Membrane proteins: From sequence to structure. *Ann. Rev. Biophys. Biomol. Struct.* 23: 167–192.

Wang, Z.L., Jones, J.D., Rizo, J. and Gierasch, L.M. (1993) Membrane-bound conformation of a signal peptide – A transferred nuclear Overhauser effect analysis. *Biochemistry* 32: 13991–13999.

Wessels, H.P. and Spiess, M. (1988) Insertion of a multispanning membrane protein occurs sequentially and requires only one signal sequence. *Cell* 55: 61–70.

Wickner, W., Driessen, A.J.M. and Hartl, F.U. (1991) The enzymology of protein translocation across the *Escherichia coli* plasma membrane. *Annu Rev Biochem* 60: 101–124.

Yang, M., Geli, V., Oppliger, W., Suda, K., James, P. and Schatz, G. (1991) The MAS-encoded processing protease of yeast mitochondria – Interaction of the purified enzyme with signal peptides and a purified precursor protein. *J. Biol. Chem.* 266: 6416–6423.

Chemistry and biochemistry

Enzymes: Chemistry and biochemistry

J. Jeffery

Department of Molecular and Cell Biology, University of Aberdeen, Marischal College, Aberdeen AB9 1AS, Scotland, UK

Summary. Basic principles underlying enzyme action are considered. Catalytic antibodies (abzymes), catalytic RNA (ribozymes), and non-biological counterparts of enzyme-catalyzed reactions are mentioned. Enzyme evolution is considered in terms of divergence, convergence, and lateral gene transfer.

Introduction

This chapter considers principles underlying enzyme function, dealing with selected aspects at the interface of chemistry and biochemistry. Comprehensive coverage is not attempted.

How enzymes work

Reaction in an enzyme-substrate complex that brings the groups that are to react into juxtaposition, with geometry appropriate for the reaction, makes the entropy of activation more favourable by 9–190 J/deg./mole (Fersht, 1985; Rawn, 1989). The energy of enzyme-substrate binding makes this possible. The substrate must bind to the enzyme in such a way that, in the transition state of the forthcoming reaction, a satisfactory overlap of orbitals is achievable. Distortion of a fully formed C-C bond by $5°-10°$ requires 2.8–11 kJ/mole, and this serves as a guide to the energy cost of imperfect alignment (Storm and Koshland, 1970; Fersht, 1985). The mechanism of HIV-1 protease provides an example (Baca and Kent, 1993).

Within an active site lined with hydrophobic amino acid side-chains, the local dielectric constant may be much lower than that of water, making electrostatic interactions stronger, and able to influence binding, conformational stability, or catalysis. In the lysozyme reaction, the β-carboxylate of Asp-52 stabilizes the carbonium ion intermediate, contributing 37.6 kJ/mole (Warshel and Levitt, 1976), corresponding to more than 10^6-fold rate enhancement. Arg-115 helps to stabilize the active site cleft (Harata et al., 1993). In several enzymes, interaction of an Arg with the 2'-group of the coenzyme adenosine ribose is a basis for discrimination between NADP(H) and NAD(H) (Hurley et al., 1991;

Miyazaki and Oshima, 1994; Sem and Kasper, 1993). The charge distribution of dihydrofolate reductase facilitates access of coenzyme and substrate to the active site (Böhm and Jaenicke, 1994). Macrodipoles of α-helices often contribute to coenzyme and substrate binding (Hol, 1985). Electrostatic interactions influence α-helix stability (Armstrong and Baldwin, 1993; Richardson and Richardson, 1988).

Hydrogen bonds are not uniform in their characteristics. Collinearity is most favourable energetically, but non-collinear arrangements with angles at hydrogen down to 165° are compatible with significant hydrogen bonding, which is therefore possible in the numerous situations in which strict collinearity is prevented by other steric constraints. However, hydrogen bonding becomes vanishingly weak if the angle at hydrogen falls below 140°. Hydrogen bonds in well-known systems have energies in the range 12–38 kJ/mole (Hagler et al., 1974). Hydrogen bonded arrays can delocalize charge (Nienaber et al., 1993). Hydrogen bonds involving sulphur occur in various enzymes, including papain, and glyceraldehyde-3-phosphate dehydrogenase, in which deprotonization of a Cys thiol is facilitated by formation of a hydrogen bond to the imidazole N of a His, the other N of which is hydrogen bonded to an oxygen (Garavito et al., 1977). In tyrosyl-tRNA synthetase, Cys-35 interacts with the carbonyl oxygen of Ile-190 and with a water molecule, (which is hydrogen bonded also to the 3′-hydroxyl of the ribose) (Brick et al., 1988).

Amino groups can form weak hydrogen bonds involving the π electrons of the benzene rings of aromatic side chains in proteins (Burley and Petsko, 1988), the energy being around 12 kJ/mol (Levitt and Perutz, 1988). Somewhat different is the interaction between aromatic side chains and Asn or Gln (Flocco and Mowbray, 1994). Interactions between Arg residues and aromatic rings show further characteristics (Flocco and Mowbray, 1994). Pancreatic lipase, and carboxypeptidase A provide examples, and a diagram illustrating the stacking arrangements of Arg-257 in glycollate oxidase is given by Flocco and Mowbray (1994).

Amino acid side-chains obstruct peptide-to-solvent hydrogen bonding in the unfolded state, and this steric blocking effect contributes about 0.5 kJ per residue to protein hydrogen bond strength, and accounts for the β-sheet propensities of the amino acids (Bai and Englander, 1994). A corresponding influence on α-helix propensities is obscured by large context-dependent effects. Where α-helices begin and end in proteins is influenced by hydrogen bonding, and is dependent upon the identity of residues at particular positions in the polypeptide chain, (Armstrong and Baldwin, 1993; Harpaz et al., 1994).

A general source of Gibbs free energy of binding is the greater number of hydrogen bonds that results from removal of the substrate from the bulk water to the hydrophobic interior of the protein. In cholesterol oxidase, the substrate binds in an internal cavity which is sealed from the solvent. In the free enzyme structure, a lattice of 13 water molecules occupies this cavity, and 12 of these water molecules are displaced when the steroid substrate binds (Li et al., 1993). *Pseudomonas* formate dehydrogenase apoenzyme has an interdomain cleft from which some water molecules are expelled when NAD^+ binds, inducing a conformational change that closes the cleft (Lamzin et al., 1994). Each mole of water released to interact optimally in the bulk phase assists binding by about 40 kJ/deg in the entropy term (Fersht, 1985). When the inhibitor zopolrestat binds in the active site of aldose reductase, at least six ordered water molecules are expelled (Wilson et al., 1993).

The milieu in which chemical reactions occur can greatly affect reaction rates. When an S_N1 reaction involves *development* of charge in the transition state, change to a solvent with better ion-solvating ability, or higher dielectric constant, can stabilize the transition state and increase the rate, sometimes as much as 10^3-fold. Also, in the case of an S_N2 reaction, change from a polar *hydroxylic* solvent (such as methanol) to a *non-hydroxylic* solvent of similar dielectric constant (such as dimethylformamide or dimethyl sulphoxide) can increase the concentration of unsolvated nucleophile, resulting in 10^4-fold or even much higher rate increases.

The clefts and cavities of an enzyme generally constitute a hydrophobic microenvironment, equivalent to low dielectric constant, and the active site brings the reactants together in appropriate orientations and in intimate proximity to suitable *polar* groups. In this way, local charge densities can be higher, or charges can be spread or stabilized. A nucleophile that is unsolvate, anchored and juxtaposed to the target electron acceptor, can be equivalent to a very high concentration of nucleophile in a favourably polar solvent. Enzymic reactions occur in "submicroenvironments", located within the hydrophobic microenvironements of the enzymes' crevices.

Many reactions are catalysed by proton donation or abstraction. General-acid catalysis by the unionised γ-carboxyl of Glu-35 occurs in lysozyme (Malcolm et al., 1989), and in many enzymes His functions as a proton donor/acceptor. In lactate dehydrogenase, His-195 acts as an acid-base catalyst, in addition to helping to orient the substrate by hydrogen bonding (Adams et al., 1973), and in glutathione reductase, His-467 acts as an acid catalyst to protonate the first molecule of glutathione (Wong et al., 1988). In glutamate dehydrogenase, the β-carboxyl of Asp-165 behaves as a general base (Stillman et al., 1993).

Imidazole can act catalytically as a general base, or as a nucleophile, while thiols can function as nucleophiles or provide good leaving groups. Pyridine facilitates the acylation of alcohols or phenols, but 4-dimethylaminopyridine is a much more effective catalyst than either pyridine or N,N-dimethylaniline. The rate of acetylation of propan-2-ol by acetic anhydride, for example, is more than 10^4-fold faster in the presence of 4-dimethylaminopyridine than in the presence of pyridine (Connors and Albert, 1973). Thus, considerable rate enhancements can be achieved in simple chemical reactions. At pH values between 7 and 8, the side-chains of Cys and His can be nucleophilic, but those of lysine, serine, and threonine require special microenvironments to achieve sufficient deprotonation to develop nucleophilic reactivity. For electronic polarizability reasons, the deprotonated thiol of a cysteine side-chain will have 10- to 100-fold higher nucleophilic reactivity than normal oxygen and nitrogen bases with the same pK_a value (Walsh, 1979). Electrophilic groups capable of functioning as electron acceptors during catalysis are often provided by *coenzymes*, or may be formed via enzyme-bound metals.

The bond energies, geometries, and other characteristics of metal ion coordination complexes show much variation. Moreover, many metal ions carry formal charges greater $+1$, and may have more than one stable valency state. Consequently, metal ions are well suited to catalytic roles. Catalysis of glycylglycine hydrolysis by a Co^{3+} complex increases the rate almost 10^{10}-fold (Buckingham et al., 1974) because of the proximity of the glycylglycine ligand to the hydroxyl ligand in the complex, the geometry, and the characteristics of the bonds to the cobalt. Hydrolytic enzymes utilizing Zn^{2+} are well known. In carboxypeptidase A, a Zn^{2+} is ligated to His-196, Glu-72, His-69 and H_2O. The substrate peptide displaces the bound water, and the zinc acts as an electrophile towards the carbonyl oxygen of the peptide bond to be cleaved (Christianson and Lipscomb, 1989). Rearrangements involving 1,2-shifts, and others involving methylation reactions are in some cases catalysed by enzymes dependent upon coenzyme B_{12}, which provides a Co^{+3} centre. Depending upon the reaction, the Co-C bond cleavage involved may occur by homolytic fisson (forming a carbon radical), heterolytic fission forming a carbonium ion, or heterolytic cleavage forming a carbanion, these alternatives being possible because the Co can transiently have a formal charge of $+2$, $+1$, or $+3$, respectively (Walsh, 1979). Class II aldolases catalyse cleavage of C-C bonds with the participation of enzyme-bound zinc. Non-enzymic, homogeneous, catalytic cleavages of various types of C-C bond have been achieved using complexes of rhodium and of iridium under harsh conditions (Gozin et al., 1993). Galactose oxidase contains one copper atom liganded in an arrangement similar to a square pyramid, which is thought to be favourable for redox change from Cu^{2+} to Cu^{1+} after addition of the D-galactose substrate (Ito et al., 1991). *Clostridium barkeri* nicotinic acid hydroxylase contains iron in iron-sulphur clusters, and selenium (which is essential for nicotinic acid hydroxylase activity) coordinated directly with molybdenum in the molybdopterin cofactor

(Gladyshev et al., 1994). In *Escherichia coli* ribonucleotide reductase, various properties of the iron appear to serve different requirements, providing a structurally frugal and ingeniously effective catalytic mechanism (Fontecave et al., 1992).

When coenzyme or substrate binds to an enzyme, part of the interaction energy that could otherwise have been available for even tighter binding may be used to distort the coenzyme or substrate. Distortion towards the geometry or electronic structure of the transition state for the enzymic reaction lowers the activation energy. An interesting example of the effect of strain upon a *non-enzymic* reaction is provided by the hydrolysis of β-lactones that are asymmetric at the β carbon (Cowdrey et al., 1937; Olson and Miller, 1938). Because the strain in the transition state for attack by water leading to acyl oxygen fission is greater than in the transition state for attack by water leading to alkyl oxygen fission, the latter is favoured in aqueous solution in the pH range of about 3–8, yielding the β-hydroxy acid of opposite configuration to that of the lactone. At more acid or more alkaline pH values, acid or alkaline hydrolysis predominates, and leads in either case to acyl oxygen fission, forming the β-hydroxyacid with the same configuration as the lactone. Possibilities for generation of favourable strain energy by substrate distortion through enzyme-substrate binding have been considered to be poor because of the nature of the forces available (Fersht, 1985). In many enzyme-substrate complexes, the enzyme is more likely to be distorted than the substrate (cf. Koshland, 1958). Nevertheless, substrate distortion contributes to catalysis by triosephosphate isomerase (Belasco and Knowles, 1980), lysozyme (Strynadka and James, 1991), influenza B virus neurominidase (Burmeister et al., 1993); and possibly serine proteases (Kraut, 1977); nicotinamide coenzyme distortion has been suggested to contribute to catalysis by dehydrogenases (Li and Goldstein, 1992; Lamzin et al., 1994).

Catalysis requires that the substrate-binding site of the enzyme should be complementary to the transition state structure because it is this to which the enzyme should bind well (Kraut, 1988). k_{cat} should be high. k_{cat}/K_M should be high, and what is generally advantageous is for the enzyme to have a structure that gives the highest value of K_M compatible with the highest value of k_{cat}/K_M. Increasingly high values of K_M imply progressively weaker substrate binding, and eventually problems of specificity, but a K_M value somewhat above the concentration at which the enzyme functions *in vivo* is adequate in practice. According to these criteria, the scope for improvement of the enzyme can be judged (Fersht, 1985; Burbaum et al., 1989; Pettersson, 1989). So assessed, carbonic anhydrase, and triosephosphate isomerase are deduced to have no scope for improvement (Fersht, 1985; but see also Pettersson, 1989). A further view suggests that for enzymes that have a detoxification role, biological advantage lies in specificity mainly for functional groups

(Jakoby and Ziegler, 1990), in terms of which aldose reductase, alcohol dehydrogenase, lactate dehydrogenase, aldehyde dehydrogenase, glutathione transferase, and O-methyltransferase have been discussed (Grimshaw, 1992).

Enzymes frequently undergo conformational changes in the course of the catalytic cycle. The movements vary in magnitude, and in character, and are usually associated with function. In the different structures and catalytic mechanisms of liver alcohol dehydrogenase, lactate dehydrogenase, and glyceraldehyde 3-phosphate dehydrogenase, different movements of helices and loops (by some 0.5 to 0.9 nm) shield the active site of the ternary complexes from the bulk solvent, and make possible the correct hydride transfers (Brändén and Eklund, 1980). An essential, large conformational change involving rotation of the NADPH domain has been proposed for thioredoxin reductase (Waksman et al., 1994). NAD^+ binding to *Pseudomonas* formate dehydrogenase results in a rotatory movement of the catalytic domain, which comes closer to the active centre, closing a cleft between the coenzyme-binding domain and the catalytic domain, and the disordered region after Ala-374 forms the C-terminal helix (Lamzin et al., 1994). Glutamate dehydrogenase undergoes a conformational change that closes the cleft between the N-terminal and C-terminal domains when glutamate binds (Stillman et al., 1993). A mobile loop envelopes the transition state in tyrosyl-tRNA synthetase (Fersht et al., 1988). A flexible loop in triosephosphate isomerase stabilizes a reaction intermediate (Pompliano et al., 1990). Glucose binding to hexokinase causes a rotation of the domains, closing the interdomain cleft and excluding water from the active site (Anderson et al., 1979). Hinge motions are a feature of the phosphoglycerate kinase mechanism (Harlos et al., 1992). Phosphorylation induces a domain rotation that produces the active form of glycogen phosphorylase (Sprang et al., 1991). The MAP kinase ERK2 is thought to undergo both extended and local conformational changes during activation (Zhang et al., 1994). During the allosteric transition in fructose-1, 6-*bis*phosphatase, conformational changes are propagated through different levels of structural organization (Liang et al., 1993). Thus, in very many cases, some of the binding energy is utilized to produce movement that is important for bringing about transformation. In this sense, the enzyme has characteristics of a *mechanical* device (Williams, 1993). The movements variously involve loops, helices, domains, entire subunits, or even all of these in a hierarchical manner. It is important to think not in terms of just a local reactive site, but to consider the whole protein. Changes introduced by site-specific mutagenesis may have effects considerably different from what might be expected from application of organic chemical theory to the immediate surroundings of the amino acid exchanged. An unexpected long-range effect of site-directed mutagenesis has been characterized in, for example, dihydrofolate reductase (Brown et al., 1993).

Catalysis can involve transient covalent bond formation. Class I aldolases utilize the ε-amino group of a lysine residue to form an iminium ion by reaction with the oxo group of the substrate. Formation of the iminium cation can facilitate acceptance of a negative charge at one of the adjacent carbon atoms, and, because the imine formation is readily reversible, the catalytic cycle can easily be completed by regeneration of the lysine ε-amino group. A reaction scheme has been given (Littlechild and Watson, 1993). Similar formation of an iminium cation between the substrate and the enzyme's active site lysine is used in acetoacetate decarboxylase. This reaction can be mimicked non-enzymically by using aniline as catalyst (Fersht, 1985).

In many enzymic reactions, a lysine ε-amino group of the enzyme forms an iminium cation with the oxo group of the coenzyme pyridoxal 5′-phosphate. Most such reactions involve an α-amino acid substrate and are (1) α-carbon reactions, achieving transamination, racemization, or decarboxylation, (2) β-carbon reactions, involving elimination or replacement, and (3) γ-carbon reactions, involving elimination or replacement (Walsh, 1979).

Aspartate aminotransferase exemplifies pyridoxal 5′-phosphate dependent transaminases (Yano et al., 1993; Tanaka et al., 1994; reaction schemes in Rawn, 1989, and Voet and Voet, 1990). Examples of other pyridoxal 5′-phosphate-dependent enzymes are tryptophan synthase (Miles, 1991), 3-ketosphinganine synthase, and serine hydroxymethyltransferase; a mechanistic scheme for each is illustrated in Rawn (1989).

The non-enzymic decarboxylation of pyridine-2-carboxylic acid occurs relatively readily because proton acquisition by the pyridine nitrogen generates an N^+ which can stabilize the development of negative charge on C-2, leading to decarboxylation via an ylid intermediate (Sykes, 1986). In enzymic reactions the formation of ylid-type intermediates with thiamine pyrophosphate coenzyme assists rate enhancement. Examples are pyruvate decarboxylase (Dyda et al., 1993), pyruvate dehydrogenase (mechanistic schemes in Rawn, 1989; Voet and Voet, 1990), bacterial pyruvate oxidase (Muller and Schultz, 1993), and transketolase (Schneider and Lindqvist, 1993). Transketolase, pyruvate oxidase, and pyruvate decarboxylase each bind the thiamine pyrophosphate in a similar way (Muller et al., 1993).

Several carboxylases utilize biotin as coenzyme. Pyruvate carboxylase, acetyl-CoA carboxylase, propionyl-CoA carboxylase, and geranyl-CoA carboxylase function this way. Methylcrotonyl-CoA carboxylase, and urea carboxylase have analogous mechanisms. Somewhat different, but utilizing the properties of biotin similarly, is methylmalonyl-CoA carboxytransferase (Wood and Zwolinski, 1976). The reaction mechanisms of these enzymes are discussed by Walsh (1979), and a mechanistic scheme for pyruvate carboxylase is illustrated in Rawn (1989). The important enzymes ribulose 1,5-*bis*phosphate carboxylase-oxygenase

(Hartman and Harpel, 1994), and Vitamin-K-dependent carboxylase (Suttie, 1980) function without biotin, pyridoxal 5'-phosphate, or thiamine pyrophosphate.

Protein structures are not limited to the 20 amino acids encoded in the genetic code

Many post-translational modifications are known (Krishna and Wold, 1993). In galactose oxidase, post-translational modification covalently links Tyr-272 to the S of Cys-228, forming an extended aromatic system which probably assists delocalization of a free radical associated with Tyr-272, Trp-290 aiding its stabilization, and the arrangement resembling an enzyme-bound coenzyme (Ito et al., 1991). In methylamine dehydrogenase, a heterotetrameric enzyme of the type H_2L_2, post-translational modifications convert Trp-57 into the orthoquinone, 6,7-dioxotryptophan, and link it covalently 2'–4 to Trp-107 (Klinman and Mu, 1994). This tryptophan tryptophanylquinone moiety is designated TTQ. The natural electron acceptor is amicyanin, a copper-containing protein, which forms with it a heterohexamer of the type $H_2L_2A_2$. The copper atom is in a hydrophobic region formed by residues from the A, L and H subunits, and including the TTQ. Electron transfer is thought to occur through Trp-107 via the amicyanin His-95 to the copper atom. Copper-containing amine oxidases have, as part of the polypeptide chain, a quinone moiety, topaquinone, formed by post-translational modification of tyrosine (Klinman and Mu, 1994).

Selenocysteine (an amino acid that differs from cysteine in having a selenium atom in place of the sulphur), occurs in several enzymes, at positions corresponding to TGA in the gene (Böck et al., 1991). In *E. coli*, biosynthesis of the selenocysteine-containing enzymes requires a special transfer RNA, designated tRNA^{Sec}, which is loaded with *serine*. The Seryl-tRNA^{Sec} is then modified in its seryl moiety, being converted into selenocysteinyl-tRNA^{Sec}. The anticodon (UCA) of selenocysteinyl-tRNA^{Sec} can recognise UGA, but incorporation into protein requires an elongation factor (SELB), GTP, and presence of the UGA in an appropriate context within the mRNA (Forchhammer et al., 1990; 1991). In eukaryotes, the formation pathway for selenocysteinyl-tRNA, and the requirements for selenocysteine incorporation are evidently different, but they are broadly equivalent insofar as serine is the amino acid ligated to a special tRNA, and UGA is the codon recognised for selenocysteine incorporation (Lee et al., 1990). Mammalian glutathione peroxidase has one selenocysteine per subunit. During the main catalytic cycle, it is believed to shuttle between the -Se$^-$ form, and the -Se-OH form (Epp et al., 1983). A -SeH is more easily deprotonated than, and is more nucleophilic than a -SH. Selenocysteine-containing formate dehydrogenase is 380-fold more efficient than the mutant in which the selenocysteine has been replaced by cysteine (Axley et al., 1991). Theories concerning the evolution of codon reassignment have been proposed (Osawa et al., 1992; Schultz and Yarus, 1994). In other known departures from the "universal" genetic code, the new codon encodes one of the 20 amino acids of the "universal" code. The selenocysteine story presently appears to be unique.

Antibodies with catalytic properties can be produced

Antibodies raised against a hapten that resembles a transition state of a reaction can catalyse that reaction. By appropriate hapten design, catalytically useful groups, such as a carboxylate (Shokat et al., 1989) or a nucleophile (Shokat alnd Schultz, 1991), have been obtained at the binding site of the antibody, and an antibody has been raised that binds metal complexes, generating specific peptidase activity (Iverson and Lerner, 1989; Blackburn and Wentworth, 1993). Model building and site-directed mutagenesis of an antibody with esterase activity have been used to investigate whether any particular residue was involved in its esterase activity, and a light-chain residue (Arg L96) was deduced to stabilize high-energy transition states during catalysis (Roberts et al., 1994). The properties of catalytic antibodies (Lerner and Schultz, 1995) are fully consistent with principles of enzyme catalysis already discussed, above.

Enzyme-type catalysis is not limited to proteins

Catalysis by RNA is well established (Cech, 1993). Reactions catalysed include RNase activity, self-splicing, self-cleaving, inverse splicing, activity as mobile elements, intron removal and 3' end formation during pre-mRNA processing, RNA editing (that is, changing an RNA sequence relative to that of the encoding DNA), and peptidyl transferase activity. A general mechanism for catalytic RNA has been proposed (Steitz and Steitz, 1993). *In vitro* evolution of catalytic RNA has been demonstrated (Bartel and Szostak, 1993; Lehman and Joyce, 1993; Piccirilli et al., 1993).

Catalytic RNA structure is closely related to function (Flor et al., 1989; Tuschl and Eckstein, 1993). In the course of reaction, complexes are formed between catalytic RNAs and their substrates. Transition state stabilization in catalysis is evident (Piccirilli et al., 1993; Tuschl and Eckstein, 1993), hydrogen bonding is extremely important for structure and function, and metal ions fulfil structural and catalytic roles (Piccirilli et al., 1993; Steitz and Steitz, 1993). The principles that underly the workings of catalytic RNA are exactly the same as those that are described for protein enzymes, above, and which apply also to catalytic antibodies.

Chemically synthesized molecules of various types and sizes have highly selective binding and catalytic properties (Mock et al., 1989; Nowick et al., 1991; Sneider et al., 1993; Johnsson et al., 1993; Bonar-Law et al., 1994; Sanders, 1994). In a porphyrin cyclic trimer, for example, acetyl transfer from acetylimidazole to 4-(hydroxy-methyl)pyridine is catalysed (Mackay et al., 1994). The enzyme-like activities observed among all these structures derive from principles described above, further demonstrating the general validity of the principles.

Enzyme specificity

Although enzymes are highly selective, most utilize a number of related substances as substrates. Sorbitol dehydrogenase, for example, utilizes sorbitol, xylitol, L-iditol, and 1-phenyl-1,2-ethanediol, (Jeffery and Jörnvall, 1988; Maret, 1991). The specificity, as usual, involves van der Waal's contacts, hydrogen bonds, and water exclusion, but, in this case, a Glu ligand to the catalytic zinc probably aids selection of the secondary hydroxyl of a 1,2-glycol for oxidation (Eklund et al., 1985). Among similar enzymes, liganding of the active site zinc appears to be broadly equivalent but not wholly uniform (Feiters and Jeffery, 1989; Karlsson et al., 1994).

In some cases, enzymes catalyse reactions that would generally be considered to be of different types. Leukotriene A_4 hydrolase catalyses the conversion of leukotriene A_4 (a 5,6-epoxide) into leukotriene B_4 (a 5,12-dihydroxy compound), and also has Cl^--stimulated peptidase activity (Wetterholm and Haeggström, 1992). Nuclear DNA uracil glycosylase is an enzyme of nucleotide excision, cleaving the base-sugar bond. It is monomeric, but forms tetramers which are glyceraldehyde-3-phosphate dehydrogenase (Meyer-Siegler et al., 1991), and the same protein is also reported to have protein kinase activity (Kawamoto and Caswell, 1986), and single-stranded DNA binding activity (Morgenegg et al., 1986).

Ribulose-1,5-*bis*phosphate carboxylase-oxidase catalyses the reaction of ribulose-1,5-*bis*phosphate with *either* carbon dioxide, to form two molecules of 3-phosphoglycerate, *or* oxygen, to form one molecule each of 3-phosphoglycerate and 2-phosphoglycollate. The first steps are common to both pathways, and lead to an enolized ribulose-1,5-*bis*phosphate. The reaction with oxygen is illustrated (Fig. 1), showing well-known mechanistic proposals (Miziorko and Lorimer, 1983; Hartman and Harpel, 1994; Rawn, 1989), and a radical mechanism proposed by Frederick C. Krebs, in which the cyclic transition state for breakdown

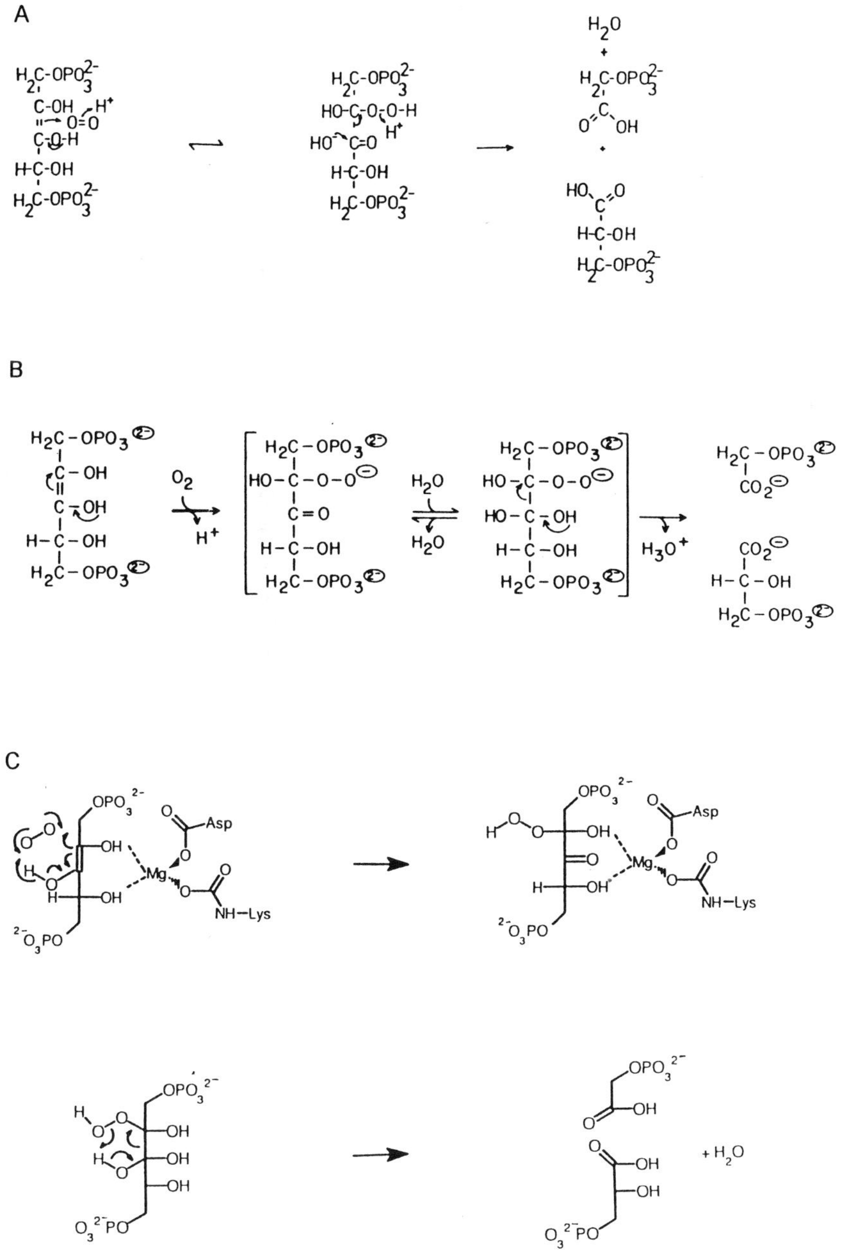

Figure 1. Oxygenase reaction of ribulose-5-phosphate carboxylase/oxygenase. (A) mechanism given by Miziorko and Lorimer (1983); (B) mechanism given by Hartman and Harpel (1994); (C) radical reactions suggested by Frederik C. Krebs for the hydroperoxide formation step (C, *top*), and, after hydration of the 3-oxo group (not shown), for formation of one molecule each of phosphoglycollic acid and D-3-phosphoglyceric acid (C, *bottom*).

of the 2-hydroperoxy-3-*gem*-diol intermediate avoids direct formation of an anion. Replacement of the Mg^{2+} by Co^{2+} in the enzyme from *Rhodospirillum rubrum* (Robison et al., 1979) and by Cu^{2+} in the spinach enzyme (Brändén et al., 1984) results in loss of the carboxylase activity, but retention of oxygenase activity.

Mammalian aldose reductase has a broad substrate specificity, catalysing reduction of the aldehyde group in molecules such as glucose, glyceraldehyde, and benzaldehyde (Morjana and Flynn, 1989), and also reduction of the 20-oxo group (to 20α-hydroxyl) in 17α-hydroxyprogesterone, progesterone, and some related steroids (Warren et al., 1993). A different enzyme, which has a similar kind of structure (Hoog et al., 1994), has 3α-hydroxysteroid dehydrogenase activity, and also utilizes various non-steroids, such as *trans*-dihydrodiols (Smithgall et al., 1986). Human placental oestradiol 17β-dehydrogenase will reduce certain 20-oxo steroids (e.g. progesterone) to the 20α-alcohols (Stricker et al., 1981), but this enzyme has a different kind of structure from mammalian aldose reductase (Peltoketo et al., 1992; Bruce et al., 1994). It does, however, have the same kind of structure as a bacterial enzyme, which has 3α-, and 20β-hydroxysteroid dehydrogenase activities in a variety of structural contexts, and catalyses a few similar reactions with non-steroid substrates (Pocklington and Jeffery, 1968; Pocklington et al., 1970; Gibb and Jeffery, 1971, 1972a–e, 1973a,b; Gibb et al., 1975; White and Jeffery, 1972, 1973a,b, 1974a,b, 1975, 1977; Ghosh et al., 1994). Octanol dehydrogenase scarcely utilizes ethanol, but 12-hydroxydodecanoate serves, and the enzyme has principally S-hydroxymethylglutathione dehydrogenase activity (glutathione-dependent formaldehyde dehydrogenase activity) (Koivusalo et al., 1989; Danielsson et al., 1994). The structure of this enzyme is different from the two kinds found among the hydroxysteroid dehydrogenases mentioned directly above, but resembles the structure of sorbitol dehydrogenase, mentioned at the beginning of this section (Eklund et al., 1985; Eklund et al., 1990).

Enzymes often form families or superfamilies

Comparisons of enzyme amino acid sequences (or corresponding nucleotide sequences) indicate that enzymes can often be sorted into groups based on residue identities, and that this is informative concerning structure, mechanism, ancestry and evolution. Among dehydrogenases dependent upon nicotinamide coenzymes, the short-chain dehydrogenase superfamily (Jörnvall et al., 1981, Jörnvall et al., 1995), the medium-chain dehydrogenase/reductase superfamily (originally called the long-chain family) (Jörnvall et al., 1981; Persson et al., 1994), and the aldo-keto reductase superfamily (Bohren et al., 1989; Bruce et al., 1994) are well characterized. Enzymes of the short-chain family are

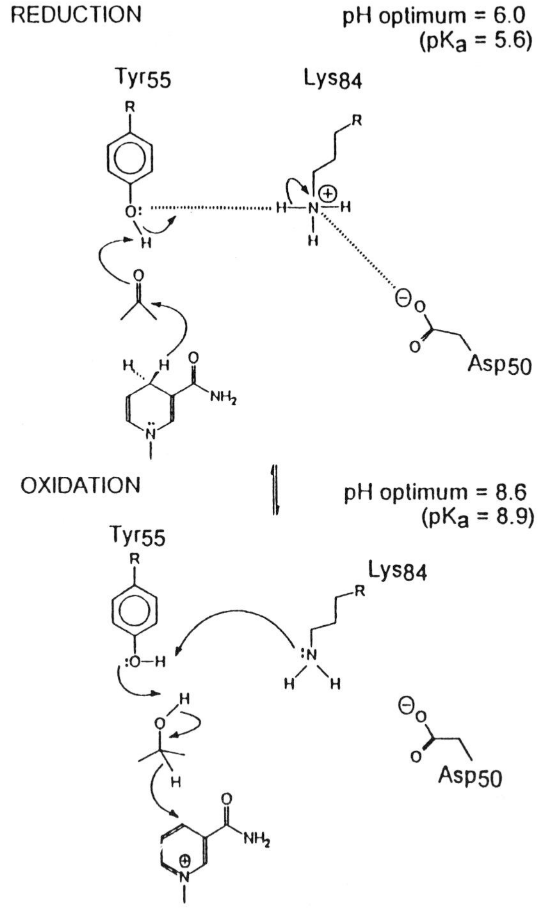

Figure 2. Rat liver 3α-hydroxysteroid/dihydrodiol dehydrogenase reaction mechanism (Hoog et al., 1994). Similar Tyr, Lys mechanisms operate in other members of the aldo-keto reductase superfamily (Wilson et al., 1992; Bruce et al., 1994), and in the structurally-different short-chain dehydrogenases (Jörnvall et al., 1995).

usually dimers or tetramers, function without a metal ion, and have subunit sizes around M_r 29 k. The subunit has an α/β structure consisting of essentially one domain (Varughese et al., 1992; Ghosh et al., 1994). In contrast, enzymes of the aldo-keto reductase superfamily are usually monomeric (though a dimeric member may be possible, Inazu, et al., 1994), and their structural type is an $(\alpha/\beta)_8$-barrel (Wilson et al., 1992; Wilson et al., 1993; Hoog et al., 1994; Bruce et al., 1994). In all cases investigated, short-chain dehydrogenases have been found to transfer the *4-pro-S* hydrogen of the nicotinaminde coenzyme, and enzymes of the aldo-keto reductase superfamily to transfer the *4-pro-R* hydrogen. In general, the interactions determining the specificity of dehydrogenases with respect to coenzyme are van der Waal's contacts and hydrogen bonds, and they differ somewhat in different cases (Jeffery, 1982). Various correlations have been noted (Jeffery, 1980, 1982; Weinhold et al., 1991).

Although the short chain dehydrogenase superfamily and the aldoketo reductase superfamily have not been proved to be related, their catalytic mechanisms both utilize a Tyr, Lys system (Fig. 2). However, the medium-chain dehydrogenase/reductase superfamily uses a different mechanism. The enzymes are dimers or tetramers, the subunits have M_r about 39 k, the structures are two domain α/β proteins, hydrogen transfer is *4-pro-R* specific with respect to the nicotinamide coenzyme, and catalysis involves an active-site zinc atom (Eklund et al., 1981; Eklund, 1989). Interestingly, somes enzymes do not belong to any presently-identified family. Glucose-6-phosphate dehydrogenases, for example, are clarly realted to one another, but show little sequence similarity to other proteins so far recorded in the Swissprot data bank (Persson et al., 1991; Jeffery et al., 1993).

Enzymes are continuously evolving

On the basis of the number and location of residue identities between the eukaryotic transcription factor TFIIS, a small subunit of eukaryotic RNA polymerase I (A12.2), the counterpart of this from RNA polymerase II (B12.6), and gene products from distantly related *Archaea*, it has been proposed that the eukaryotic polymerase subunits are the counterparts of the archaeal gene products, RNA polymerase I being closer in structure to the common ancestor, with functional divergence of RNA polymerase II in a 45 residue terminal segment, the aboriginal function of which may now be fulfilled by TFIIS (Kaine et al., 1994).

Methionyl aminopeptidase, creatine amidinohydrolase, and prolidase have similar primary and tertiary structures, divergence having produced amido-, imido- and amidino-hydrolytic activities. Similarly, divergence from common ancestors has been deduced for asparagine synthetase and aspartly tRNA synthetase (Furukawa et al., 1992); for phosphoribosyltransferases and thymidine phosphorlyases (Mushegian and Koonin, 1994); and for AMP hydrolase, uridine phosphorylase, purine phosphorylase, certain uncharacterized bacterial proteins, and a group of plant stress-inducible proteins of unknown function (Mushegian and Koonin, 1994).

Divergent evolution is well characterized in the medium-chain dehydrogenase/reductase superfamily (Persson et al., 1994), within which vertebrate alcohol dehydrogenases have been divided into five classes (with perhaps a sixth class) (Jörnvall and Höög, 1994). Class III (which has S-hydroxymethylglutathione dehydrogenase activity) is closest to the last common ancestor; Class I (the classical liver alcohol dehydrogenase) evidently arose from a gene duplication among very early vertebrates (Danielsson et al., 1994); and Class IV (characteristically expressed in epithelial cells) apparently separated from Class I later in early vertebrate evolution (Parés et al., 1994).

Bacterial mandelate racemase is structurally very similar to muconate lactonizing enzyme; S-mandelate dehydrogenase is related to glycollate oxidase and flavocytochrome b_2 (a lactate dehydrogenase); benzoylformate dehydrogenase is related to thiamine pyrophosphate-dependent decarboxylases; and benzaldehyde dehydrogenase has similarities to NAD^+-dependent aldehyde dehydrogenases (Petsko et al., 1993). Relatively nonspecific ancestral enzymes with appropriate catalytic activities are suggested to have undergone gene duplications; divergent evolution leading to enzymes with changed substrate specificities and modified catalytic capabilities, the R-mandelate to succinate pathway evolving "backwards", because the enzymes of the first part of the pathway would be unlikely to confer competitive advantage on the organism until it possessed the later enzymes (Petsko et al., 1993). The work extends and reinforces with impressive data for the mandelate pathway, findings and interpretations made on other systems (Horowitz, 1945; Jeffery and Jörnvall, 1983; Parsot, 1986; Cairns et al., 1988; Crawford, 1989).

Functional convergence is exemplified by the short-chain and medium-chain dehydrogenases; each superfamily has an enzyme that converts ribitol into ribulose, and each has one that converts ethanol into acetaldehyde (Jörnvall et al., 1981). Similarly, 3α-hydroxysteroid dehydrogenase activity is found in the short-chain dehydrogenase and in the aldo-keto reductase superfamilies (Gibb and Jeffery, 1973a; Ghosh et al., 1994; Smithgall et al., 1986; Hoog et al., 1994).

Convergence of *catalytic mechanism* is seen in the subtilisin proteases and the chymotrypsin proteases, which have different structures, but both employ an Asp, His, Ser triad for catalysis (Kraut, 1977). The Tyr, Lys catalytic systems of the structurally-dissimilar short-chain dehydrogenase and aldo-keto reductase superfamilies (Fig. 2) suggest possible convergence, but although glucose-6-phosphate dehydrogenase, which differs structurally from both, has a conserved sequence **RIDHYLGKE-XVXXXXXXRF**XN that includes Lys, Asp, His, Tyr, Glu, and two Arg (shown bold) (Jeffery et al., 1993), it is not evident that this enzyme has a corresponding catalytic Asp/Glu, Tyr, Lys/(Arg) arrangement. Many kinases contain a mechanistically important motif GXXGXGK which forms a phosphate-binding cavity, and a protein kinase that is different in both primary structure and overall fold has a Gly-dependent, Lys-containing, phosphate-binding cavity formed in a different way (Knighton et al., 1991). An evolutionary tree of selected kinase subfamilies is illustrated in Alberts et al. (1994). Thioredoxin reductase and glutathione reductase both catalyse the reduction of disulphides by NADPH via an enzyme disulphide and a flavin, but have significant structural differences, and it has been suggested that these enzymes diverged from an ancestral nucleotide-binding protein, and then acquired their mechanistically very similar disulphide reductase activities by convergent evolution (Kuriyan et al., 1991).

Convergence may sometimes be towards structural motifs that are particularly stable, or easily formed. Many enzymes contain an $(\alpha/\beta)_8$-*barrel* structure, and some of them evidently have a common ancestor (Wilmanns et al., 1991), but the primary structures of numerous others show no evidence of common ancestry, and convergent evolution has been discussed (Lesk et al., 1989; Rice et al., 1990). Several hydrolytic enzymes of widely differing phylogenetic origins and catalytic functions have an $(\alpha/\beta)_8$-*sheet* structure (Ollis et al., 1992; Verschueren et al., 1993), and are thought to have resulted from convergent evolution towards a catalytic triad, thereafter diverging with conservation of principal tertiary structure features because of their stability and functional utility, though little sequence similarity was retained, and the catalytic triads came to consist of a nucleophile (which was Cys, Ser, or Asp), a His (which was invariant), and an acid (Asp or Glu). The related structure of *Candida antarctica* lipase B has a Ser, His, Asp catalytic triad, but only seven strands, and a TXSXG segment instead of the GXSXG motif characteristic of many lipases (Uppenberg et al., 1994).

Kinases that catalyse phosphorylation of sugars comprise at least three nonhomologous families, each of which has a distinct three-dimensional fold. Fructokinase specificity has evolved independently in both the hexokinase and ribokinase families, and glucose specificity has evolved independently in different branches of the hexokinase family (Bork et al., 1993). In various dehydrogenases and kinases, nucleotide binding structures appear to be homologous, and to have been incorporated into dissimilar structures that provide the requirements for catalysis, often closely associated with substrate binding. Liver alcohol dehydrogenase, lactate dehydrogenase, and glyceraldehyde-3-phosphate dehydrogenase exemplify this (Brändén and Eklund, 1980). Human hormone-sensitive lipase has one segment apparently of common ancestry with bacterial hydrolytic enzymes, while other parts of the molecule are of different lineage (Langin and Holm, 1993). Enzymes with several domains or segments that appear to have markedly different origins are also known; urokinase contains epidermal growth factor, kringle, and serine protease domains, and prostaglandin H_2 synthase-1 has epidermal growth factor, membrane-binding and peroxidase domains (Picot et al., 1994).

Anomalously high similarity between the primary structures of certain archaeal glutamine synthetase I enzymes and the eubacterial enzymes suggests one or more lateral gene transfers (Smith et al., 1992). Evidence for transfer of a glyceraldehyde-3-phosphate dehydrogenase gene, and of a glucose-6-phosphate isomerase gene from a eukaryote to an ancestor of *E. coli* has been considered to be good, and transfer of a glutamine synthetase II gene from a eukaryote to a eubacterium may have occurred (Smith et al., 1992). The sequence of the isopenicillin N synthetase genes for two bacteria and three fungi are all very similar, and the bacterial – fungal similarities are only a little less than those among the bacteria, or among the fungi (Miller and Ingolia, 1989). Similar conclusions were reached for the deacetoxycephalosporin C synthetase genes, and both were suggested to have arisen first in bacteria, and to have transferred to fungi after the separation of the fungal line from the yeast line (Miller and Ingolia, 1989), but other evidence suggests that lateral gene transfer need not be proposed for these genes (Smith et al., 1992). Iron-superoxide dismutase from the amitochondrial eukaryote *Entamoeba histolytica* has a primary structure anomalously similar to the eubacterial enzyme structures, suggesting its acquisition by lateral transfer from a prokaryote (Smith et al., 1992). The same eukaryote also has an alcohol dehydrogenase homologous to the multifunctional *adhE* gene product of *E. coli* (Yang et al., 1994). The anomalous similarity of the *Saccharomyces cerevisiae* Class II aldolase to that of *E. coli* also suggests a prokaryote to eukaryote gene transfer (eubacterium to yeast) (Smith et al., 1992). Studies on chitinases (Gooday, 1994) showed that infection of cultured insect cells with *Autographa californica* nuclear polyhedrosis virus caused a great increase in virally-encoded chitinase. Because of primary structure similarities to chitinase A of *Serratia marcescens* (an insect gut pathogenic bacterium) it was thought likely that the virus acquired the chitinase gene relatively recently by transfer from the bacterium (Gooday, 1994).

Conclusion

Chemical principles underlie biological catalysis, while principles of inheritance, genetic unrest, and natural selection govern enzyme emergence, and the characteristics of enzyme families. For the future, metal-ion binding sites provide opportunities to engineer novel enzymes for use *in vitro*, especially when the metal can be inserted into the apo-enzyme *in vitro*, and choice of metal is not limited, as it is during natural evolution *in vivo*, by toxicity to the organism. Refinement of the understanding of three-dimensional structural units (Orengo and Thornton, 1993) will facilitate protein engineering. Incorporation of chemically synthesized segments into designer proteins permits utilization of structures outside the scope of the genetic code, and regulatory responses to desired signals might be contrived for industrial purposes, perhaps borrowing ideas from work on molecules that perform logic operations (de Silva et al., 1993).

Inhibitors of HIV protease, HIV reverse transcriptase, and HIV integrase are currently major targets for industry. So, too are inhibitors of prostaglandin H_2 synthase-2 that do not also inhibit prostaglandin H_2 synthase-1, and appropriately selective inhibitors of 5-lipoxygenase, leukotriene A_4 hydrolase, and steroid 5α-reductase. Paradigms from the past include inhibition of prostaglandin H_2 synthase-1 by O-acetylsalicylic acid, through irreverisible acetylation of Ser-530 (Roth et al., 1975; Picot et al., 1994), and by non-covalent binding of the carboxyl group of S-flurbiprofen to the guanidinium group of Arg-120 (Picot et al., 1994); the inhibition of β-lactamase by clavulanic acid through highly specific covalent modification (Charnas and Knowles, 1981); and the inhibition of aldose reductase by zopolrestat, through non-covalent, tight binding in the active site of the holoenzyme (Wilson et al., 1993). Molecular enzymology, at the interface of chemistry and biochemistry, is a hugely productive field in which much remains to be done.

References

Adams, M.J., Buehner, M., Chandrasekhar, K., Ford, G.C., Hackert, M.L., Liljas, A., Rossmann, M.G., Smiley, I.E., Allison, W.S., Everse, J., Kaplan, N.O. and Taylor, S.S. (1973) Structure – function relationships in lacetate dehydrogenase. *Proc. Natl. Acad. Sci. USA* 70: 1968–1972.

Alberts, B., Bray, D., Lewis, J., Raff, M., Roberts, K. and Watson, J.D. (1994) *Molecular Biology of the Cell*, Third Edition, Garland, New York.

Anderson, C.M., Zucker, F.H. and Steitz, T.A. (1979) Space-filling models of kinase clefts and conformation changes. *Science* 204: 375–380.

Armstrong, K.M. and Baldwin, R.L. (1993) Charged histidine affects alpha-helical stability at all positions in the helix by interacting with the backbone charges. *Proc. Natl. Acad. Sci. USA* 90: 11337–11340.

Axley, M.J., Böck, A. and Stadtman, T.C. (1991) Catalytic properties of an *Escherichia coli* formate dehydrogenase mutant in which sulphur replaces selenium. *Proc. Natl. Acad. Sci. USA* 88: 8450–8454.

Baca, M. and Kent, S.B.H. (1993) Catalytic contribution of flap-substrate hydrogen bonds in "HIV-1 protease" explored by chemical synthesis. *Proc. Natl. Acad Sci. USA* 90: 11638–11642.

Bai, Y. and Englander, S.W. (1994) Hydrogen bond strength and β-sheet propensities: the role of a chain blocking effect. *Proteins Struct. Funct. Genet.* 18: 262–266.

Bartel, D.P. and Szostak, J.W. (1993) Isolation of new ribozymes from a large pool of random sequences. *Science* 261: 1411–1418.

Belasco, J.G. and Knowles, J.R. (1980) Direct observation of substrate distortion by triosephosphate isomerase using Fourier transformation. *Biochemistry* 19: 472–477.

Blackburn, G.M. and Wentworth, P. (1993) Catalytic antibodies. *In*: J. Gosling and D.E. Rean (eds): *Immunotechnology*, Portland Press, Colchester, pp 107–125.

Böck, A., Forchhammer, K., Heider, J. and Baron, C. (1991) Selenoprotein synthesis: an expansion of the genetic code. *Trends Biochem. Sci.* 16: 463–467.

Böhm, G. and Jaenicke, R. (1994) A structure-based model for the halophilic adaptation of dihydrofolate reductase from *Halobacterium volcanii*. *Protein Engng.* 7: 213–220.

Bohren, K.M., Bullock, B., Wermuth, B. and Gabbay, K.H. (1989) The aldo-keto reductase superfamily. cDNAs and deduced amino acid sequences of human aldehyde and aldose reducatase. *J. Biol. Chem.* 264: 9547–9551.

Bonar-Law, R.P., Mackay, L.G., Walter, C.J., Marvaud, V. and Sanders, J.K.M. (1994) Towards synthetic enzymes based on porphyrins and steroids. *Pure and Appl. Chem.* 66: 803–810.

Bork, P., Sander, C. and Valencia, A. (1993) Convergent evolution of similar enzymatic function on different protein folds: the hexokinase, ribokinase, and galactokinase families of sugar kinases. *Protein Sci.* 2: 31–40.

Brändén, C.-I. and Eklund, H. (1980) Structure and mechanism of liver alcohol dehydrogenase, lactate dehydrogenase and glyceraldehyde-3-phosphate dehydrogenase. *In*: J. Jeffery (ed): *Dehydrogenases Requiring Nicotinamide Coenzymes*, Birkhäuser Verlag, Basel, pp 41–84.

Brändén, R., Nilson, T. and Styring, S. (1984) An intermediate formed by the Cu^{2+}-activated ribulose-1,5-bisphosphate carboxylase/oxygenase in the presence of ribulose 1,5-bisphosphate and O2. *Biochemistry* 23: 4378–4382.

Brick, P., Bhat, T.N. and Blow, D.M. (1988) Structure of tyrosyl-tRNA synthetase refined at 2.3 Å resolution. Interaction of the enzyme with the tyrosyl adenylate intermediate. *J. Mol. Biol.* 208: 83–98.

Brown, K.A., Howell, E.E. and Kraut, J. (1993) Long-range structural effects in a second-site revertant of a mutant dihydrofolate reductase. *Proc. Natl. Acad. Sci. USA* 90: 11753–11756.

Bruce, N.C., Willey, D.L., Coulson, A.F.W. and Jeffery, J. (1994) Bacterial morphine dehydrogenase further defines a distinct superfamily of oxidoreductases with diverse functional activities. *Biochem. J.* 299: 805–811.

Buckingham, D.A., Keene, F.R. and Sargeson, A.M. (1974) Facile intramolecular hydrolysis of dipeptides and glycinamide. *J. Am. Chem. Soc.* 96: 4981–4983.

Barbaum, J.J., Raines, R.T., Albery, W.J. and Knowles, J.R. (1989) Evolutionary optimization of the catalytic effectiveness of an enzyme. *Biochemistry* 28: 9293–9305.

Burley, S.K. and Petsko, G.A. (1988) Weakly polar interactions in proteins. *Adv. Protein Chem.* 39: 125–189.

Burmeister, W.P., Henrissat, B., Bosso, C., Cusack, S. and Ruigrok, R.W.H. (1993) Influenza B virus neurominidase can synthesize its won inhibitor. *Structure* 1: 19–26.

Cairns, J., Overbaugh, J. and Miller, S. (1988) The origin of mutants. *Nature* 335: 142–145 and 336: 525–528.

Cech, T.R. (1993) Catalytic RNA: structure and mechanism. *Biochem. Soc. Trans.* 21: 229–234.

Charnas, R.L. and Knowles, J.R. (1981) Inactivation of RTEM β-lactamase from *Escherichia coli* by clavulanic acid and 9-deoxyclavulanic acid. *Biochemistry* 20: 3214–3219.

Christianson, D.W. and Lipscomb, W.N. (1989) Carboxypeptidase A. *Accounts Chem. Res.* 22: 62–69.

Connors, K.A. and Albert, K.S. (1973) Determination of hydroxy compounds by 4-dimethylaminopyridine-catalyzed acetylation. *J. Pharmaceut. Sci.* 62: 845–846.

Cowdrey, W.A., Hughes, E.D., Ingold, C.K., Mastermann, S. and Scott, A.D. (1937) Reaction kinetics and Walden inversion. Part VI. Relation of steric orientation to mechanism in substitutions involving halogen atoms and simple or substituted hydroxyl groups. *J. Chem. Soc.* 1937: 1252–1271.

Cawford, I.P. (1989) Evolution of a biosynthetic pathway: the tryptophan paradigm. *Annu. Rev. Microbiol.* 43: 567–600.

Danielsson, O., Atrian, S., Luque, T., Hjelmqvist, L., Gonzàlez-Duarte, R. and Jörnvall, H. (1994) Fundamental molecular differences between alcohol dehydrogenae classes. *Proc. Natl. Acad. Sci. USA* 91: 4980–4984.

de Silva, A.P., Gunaratne, H.Q.N. and McCoy, C.P. (1993) A molecular photoionic AND gate based on fluorescent signalling. *Nature* 364: 42–44.

Dyda, F., Furey, W., Swaminathan, S., Sax, M., Farrenkopf, B. and Jordan, F. (1993) Catalytic centers in the thiamin diphosphate enzyme pyruvate decarboxylase at 2.4 Å resolution. *Biochemistry* 32: 6165–6170.

Eklund, H., Samama, J.-P., Wallén, L., Brändén, C.-I. Åkeson, Å. and Jones, T.A. (1981) Structure of a triclinic ternary complex of horse liver alcohol dehydrogenase at 2.9 Å resolution. *J. Mol. Biol.* 146: 561–587.

Eklund, H., Horjales, E., Jörnvall, H., Brändén, C.-I. and Jeffery, J. (1985) Molecular aspects of differences between alcohol and sorbitol dehydrogenases. *Biochemistry* 24: 8005–8012.

Eklund, H. (1989) Coenzyme binding in alcohol dehydrogenase. *Biochem. Soc. Trans.* 17: 293–296.
Eklund, H., Müller-Wille, P., Horjales, E. Futer, O., Holmquist, B., Vallee, B.L., Höög, J.-O., Kaiser, R. and Jörnvall, H. (1990) Comparison of three classes of human liver alcohol dehydrogenase. Emphasis on different substrate-binding pockets. *Eur. J. Biochem.* 193: 303–310.
Epp, O., Ladenstein, R. and Wendel, A. (1983) The refined structure of the selenoenzyme glutathione peroxidase at 0.2 nm resolution. *Eur. J. Biochem.* 133: 51–69.
Feiters, M.C. and Jeffery, J. (1989) Zinc environment in sheep liver dehydrogenase. *Biochemistry* 28: 7257–7262.
Fersht, A. (1985) *Enzyme Structure and Mechanism*, Second Edition, Freeman, New York.
Fersht, A.R., Knill-Jones, J.W., Bedouelle, H. and Winter, G. (1988) Reconstruction by site-directed mutagenesis of the transition state for the activation of tryosine by the tyrosyl-tRNA synthetase: a mobile loop envelopes the transition state in an induced-fit mechanism. *Biochemistry* 27: 1581–1587.
Flocco, M.M. and Mowbray, S.L. (1994) Planar stacking of arginine and aromatic side-chains in protein. *J. Mol. Biol.* 235: 709–717.
Flor, P.J., Flanegan, J.B. and Cech, T.R. (1989) A conserved base pair within helix P4 of the *Tetrahymena* ribozyme helps to form the tertiary structure required for self-splicing. *EMBO J.* 1989: 3391–3399.
Fontecave, M., Nordlund, P., Eklund, H. and Reichard, P. (1992) The redox centres of ribonucleotide reductase of *Escherichia coli*. *Adv. Enzymol. Rel. Areas Mol. Biol.* 65: 147–183.
Forchhammer, K., Rucknagel, K.-P. and Böck, A. (1990) Purification and biochemical characterization of SELB, a translation factor involved in selenoprotein synthesis. *J. Biol. Chem.* 265: 9346–9350.
Forchhammer, K., Leinfelder, W., Boesmiller, K., Verprek, B. and Böck, A. (1991) Selenocysteine synthase from *Escherichia coli*. Nucleotide sequence of the gene (selA) and purification of the protein. *J. Biol. Chem.* 266: 6318–6323.
Furukawa, T., Yoshimura, A., Sumizawa, T., Haraguchi, M., Akiyama, S.I., Fukui, K., Hinchman, S.K., Henikoff, S. and Schuster, S.M. (1992) A relationship between asparagine synthetase A and aspartyl tRNA synthetase. *J. Biol. Chem.* 167: 144–149.
Garavito, R.M., Rossmann, M.G., Argos, P. and Eventoff, W. (1977) Convergence of active center geometries. *Biochemistry* 16: 5065–5071.
Ghosh, D., Wawrzak, Z., Weeks, C.M., Duax, W.L. and Erman, M. (1994) The refined three-dimensional structure of $3\alpha,20\beta$-hydroxysteroid dehydrogenase and possible roles of the residues conserved in short-chain dehydrogenase. *Structure* 2: 629–640.
Gibb, W. and Jeffery, J. (1971) Relationship between 3α and 20β-hydroxysteroid: NAD oxidoreductase activity of a crystalline enzyme preparation. *Eur. J. Biochem.* 23: 336–342.
Gibb, W. and Jeffery, J. (1972a) Some effects of ring A structure on the interaction of 3-oxo steroids of the androstane series with cortisone reductase. *Eur. J. Biochem.* 25: 136–140.
Gibb, W. and Jeffery, J. (1972b) Steric, chiral and conformational aspects of the 3-hydroxy and 20-hydroxy steroid dehydrogenase activities of cortisone reductase preparations. *Biochim. Biophys. Acta* 268: 13–20.
Gibb, W. and Jeffery, J. (1972c) Reduction of the non-steroid adamantanone by crystalline preparations of cortisone reductase. *Biochem. J.* 126: 443.
Gibb, W. and Jeffery, J. (1972d) The different effects of an epoxy and a methylene group on enzymic reductions of a vicinal oxo group. *Bioorg. Chem.* 2: 73–76.
Gibb, W. and Jeffery, J. (1972e) 5α-Dihydrotestosterone sulphate and cortisone reductase. *Biochim. Biophys. Acta* 280: 646–651.
Gibb, W. and Jeffery, J. (1973a) Steric course with respect to the reduced nicotinamide-adenine dinucleotide of the reduction of 3-oxo steroids catalysed by cortisone reductase. *Eur. J. Biochem.* 34: 395–400.
Gibb, W. and Jeffery, J. (1973b) 3-Hydroxy steroid dehydrogenase activities of cortisone reductase. *Biochem. J.* 135: 881–888.
Gibb, W. and Jeffery, J., Kirk, D.N. and Mahdi, H. (1975) The altered specificity of cortisone reductase with certain retroandrostan-3-one substrates. *Biochem. J.* 145: 483–489.
Gladyshev, V.N., Khangulov, S.V. and Stadman, T.C. (1994) Nicotinic acid hydroxylase from *Clostridium barkeri*: Electron paramagnetic resonance studies show that selenium is coordi-

nated with molybdenum in the catalytically active selenium-dependent enzyme. *Proc. Natl. Acad. Sci. USA* 91: 232–236.
Gooday, G.W. (1994) Physiology of microbial degradation of chitin and chitosin. *In*: C. Ratledge (ed): *Biochemistry of Microbial Degradation*, Kluwer, Dordrecht, pp 279–312.
Gozin, M., Weisman, A., Ben-David, Y. and Milstein, D. (1993) Activation of a carbon-carbon bond in solution by transition-metal insertion. *Nature* 364: 699–701.
Grimshaw, C.E. (1992) Aldose reductase: model for a new paradigm of enzymic perfection in detoxification catalysts. *Biochemistry* 31: 10139–10145.
Hagler, A.T., Lifson, S. and Huler, E. (1974) *Peptides, Polypeptides, and Proteins*, John Wiley, New York.
Harata, K., Muraki, M. and Jigami, Y. (1993) Role of Arg115 in the catalytic action of human lysozyme. X-ray structure of His115 and Glu115 mutants. *J. Mol. Biol.* 233: 524–535.
Harlos, K., Vas, M. and Blake, C.F. (1992) Crystal structure of the binary complex of pig muscle phosphoglycerate kinase and its substrate 3-phospho-D-glycerate. *Proteins Struct. Funct. Genet.* 12: 133–144.
Harpaz, Y., Elmasry, N., Fersht, A.R. and Henrick, K. (1994) Direct observation of better hydration at the N terminus of an α-helix with glycine rather than alanine as the N-cap residue. *Proc. Natl. Acad. Sci. USA* 91: 311–315.
Hartman, F.C. and Harpel, M.R. (1994) Structure, function, regulation, and assembly of D-ribulose-1,5-bisphosphate carboxylase/oxygenase. *Annu. Rev. Biochem.* 63: 197–234.
Hol, W.G.J. (1985) The role of the α-helix dipole in protein function and structure. *Prog. Biophys. Mol. Biol.* 45: 149–195.
Hoog, S.S., Pawlowski, J.E., Alzari, P.M., Penning, T.M. and Lewis, M. (1994) Three-dimensional structure of rat liver 3α-hydroxysteroid/dihydrodiol dehydrogenase: a member of the aldo-keto reductase superfamily. *Proc. Natl. Acad. Sci. USA* 91: 2517–2521.
Horowitz, N.H. (1945) The evolution of biochemical systems. *Proc. Natl. Acad. Sci. USA* 31: 153–157.
Hurley, J.H., Anthony, M.D., Koshland, D.E. and Stroud, R.M. (1991) Catalytic mechanism of $NADP^+$-dependent isocitrate dehydrogenase: implications from the structures of magnesium-isocitrate and $NADP^+$ complexes. *Biochemistry* 30: 8671–8678.
Inazu, A., Sato, K., Nakayama, T., Deyashiki, Y., Hara, A. and Nozawa, Y. (1994) Purification and characterization of a novel dimeric 20α-hydroxysteroid dehydrogenase from *Tetrahymena pyriformis*. *Biochem. J.* 297: 195–200.
Ito, N., Phillips, S.E.V., Stevens, C., Ogel, Z.B., McPherson, M.J., Keen, J.N., Yadav, K.D.S. and Knowles, P.F. (1991) Novel thioether bond revealed by a 1.7 Å crystal structure of galactose oxidase. *Nature* 350: 87–90.
Iverson, B.C. and Lerner, R.A. (1989) Sequence-specific peptide cleavage catalyzed by an antibody. *Science* 243: 1184–1188.
Jakoby, W.B. and Ziegler, D.M. (1990) The enzymes of detoxication. *J. Biol. Chem.* 265: 20715–20718.
Jeffery, J. (1980) The specificity of dehydrogenases. *In*: J. Jeffery (ed): *Dehydrogenases Requiring Nicotinamide Coenzymes*, Birkhäuser Verlag, Basel, pp 85–125.
Jeffery, J. (1982) Stereochemistry of dehydrogenases. *In*: C. Tamm (ed): *Stereochemistry*, Elsevier, Amsterdam, pp 113–160.
Jeffery, J. and Jörnvall, H. (1983) Enzyme relationships in a sorbitol pathway that bypasses glycolysis and pentose phosphates in glucose metabolism. *Proc. Natl. Acad. Sci. USA* 80: 901–905.
Jeffery, J. and Jörnvall, H. (1988) Sorbitol dehydrogenase. *Adv. Enzymol. Rel. Areas Mol. Biol.* 61: 47–106.
Jeffery, J., Persson, B., Wood, I., Bergman, T., Jeffery, R. and Jörnvall, H. (1993) Glucose-6-phosphate dehydrogenase. Structure-function relationships and the Pichia jadinii enzyme structure. *Eur. J. Biochem.* 212: 41–49.
Jörnvall, H., Persson, M. and Jeffery, J. (1981) Alcohol and polyol dehydrogenases are both divided into two protein types, and structural properties cross-relate the different enzyme activities within each group. *Proc. Natl. Acad. Sci. USA* 78: 4226–4230.
Jörnvall, H. and Höög, J.-O. (1995) Nomenclature of alcohol dehydrogenases. *Alcohol and Alcoholism*; in press.
Jörnvall, H., Persson, B., Krook, M., Atrian, S., Gonzàlez-Duarte, R., Jeffery, J. and Ghosh, D. (1995) Short-chain dehydrogenase/reductases (SDR). *Biochemistry*; in press.

Johnsson, K., Alleman, R.K., Widmer, H. and Benner, S.A. (1993) Synthesis, structure and activity of artificial, rationally designed catalytic polypeptides. *Nature* 365: 530–532.

Kaine, B.P., Mehr, I.J. and Woese, C.R. (1994) The sequence, and its evolutionary implications, of a *Thermococcus celer* protein associated with transcription. *Proc. Natl. Acad. Sci. USA* 91: 3854–3856.

Karlsson, C., Jörnvall, H. and Höög, J.-O. (1995) Zinc binding of alcohol and sorbitol dehydrogenases. *In*: H. Weiner, B. Wermuth and R. Holmes (eds): *Enzymology and Molecular Biology of Carbonyl Metabolism 5*, Plenum Press, New York; *in press*.

Kawamoto, R.M. and Caswell, A.H. (1986) Autophosphorylation of glyceraldehydephosphate dehydrogenase and phosphorylation of protein from skeletal muscle microsomes. *Biochemistry* 25: 656–661.

Klinman, J.P. and Mu, D. (1994) Quinoenzymes in biology. *Annu. Rev. Biochem.* 63: 299–344.

Knighton, D.R., Zheng, J., Eyck, L.F.T., Ashford, V.A., Xuong, N.-H., Taylor, S.S. and Sowadski, J.M. (1991) Crystal structure of the catalytic subunit of cyclic adenosine monophosphate-dependent protein kinase. *Science* 253: 407–414.

Koivusalo, M., Baumann, M. and Uotila, L. (1989) Evidence for the identity of glutathionedependent formaldehyde dehydrogenase and Class III alcohol dehydrogenase. *FEBS Lett.* 257: 105–109.

Koshland, D.E. (1958) Application of a theory of enzyme specificity to protein synthesis. *Proc. Natl. Acad. Sci. USA* 44: 98–104.

Kraut, J. (1977) Serine proteases: structure and mechanism of catalysis. *Annu. Rev. Biochem* 46: 331–358.

Kraut, J. (1988) How do enzymes work? *Science* 242: 533–540.

Krishna, R.G. and Wold, F. (1993) Post-translational modification of proteins. *Adv. Enzymol. Rel. Areas Mol. Biol.* 67: 265–298.

Kuriyan, J., Krishna, T.S.R., Wong, L., Guenther, B., Pahler, A., Williams, C.H. and Model, P. (1991) Convergent evolution of similar function in two structurally divergent enzymes. *Nature* 352: 172–174.

Lamzin, V.S., Dauter, Z., Popov, V.O., Harutyunyan, E.H. and Wilson, K.S. (1994) High resolution structures of holo and apo formate dehydrogenases. *J. Mol. Biol.* 236: 759–785.

Langin, D. and Holm, C. (1993) Sequence similarities between hormone-sensitive lipase and five prokaryotic enzymes. *Trends Biochem. Sci.* 18: 466–467.

Lee, B.J., Rajagopalan, M., Kim, Y.S., You, K.H., Jacobson, B.J. and Hatfield, D. (1990) Selenocysteine tRNA[Ser]Sec gene is ubiquitous within the animal kingdom. *Mol. Cell Biol.* 10: 1940–1949.

Lehman, N. and Joyce, G.F. (1993) Evolution *in vitro*: analysis of a lineage of ribozymes. *Current Biology* 3: 723–734.

Lesk, A., Brändén, C.-I. and Chothia, C. (1989) Structural principles of α/β barrel proteins: the packing of the interior of the sheet. *Proteins: Struct. Funct. Genet.* 5: 139–148.

Levitt, M. and Perutz, M.F. (1988) Aromatic rings act as hydrogen bond acceptors. *J. Mol. Biol.* 55: 379–400.

Li, H. and Goldstein, B.M. (1992) Carboxamide group conformation in the nucotinamide and thiazole-4-carboxamide rings: implications for coenzyme binding. *J. Med. Chem.* 35: 3560–3567.

Li, J., Vrielink, A., Brick, P. and Blow, D.M. (1993) Crystal structure of cholesterol oxidase complexed with a steroid substrate: implications for flavin adenine dinucleotide dependent alcohol oxidases. *Biochemistry* 32: 11507–11515.

Liang, J.-Y., Zhang, Y., Huang, S. and Lipscomb, W.N. (1993) Allosteric transition of fructose-1,6-bisphosphatase. *Proc. Natl. Acad. Sci. USA* 90: 2132–2136.

Littlechild, J.A. and Watson, H.C. (1993) A data-based reaction mechanism for type I fructose bisphosphate aldolase. *Trends Biochem. Sci.* 18: 36–39.

Mackay, L.G., Wylie, R.S. and Sanders, J.K.M. (1994) Catalytic acyl transfer by a cyclic porphyrin trimer: efficient turnover without product inhibition. *J. Am. Chem. Soc.* 116: 3141–3142.

Malcolm, B.A., Rosenberg, S., Corey, M.J., Allen, J.S., de Baetselier, A. and Kirsh, J.F. (1989) Site-directed mutagenesis of the catalytic residues Asp-52 and Glu-35 of chicken egg white lysozyme. *Proc. Natl. Acad. Sci. USA* 86: 133–137.

Maret, W. (1991) Novel substrates and inhibitors of human liver sorbitol dehydrogenase. In: H. Weiner, D.W. Crabb and T.G. Flynn (eds): *Enzymology and Molecular Biology of Carbonyl Metabolism 3*, Plenum Press, New York, pp 327–336.

Meyer-Siegler, K., Mauro, D.J., Seal, G., Wurzer, J., de Riel, J.K. and Sirover, M.A. (1991) A human nuclear uracil DNA glycosylase is the 37-kD subunit of glyceraldehyde-3-phosphate dehydrogenase. *Proc. Natl. Acad. Sci. USA* 88: 8460–8464.

Miles, E.W. (1991) Structural basis for catalysis by tryptophan synthase. *Adv. Enzymol. Rel. Areas Mol. Biol.* 64: 93–172.

Miller, J.R. and Ingolia, T.D. (1989) Cloning and characterization of β-lactam biosynthetic genes. *Mol. Microbiol.* 3: 689–695.

Miyazaki, K. and Oshima, T. (1994) Co-enzyme specificity of 3-isopropylmalate dehydrogenase from *Thermus thermophilus* HB8. *Protein Engng.* 7: 401–403.

Miziorko, H.M. and Lorimer, G.H. (1983) Ribulose-1,5-bisphosphate carboxylase-oxygenase. *Annu. Rev. Biochem.* 52: 507–535.

Mock, W.L., Irra, T.A., Wepsiec, J.P. and Adhya, M. (1989) Catalysis by curcurbituril. The significance of bound-substrate destabilization for induced trizole formation. *J. Org. Chem.* 54: 5302–5308.

Morgenegg, G., Winkler, G.C., Hubscher, U., Heizmann, C.W., Mous, J. and Kuenzle, C.C. (1986) Glyceraldehyde-3-phosphate dehydrogenase is a nonhistone protein and a possible activator of transcription in neurons. *J. Neurochem.* 47: 54–62.

Morjana, N.A. and Flynn, T.G. (1989) Aldose reductase from human psoas muscle. Purification, substrate specificity, immunological characterization, and effect of drugs and inhibitors. *J. Biol. Chem.* 264: 2906–2911.

Muller, Y.A. and Schulz, G.E. (1993) Structure of the thiamine and flavine-dependent enzyme pyruvate oxidase. *Science* 259: 965–967.

Muller, Y.A., Lindqvist, Y., Furey, W., Schulz, G.E., Jordan, F. and Schneider, G. (1993) A thiamin diphosphate binding fold revealed by comparison of the crystal structures of transketolase, pyruvate oxidase and pyruvate decarboxylase. *Structure* 1: 95–103.

Mushegian, A.R. and Koonin, E.V. (1994) Unexpected sequence similarity between nucleotidases and phosphoribosyltransferases of different specificity. *Protein Sci.* 3: 1081–1088.

Nienaber, V.L., Breddam, K. and Birktoft, J.J. (1993) A glutamic acid specific protease utilizes a novel histidine triad in substrate binding. *Biochemistry* 32: 11469–11475.

Norwick, J.S., Feng, Q., Tjivikua, T., Ballester, P. and Rebek, J. (1991) Kinetic studies and modeling of a self-replicating system. *J. Am. Chem. Soc.* 113: 8831–8839.

Ollis, D.L., Cheah, E., Cygler, M., Dijkstra, B., Frolow, F., Franken, S.M., Harel, M., Remington, S.J., Silman, I., Schrag, J., Sussman, J.L., Verschueren, K.H.G. and Goldman, A. (1992) The α/β hydrolase fold. *Protein Engng.* 5: 197–211.

Olson, A.R. and Miller, R.J. (1938) The mechanism of the aqueous hydrolysis of β-butyrolactone. *J. Am. Chem. Soc.* 60: 2687–2692.

Orengo, C.A. and Thornton, J.M. (1993) Alpha plus beta folds revisited: some favoured motifs. *Structure* 1: 105–120.

Osawa, S., Jukes, T.H., Watanabe, K. and Muta, A. (1992) Recent evidence for evolution of the genetic code. *Microbiol. Rev.* 56: 229–264.

Parés, X., Cederlund, E., Moreno, A., Hjelmqvist, L., Farres, J. and Jörnvall, H. (1994) Mammalian class IV alcohol dehydrogenase (stomach alcohol dehydrogenase): structure, origin, and correlation with enzymology. *Proc. Natl. Acad. Sci. USA* 91: 1893–1897.

Parsot, C. (1986) Evolution of biosynthetic pathways: a common ancestor for threonine synthase, threonine dehydratase and D-serine dehydratase. *EMBO J.* 5: 3013–3019.

Peltoketo, H., Isomaa, V. and Vihko, R. (1992) Genomic organization and DNA sequences of human 17β-hydroxysteroid dehydrogenase genes and flanking regions. *Eur. J. Biochem.* 209: 459–466.

Persson, B., Jeffery, J. and Jörnvall, H. (1991) Different segment similarities in long-chain dehydrogenases. *Biochem. Biophys. Res. Commun.* 177: 218–223.

Persson, B., Zigler, J.S. and Jörnvall, H. (1994) A superfamily of medium-chain dehydrogenases/reductases (MDR): sub-lines including ζ-crystallin, alcohol and polyol dehydrogenases, quinone oxidoreductases, enoyl reductases, VAT-1 and other proteins. *Eur. J. Biochem.* 226: 15–22.

Petsko, G.A., Kenyon, G.L., Gerlt, J.A., Ringe, D. and Kozarich, J.W. (1993) On the origin of enzymatic species. *Trends Biochem. Sci.* 18: 372–376.

Pettersson, G. (1989) Effect of evolution on the kinetic properties of enzymes. *Eur. J. Biochem.* 184: 561–566.

Piccirilli, J.A., Vyle, J.S., Caruthers, M.H. and Cech, T.R. (1993) Metal ion catalysis in the *Tetrahymena* ribozyme reaction. *Nature* 361: 85–88.

Picot, D., Loll, P.J. and Garavito, R.M. (1994) The X-ray crystal structures of the membrane protein prostaglandin H2 synthase-1. *Nature* 367: 243–249.

Pocklington, T. and J. Jeffery, J. (1968) 3α-Hydroxysteroid: NAD oxidoreductase activity in crystalline preparations of 20β-hydroxysteroid: NAD oxidoreductase. *Eur. J. Biochem.* 7: 63–67.

Pocklington, T., Jeffery, J., Middleditch, B.S. and Brooks, C.J.W. (1970) Reduction of 5α-androstane-3,16-dione by a crystalline 20β-hydroxysteroid-nicotinamide-adenine dinucleotide oxidoreductase preparation. *Biochem. J.* 119: 803–804.

Pompliano, D.L., Peyman, A. and Knowles, J.R. (1990) Stabilization of a reaction intermediate as a catalytic device: definition of the functional role of the flexible loop in triosephosphate isomerase. *Biochemistry* 29: 3186–3194.

Rawn, J.D. (1989) *Biochemistry*, Second Edition, Patterson, Burlington, North Carolina.

Rice, P.R., Goldman, A. and Steitz, T.A. (1990) A helix-turn-strand motif common in α-β proteins. *Proteins: Struct. Funct. Genet.* 8: 334–341.

Richardson, J.S. and Richardson, D.C. (1988) Amino acid preferences for specific locations at the ends of α helices. *Science* 240: 1648–1652.

Roberts, V.A., Stewart, J., Benkovic, S.J. and Getzoff, E.D. (1994) Catalytic antibody model and mutagensis implicate arginine in transition-state stabilization. *J. Mol. Biol.* 235: 1098–1116.

Robison, P.D., Martin, M.N. and Tabita, F.R. (1979) Differential effects of metal ions on *Rhodospirillum rubrum* ribulose bisphosphate carboxylase/oxygenase and stoichiometric incorporation of HCO3- into a cobalt(III)-enzyme complex. *Biochemistry* 18: 4453–4458.

Roth, G.J., Stanford, N. and Majerus, P.W. (1975) Acetylation of prostaglandin synthase by aspirin. *Proc. Natl. Acad. Sci. USA* 72: 3073–3076.

Sanders, J.K.M. (1994) Enzyme Mimics. *Proc. Indian Acad. Sci. (Chem. Sci.)* 106: 983–988.

Schneider, G. and Lindqvist, Y. (1993) Enzymatic thiamine catalysis: mechanistic implications from the three-dimensional structure of transketolase. *Bioorgan. Chem.* 21: 109–117.

Schultz, D.W. and Yarus, M. (1994) Transfer RNA mutation and the malleability of the genetic code. *J. Mol. Biol.* 235: 1377–1380.

Sem, D.S. and Kasper, C.B. (1993) Interaction with arginine 597 of NADPH-cytochrome P-450 oxidoreductase is a primary source of the uniform binding energy used to discriminate between NADPH and NADH. *Biochemistry* 32: 11548–11558.

Shokat, K.M., Leumann, C.J., Sugasawara, R. and Shultz, P.G. (1989) A new strategy for the generation of catalytic antibodies. *Nature* 338: 269–271.

Shokat, K.M. and Shultz, P.G. (1991) The generation of antibody combining sites containing catalytic residues. *CIBA Foundation Symposium* 159: 118–134.

Smith, M.W., Feng, D.-F. and Doolittle, R.F. (1992) Evolution by acquisition: the case for the horizontal gene transfers. *Trends Biochem. Sci.* 17: 489–493.

Smithgall, T.E., Harvey, R.G. and Penning, T.M. (1986) Regio- and Stereospecificity of homogeneous 3α-hydroxysteroid-dihydrodiol dehydrogenase for trans-dihydrodiol metabolites of polycyclic aromatic hydrocarbons. *J. Biol. Chem.* 261: 6184–6191.

Sneider, H.-J., Kramer, R. and Rammo, J. (1993) Kinetic and spectroscopic characterization of ternary complexes by numerical fitting methods. Catalysis of acyl-transfer reactions by a macrocyclic azoniacyclophan. *J. Am. Chem. Soc.* 115: 8980–8984.

Sprang, S.R., Withers, S.G., Goldsmith, E.J., Fletterick, R.J. and Madsen, N.B. (1991) Structural basis for the activation of glycogen phosphorylase b by adenosine monophosphate. *Science* 254: 1367–1371.

Steitz, T.A., and Steitz, J.A. (1993) A general two-metal-ion mechanism for catalytic RNA. *Proc. Natl. Acad. Sci. USA* 90: 6498–6502.

Stillman, T.J., Baker, P.J., Britton, K.L. and Rice, D.W. (1993) Conformational flexibility in glutamate dehydrogenase. Role of water in substrate recognition and catalysis. *J. Mol. Biol.* 234: 1131–1139.

Storm, D.R. and Koshland, D.E. (1970) A source of the special catalytic power of enzymes: orbital steering. *Proc. Natl. Acad. Sci. USA* 66: 445–452.

Strickler, R.C., Tobias, B. and Covey, D.F. (1981) Human placental 17β-estradiol dehydrogenase and 20α-hydroxysteroid dehydrogenase. Two activities at a single enzyme active site. *J. Biol. Chem.* 256: 316–321.

Strynadka, N.C.J. and James, M.N.G. (1991) Lysozyme revisited: crystallographic evidence for distortion of an N-acetylmuramic acid residue bound in site D. *J. Mol. Biol.* 220: 401–424.
Suttie, J.W. (1980) Vitamin-K-dependent carboxylation. *Trends Biochem. Sci.* 5: 302–304.
Sykes, P. (1986) *A Guidebook to Mechanism in Organic Chemistry.* Sixth Edition, Longman, London.
Tanaka, T., Yamamoto, S., Moriya, T., Taniguchi, M., Hayashi, H., Kagamiyama, H. and Oi, S. (1994) Aspartate transaminase from a thermophilic formate-utilizing methanogen. *Methanobacterium thermoformicicum* strain SF-4: relation to serine and phosphoserine aminotransferases, but not to the aspartate aminotransferase family. *J. Biochem.* 115: 309–317.
Tuschl, T. and Eckstein, F. (1993) Hammerhead ribozymes: importance of stem-loop II for activity. *Proc. Natl. Acad. Sci. USA* 90: 6991–6994.
Uppenberg, J., Hansen, M.T., Patkar, S. and Jones, T.A. (1994) The sequence, crystal structure determination and refinement of two crystal forms of lipase B from *Candida antarctica. Structure* 2: 293–308.
Verughese, K.I., Skinner, M.M., Whiteley, J.M., Matthews, D.A. and Xuong, N.H. (1992) Crystal structure of rat liver dihydropteridine reductase. *Proc. Natl. Acad. Sci. USA* 89: 6080–6084.
Verschueren, K.H.G., Seljee, F., Rozeboom, H.J., Kalk, K.H. and Dijkstra, B.W. (1993) Crystallo-graphic analysis of the catalytic mechanism of haloalkane dehydrogenase. *Nature* 363: 693–698.
Voet, D. and Voet, J.G. (1990) *Biochemistry.* Wiley, New York.
Waksman, G., Krishna, T.S.R., Williams, C.H. and Kuriyan, J. (1994) Crystal structure of *Escherichia coli* thioredoxin reductase refined at 2 Å resolution. Implications for a large conformational change during catalysis. *J. Mol. Biol.* 236: 800–816.
Walsh, C. (1979) *Enzymatic Reaction Mechanisms.* Freeman, San Francisco.
Warren, J.C., Murdock, G.L., Ma, Y., Goodman, S.R. and Zimmer, W.E. (1993) Molecular cloning of testicular 20α-hydroxysteroid dehydrogenase: identity with aldose reductase. *Biochemistry* 32: 1401–1406.
Warshel, A. and Levitt, M. (1976) Theoretical studies of enzymatic reactions: dielectric electrostatic and steric stabilization of the carbonium ion in the reaction of lysozyme. *J. Mol. Biol.* 103: 227–249.
Weinhold, E.G., Glasfeld, A., Ellington, A.D. and Benner, S.A. (1992) Structural determinants of stereospecificity in yeast alcohol dehydrogenase. *Proc. Natl. Acad. Sci. USA* 88: 8420–8424.
Wetterholm, A. and Haeggström, J.Z. (1992) Leukotriene A4 hydrolase: an anion activated peptidase. *Biochim. Biophys. Acta* 1123: 275–281.
White, I.H. and Jeffery, J. (1972) Structural features of ring A and the interaction of 20-oxo steroids with cortisone reductase. *Eur. J. Biochem.* 25: 409–414.
White, I.H. and Jeffery, J. (1973a) Identification of the product of the reaction of progesterone with cortisone reductase and reduced nicotinamide-adenine dinucleotide. *Biochem. Soc. Trans.* 1: 767–769.
White, I.H. and Jeffery, J. (1973b) Structural features of ring B and the interaction of 20-oxosteroids with cortisone reductase. *Biochim. Biophys. Acta* 296: 604–614.
White, I.H. and Jeffery, J. (1974a) Structural features of ring C of 20-oxo steroids and the interaction with cortisone reductase. *Biochem. J.* 137: 349–354.
White, I.H. and Jeffery, J. (1974b) Influence of the structure of the D ring of various 20-oxo steroids on their behaviour as substrates for cortisone reductase. *Biochem. Soc. Trans.* 2: 984.
White, I.H. and Jeffery, J. (1975) The functioning of a nicotinamide-adenine dinucleotide-dependent dehydrogenase and the structure adjacent to the reacting carbon atom of the substrate. *Biochem. Soc. Trans.* 2: 674–675.
White, I.H. and Jeffery, J. (1977) Cortisone reductase and some non-steroid analogues of its steroid substrates. *Biochem. Soc. Trans.* 5: 723–724.
Williams, R.J.P. (1993) Are enzymes mechanical devices? *Trends Biochem. Sci.* 18: 115–117.
Wilmanns, M., Hyde, C.C., Davies, D.R., Kirschner, K. and Jansonius, J.N. (1991) Structural conservation in parallel β/α-barrel enzymes that catalyze three sequential reactions in the pathway of tryptophan biosynthesis. *Biochemistry* 30: 9161–9169.

Wilson, D.K., Bohren, K.M., Gabbay, K.H. and Quiocho, F.A. (1992) An unlikely sugar substrate site in the 1.65 Å structure of the human aldose reductase holenzyme implicated in diabetic complications. *Science* 257: 81–84.

Wilson, D.K., Tarle, I., Petrash, J.M. and Quiocho, F.A. (1993) Refined 1.8 Å structure of human aldose reductase complexed with the potent inhibitor zopolrestat. *Proc. Natl. Acad. Sci. USA.* 90: 9847–9851.

Wong, D.J., Vanoni, M.A. and Blanchard, J.S. (1988) Glutathione reductase: solvent equilibrium and kinetic isotope effects. *Biochemistry* 27: 5232–5236.

Wood, H.G. and Zwolinski, G.K. (1976) Transcarboxylase: role of biotin, metals, and subunits in the reaction and its quaternary structure. *Critical Reviews in Biochemistry* 4: 47–122.

Yang, W., Li, E., Kairong, T. and Stanley, S.L. (1994) *Entamoeba histolytica* has an alcohol dehydrogenase homologous to the multifuntional *adhE* gene product of *Escherichia coli*. *Mol. Biochem. Parasitol.* 64: 253–260.

Yano, T., Hinoue, Y., Chen, V.J., Metzler, D.E., Mihyahara, I., Hirotsu, K. and Kagamiyama, H. (1993) Role of an active site residue analyzed by combination of mutagènsis and coenzyme analogue. *J. Mol. Biol.* 234: 1218–1229.

Zhang, F., Strand, A., Robbins, D., Cobb, M.H. and Goldsmith, E.J. (1994) Atomic structure of the MAP kinase ERK at 2.3 Å resolution. *Nature* 367: 704–711.

Analysis of the structure of naturally processed peptides bound by Class I and Class II major histocompatibility complex molecules

E. Appella[1], E.A. Padlan[2] and D.F. Hunt[3]

[1]*Laboratory of Cell Biology, National Cancer Institute, and*
[2]*Laboratory of Molecular Biology, National Institute of Diabetes and Digestive and Kidney Diseases, NIH, Bethesda, MD 20892;*
[3]*Departments of Chemistry and Pathology, University of Virginia, Charlottesville, VA 22901, USA*

Summary. Antigen-specific T-cell responses require antigenic peptides presented on the cell surface by the major histocompatibility complex (MHC) molecules. The structural characteristics of these peptides are being defined. It is now known that the majority of peptides that associate with MHC Class I are 8–10 residues long, with allotype-specific binding motifs containing up to three anchor positions. This is consistent with the presence of six pockets and the close-ended structure of the MHC Class I peptide binding groove. In contrast to peptides associated with MHC Class I, those associated with MHC Class II are 10–34 residues in length and are commonly presented in nested sets with extensions or truncations at the N- or C-terminal ends. Binding motifs for MHC Class II appear to contain up to four anchor positions, with more loosely defined amino acid preferences. The expression of histocompatibility proteins in cells that do not load peptides but fold them correctly has permitted the X-ray analysis of the three-dimensional structure of Class I and Class II complexes with single defined peptides. The differences in length between the Class I and Class II sequences can be ascribed to small structural differences between the two binding sites and the positioning of key residues that make hydrogen bonds to the bound peptides. Recent advances in mass spectrometry are making possible the analysis and sequencing of subpicomolar quantities of MHC-bound peptides and the estimation of the size of the total population of peptides. The sequence information derived from this technique, in conjunction with X-ray crystallographic analysis of MHC complexes involving single, defined peptides, may provide a new approach towards the development of useful reagents for therapeutic intervention.

Introduction

The immune system has developed a variety of mechanisms to detect and eliminate invading microorganisms or any foreign material. Two types of immune recognition systems have evolved. The first is based on the specific recognition properties of immunoglobulin molecules, which constitute both the receptor and effector molecules of B lymphocytes. Immunoglobulins are comprised of heavy (H) and light (L) polypeptide chains connected by disulfide bonds. Each chain has a variable region (V) and a constant region (C). The V domain of the H chain is coded by three partial genes and that of the L chain by two genes which rearrange somatically to generate the diversity of the receptor/antibody molecules. This somatic rearrangement is crucial in developing the combinatorial diversity of the immune system's antibody repertoire. The second system, responsible for cell-mediated immunity, involves a second type of lymphocyte called a T-cell. Each T-cell has a unique receptor (TCR) that recognizes peptide fragments of antigens. These peptides, by-products of endocellular degradation steps, are transferred to the cell surface in association with major histocompatibility complex (MHC) molecules. Two different MHC-molecules, Class I and Class II, have evolved to recognize peptide and further divide the cell-mediated immunity response into two systems.

Class I molecules are present on most cell types of the body and consist of a glycoprotein of 45 kDa non-covalently associated with another protein, $\beta 2$ microglobulin. Class II molecules are expressed on B cells, macrophages, monocytes, dendritic cells and induced by interferons in other cell types and are composed of a heterodimer of transmembrane subunits, α and β, 33 kDa and 29 kDa, respectively. The structures of both Class I and Class II MHC molecules have been determined by X-ray crystallography and have been found to be rather similar in three dimensions. In both Class I and Class II MHC, the peptide-binding site is comprised of a floor of eight β-strands flanked by two helical walls. In Class I MHC, the peptide-binding site is formed by the first two domains of the heavy chains ($\alpha 1$ and $\alpha 2$) (Fig. 1a,b); in Class II MHC, the site is formed by the first two domains of each chain ($\alpha 1$ and $\beta 1$) (Fig. 1c,d). The peptide-binding platform is supported underneath by two immunoglobulin-like domains, by the third domain of the heavy chain ($\alpha 3$) and β-2 microglobulin in Class I and by the two second domains of the α and β chains ($\alpha 2$ and $\beta 2$) in Class II. In both classes, the peptide binding groove is lined by polymorphic residues (Fig. 1e,f). Studies using mutant cell lines that do not interact significantly with peptides have suggested that peptide binding is required for assembly of the Class I molecule and for stable cell surface expression (Townsend and Bodmer, 1989). Class II molecules show great similarity in the structure of the binding site with Class I. Peptides in the Class II binding groove show an extended conformation, projecting out of both

Figure 1. Stereodrawings of the Cα backbone structure of the peptide-binding sites of human MHC molecules. (a) and (b): two perpendicular views of the peptide-binding site of the Class I HLA-A2 with bound LLFGYPVYV peptide (Madden et al., 1993); the peptide positions 2 and 9 are indicated in (b). (c) and (d): similar views of the peptide-binding site of the Class II HLA-DR1 with bound PKYVKQNTLKLAT peptide (Stern et al., 1994). The α chain is shown in yellow and the β chain is shown in blue. The dotted molecular surfaces lining the binding grooves were computed using program MS of M. L. Connolly (1983). Atomic coordinates were from the Protein Data Bank (Bernstein et al., 1977): Entry 1HHk for the HLA-A2 complex and entry 1DLH for the HLA-DR1 complex. The polymorphism in human Class I and Class II MHC molecules is depicted in e and f, respectively, in the form of dotted spheres; the radius of each sphere is proportional to the amino acid variability at that position computed according to Padlan (1977).

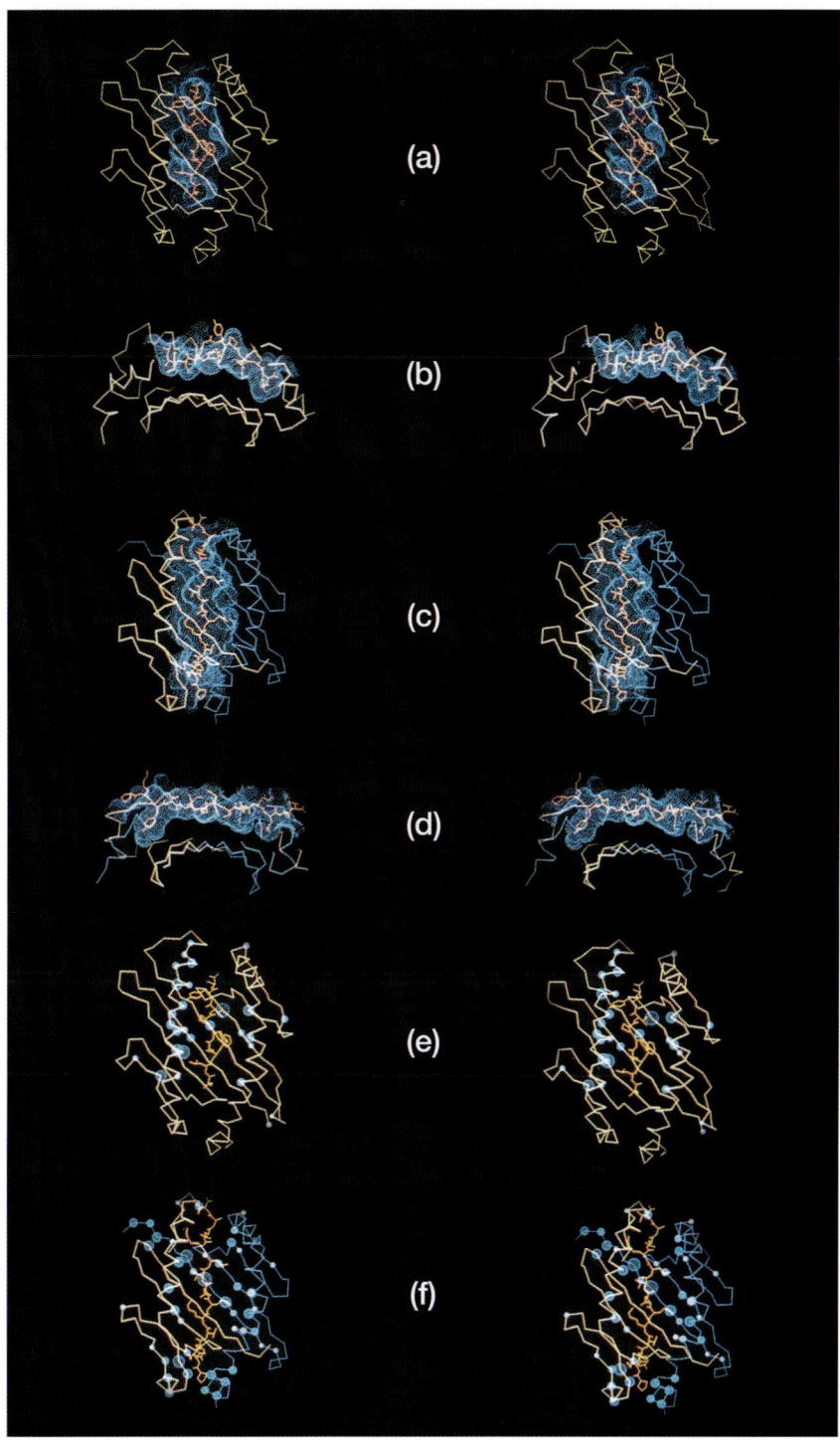

ends of the site. Several studies have demonstrated that less than 20% of Class II molecules will bind peptides derived from foreign antigens (Buus et al., 1986; Roche and Cresswell, 1990); the peptide binding sites on the remainder of the proteins are occupied by endogenous peptides which interfere with the binding of exogenous peptides. Class II molecules have been expressed in insect cell cultures and in this case the antigen binding site appears to be "empty". In contrast to Class I molecules, the subunits of Class II will assemble in the absence of peptides. Peptide binding to Class II molecules, however, does stabilize the molecules against aggregation *in vitro* (Stern and Wiley, 1992; Germain and Rinker, 1993).

Exploration of the structural motifs in naturally-occurring peptides bound to MHC molecules was initially carried out by studies with synthetic peptides (Townsend et al., 1986; Rotzscheke et al., 1990; Corr et al., 1992, 1993). Cytotoxic T-cells were shown to recognize virally-infected cells by virtue of their specific recognition of these cells' Class I MHC molecules bearing virus-derived peptides. These peptides could be gently extracted from partially purified MHC-complexes by acid treatment and subjected to extensive biochemical and immunological comparison with synthetic peptides. The natural peptides behaved chemically like their synthetic analogues, which are predominantly of nine residues length. Absolute amounts of the naturally-bound peptides were estimated to be between 200 and 500 copies per cell. (Falk et al., 1991). These numbers were consistent with the minimal number of peptide-MHC complexes believed to be necessary for stimulating MHC Class II-restricted T cells (Demotz et al., 1990; Harding and Unanue, 1992). In contrast to Class I bound peptides, most of the naturally-occurring peptides extracted from Class II molecules are 12–19 residues in length and many share a core sequence with differences localized predominantly at the amino and carboxyl termini (Rudensky et al., 1991; Hunt et al., 1992; Chicz et al., 1993).

HPLC fractionation and automated Edman degradation have been employed for detailed structural analysis of peptides extracted from some MHC molecules. In a few cases complete amino acid sequences were obtained (Jardetzky et al., 1991; Rudensky et al., 1991). Given the potential diversity of MHC-bound peptides, however, it became apparent that detailed analysis could only be successfully applied if the protein extracts contained one or two dominant peptides by HPLC. In the case of more complex mixtures one could define statistical preferences for amino acids with respect to position in the bound peptide (Falk et al., 1991).

The task of sequencing a highly complex mixture of peptides with a concentration range of the order of 10–1000 femtomoles per 10^8 cells is formidable. The development of microcapillary HPLC in combination with electrospray ionization/tandem mass spectrometry made it possible to sequence peptides presented by both Class I and Class II molecules (Hunt et al., 1992a, 1992b). In addition, the approximate number and quantity of individual peptides can be determined, their molecular mass defined, and the analysis of peptides in mixtures can be performed at the sub-picomole level.

One technique which permits a direct visualization of molecular interactions is X-ray crystallography. As applied to the study of MHC molecules, it has provided not only the three-dimensional structures of both Class I and Class II molecules, but also the mode of binding of peptide and the conformation of the bound peptide. In the beginning, the complexes studied involved peptide mixtures. More recently, advances in the expression of MHC molecules in non-mammalian systems have permitted the formation of complexes involving single, defined peptides. The MHC-peptide complexes studied by crystallography include the human Class I HLA-A2 molecule complexed with an endogenous mixture of peptides (Bjorkman et al., 1987; Saper et al., 1991) and more recently complexed with five different viral peptides (Madden et al., 1993); the human Class I HLA-Aw68 complexed with an endogenous mixture (Garrett et al., 1989; Guo et al., 1992) and also with a viral peptide (Silver et al., 1992); the human Class I HLA-B27 complexed with an endogenous mixture (Madden et al., 1991, 1992), the murine Class I H-2K^b complexed with two viral peptides (Fremont et al., 1992; Zhang et al., 1992), the murine Class I H-2D^b complexed with a viral peptide (Young et al., 1994), and the human Class II HLA-DR1 initially complexed with a mixture of peptides (Brown et al., 1993), and very recently complexed with an influenza hemagglutinin peptide (Stern et al., 1994). The pictures of the MHC-peptide interactions obtained from these studies provide a structural basis for the peptide selectivity of the different MHC molecules.

The identification of peptide-binding motifs has important implications in medicine. Indeed, the analysis of individual MHC-associated peptides and their sequences not only allows meaningful comparisons of the binding specificity between the different alleles of each MHC class but, very importantly, also assists in the design of potential epitopes for vaccine development.

Results

Microcapillary HPLC/electrospray ionization mass spectrometry

Naturally processed peptides associated with Class I and Class II MHC molecules are co-isolated with immuno-affinity purified MHC-complexes and separated from them by filtration after dissociation at low pH. An alternative method has involved the extraction of whole cells

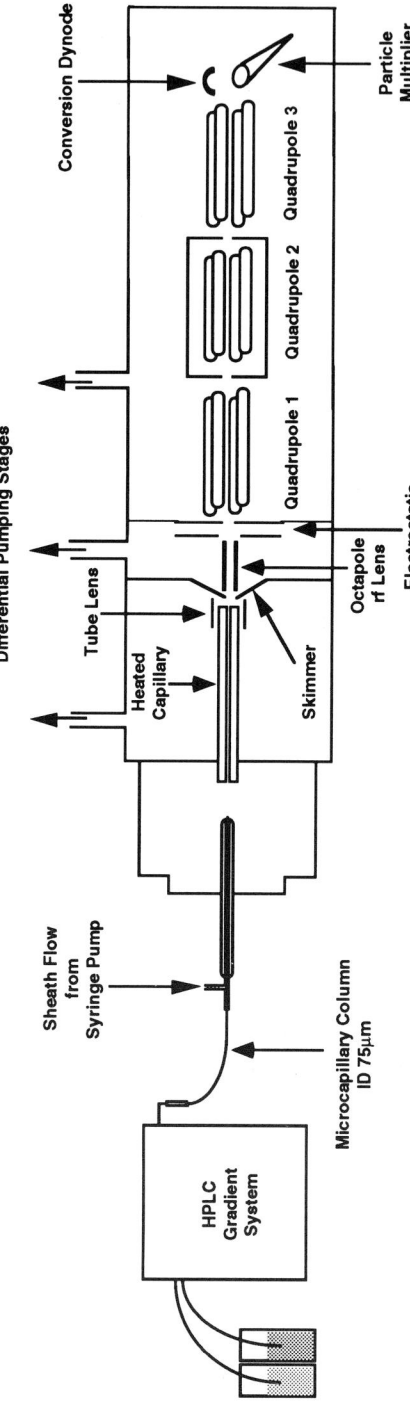

Figure 2. Triple quadrupole mass spectrometer interfaced to microcapillary HPLC via an electrospray ion source.

with trifluoroacetic acid; however, this method is used less often due to high background and somewhat lower yields for sequencing. The peptide extracts, which represent a complex mixture in low abundance, were analyzed by microcapillary HPLC in conjunction with electrospray ionization/tandem mass spectrometry. Individual species can then be directly sequenced in the mass spectrometer by collision activated dissociation analysis.

Figure 2 shows a triple quadrupole spectrometer. The instrument is interfaced with a microcapillary HPLC column via an electrospray ion source. Microcapillary HPLC columns are made of fused silica tubing with an inside diameter of 75 µm, with the last 10 cm of the column packed with Poros R/H II reverse phase absorbent (Perceptive Biosystems, Cambridge, MA, USA). Peptides eluted with a 12-min linear gradient of 0–80% acetonitrile (A), 0.5% acetic acid (B), at 500 nanoliters per min, enter the mass spectrometer as protonated ions and are analyzed by the third quadrupole, Q3. A spectrum is generated which displays the molecular mass and relative abundance of all the peptides in the mixture. More than 10 000 different peptides were estimated to be associated with Class I molecules isolated from 10^9 cells. Ninety percent of these peptides were present at 0.01–0.1% of the total extract, or 100–1000 copies/cell. Roughly the same number of peptides was also estimated to be associated with Class II molecules. Again, most of these peptides were present at levels considerably below 1% of the total sample. In some cases, viral peptides associated with murine Class II molecules represented the bulk of the total peptide material recovered; it is unclear, however, if these results are unique to the cell line from which the molecules were isolated (Rudensky et al., 1991). Sequence information on individual peptides present in the above mixtures was obtained by repeating the microcapillary HPLC and subjecting single ions of known mass to collision activated dissociation (CAD). To obtain this data the first mass filter (Q1, Fig. 2) is set to pass all ions within a 6 mass unit window centered around the selected ion; these ions are transmitted to quadrupole 2 (Q2), all the others are rejected. Quadrupole 2 is pressurized with argon and operated to transmit all ions of all masses. Protonated peptide molecules that enter Q2 suffer 10–100 collisions with argon atoms, become energized and fragment at one of the amide bonds in the molecule more or less randomly. A collection of neutral species and ions is produced by the above collisional process; neutrals are pumped away by the vacuum system, and the charged fragments, all derived from a particular peptide component, are transmitted to the third quadrupole (Q3) that separates them according to mass. The result is a ladder of fragments that differ in length (mass) by single amino acid residues. Mass measurements in Q3 generate a mass spectrum from which the order of amino acids in a particular peptide can be deduced. This process can be repeated every

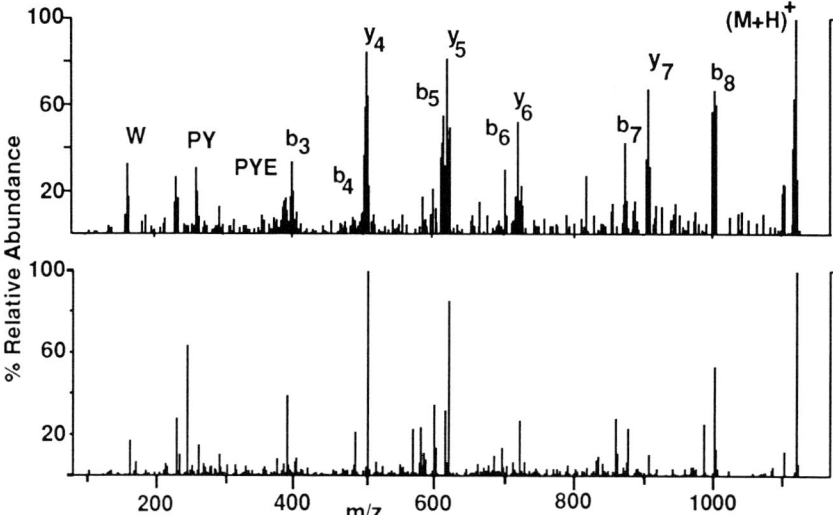

Figure 3. CAD mass spectra of the $(M+H)^+$ ion at m/z 1121. The top spectrum was recorded on a sample quantity of 100–300 femtomoles; the bottom spectrum required 2 picomoles and the spectrum is recorded at approximately unit resolution.

2–8 seconds and additional peptides in the original mixture can be analyzed. Figure 3 shows a spectrum of a $(M+H)^+$ ion at m/z 1121 recovered from the mixture of Class I extracted peptides. The sample quantity is 100–300 femtomoles. To generate the spectrum in Figure 3, the mass range from 50 to 1221 Da was scanned six times and the resulting spectra were then summed. Signals for each of the fragment ions in the spectrum appear in clusters due to the contribution of mass iostopes such as ^{13}C and because the mass resolution of the system is high in order to increase sensitivity. The sequence of amino acids in the sample is deduced by subtracting m/z values for peaks that occupy the same relative positions within two cluster patterns and are separated by at least 57 and no more than 186 mass units, the residue masses of the lightest and heaviest amino acids. predicted masses (m/z values) for ions of type y are shown below the sequence in Figure 3. These ions all contain the carboxyl-terminal residue plus one or more additional residues. Predicted mass (m/z values) for ions of type b are shown above the sequence. These ions all contain the amino terminal residue plus one or more residues. The residues observed in the spectrum are underlined. Fragments which result from a cleavage at Pro are labeled with the single letter code for the amino acids contained within the fragment. The amino acids Leu/Ile have identical masses and they cannot be differentiated on the triple quadrupole instrument and are specified in the structure as Lxx. Assignment of the residue 2 as Leu in this peptide was established by the use of synthetic peptides containing Leu or Ile at this position and by demonstrating that only the Leu containing peptide co-eluted from a microcapillary HPLC column with the naturally processed peptide.

Allele-specific consensus motifs

Table 1 summarizes the sequences of Class I HLA-associated peptides extracted from HLA-A, B and C alleles (Engelhard, 1993; Kubo et al., 1994; Falk et al., 1993). The predominance of peptides of nine amino acids allowed us to align them to look for conservation in amino acids at each position. Peptide position 2 (P2) shows the strongest residue preferences among the MHC class I allele-associated peptides. For peptides associated with HLA-A2.1, P2 is either Leu or Ile, with two exceptions to date, Val and Met. Peptide isolated from HLA-A3, A11, and Aw68 also contain predominantly hydrophobic amino acids at P2, whereas HLA-A1 has preferably Thr or Ser at this position. In the case of HLA-B7, the predominant residue at P2 is Pro, but for HLA-B27 the

Table 1. Structural characteristics and proposed motifs of peptides associated with human Class I MHC molecules

Isoform	Length	Side chain position[a]								C-Term
		1	2	3	4	5	6	7	8	
HLA-A1	9–12		**T,S**	**D,**E						**Y**
HLA-A2.1	9–14		**L,**M,I							**V,L,**I,A
HLA-A3	9		**V,L,**M							**K,**Y
HLA-A11	9–12		**T,**V							**K**
HLA-A24	9		**Y**							**L,F**
HLA-Aw68	9–11		**V,**T							**R,K**
HLA-B7	9	A,R	**P**	**R,K**						**L,**I,A,N,
HLA-B8	8–9			**K,**R		**K,**R				**L,**I
HLA-B27	9	R,K	**R**							**K,R**
HLA-Bw37	8–9		**D,**E			**V**I				**F,**M,I,L
HLA-Cw3	8			M,Y	P		**F,Y**			**L**
HLA-Cw4	9		**Y,**P			**A,**M	**V,**L			**L,**M,F
HLA-Cw6	9					**I,F,**M	**V,**I			**L**
HLA-Cw7	8		**Y**			**F,**M,Y	**V,**I			**Y,L**

[a]Letters in bold indicate amino acids in single-letter code used as anchor residues, and found in most peptide sequences. Italicized letters indicate amino acids infrequently found and some interpreted as auxiliary anchor residues.

residue is Arg. For HLA-C, P2 is variable and different for each allelic product. These results emphasize that each allelic product of the MHC-molecules binds a distinct set of peptides. Allelic variants of MHC molecules bind peptides based on selectivity for classes of amino acids predominantly at position 2. Allele by allele comparison of MHC molecules with their cognate peptides reveals wide chemical diversity in the preference of amino acids at position 2.

The residues found at position 9 (P9) are mostly those with aliphatic hydrocarbon side chains (Ala, Val, Leu and Ile) with one exception: an analysis of HLA-B27 associated peptides indicated a variety of amino acid side chains at P9, but with a bias toward positively charged amino acids. In all HLA-bound peptides the variation at P9 is higher than that observed at P2.

In contrast to the results with HLA-A2.1 derived peptides, those from HLA-B7 showed a high conservation at both positions P2 and P3 (Huczko et al., 1993). P3 was found to be Arg in 7/12 sequences and binding experiments have suggested that the P3 position plays a role in the specificity of peptide binding. An analysis of the rest of the sequences in peptide positions shows that no other residues are present at any position in greater than 50% of the HLA associated peptides. One characteristic of HLA-A2.1 peptides is the predominance of Gly and Pro at P4 which may cause a kink in the middle of the peptide backbone. This structural feature may play a role in the binding to HLA-A2.1.

Taken together, these data indicate a common mode of interaction of peptides with human MHC class I molecules based on the positioning of the amino and carboxyl termini, together with complementary interactions at P2, P3 and P9. This model, however, does not account for the binding of all peptides to these molecules. In fact, peptides longer than 9 residues can bind with affinities comparable to those of 9 residue sequences. In one rare case, for HLA-A2.1, a peptide of 14 residues was identified (Chen et al., 1994). This peptide has Leu at P2, but the last 5 residues at the carboxyl terminus all could probably be accommodated in the groove through additional kinking in the center of the peptide. Another possibility indicated by binding experiments is that the longer peptides bind in several different registers at the carboxyl terminal end, with an extension of one or more residues beyond the groove. Clearly, much remains to be learned about the factors that allow the binding of different size peptides.

Characteristics of MHC molecule peptide binding

The crystallographic analyses of peptide-class I complexes provide a structural basis for some of the observed peptide binding preferences.

For example, the X-ray studies clearly show that the amino and carboxyl terminal ends of class I associated peptides participate in the extensive set of hydrogen bonds to conserved residues that are located in two pockets, A and F, at opposite ends of the binding site. Polymorphic residues in four more centrally located pockets (B, C, D, E) may play a role in determining which peptides are bound to the MHC molecules.

The P2 side chain binds in the B pocket, which varies in structure in different class I molecules. The selectivity of the B pocket is quite well demonstrated both by crystal structure (Guo et al., 1993) and construction of models (Corr et al., 1992, 1993; Huczko et al., 1993); however, predictions based on the inspection of the side chains that form the pocket are not straightforward.

The P5 side chain can be found to interact with the C pocket of the class I binding site (Freemont et al., 1992; Zhang et al., 1992); however, this pocket is also variable in different Class I molecules and the basis for its selectivity is not well documented. For the P7 side chain, there is no single binding pocket, but several subsets of side chains from MHC molecules may or may not contact the P7 peptide side chain. The P3 side chain binds in the D pocket in the case of HLA-B7. Modeling studies have suggested that the Arg residue at P3 contributes to binding through van der Waals and electrostatic interactions in the D pocket (Huczko et al., 1993). In summary, pockets B, C and D may be largely responsible for the motifs observed for most class I molecules. On the other hand, naturally occurring peptides often do not contain all the elements of a defined motif and this suggests that other highly variable residues in the peptide also play a role in binding: the peptide residues which are not buried upon binding to MHC are available for interacting with TCR. Recognition of the presented peptide by the TCR could be via the nature of the side chains and their relative disposition; the latter would be dictated by the conformation of the peptide backbone which can be influenced by the MHC side chains lining the peptide-binding groove.

Peptide binding is different for MHC Class II. The peptides associated with MHC class II are of 10–34 residues in length with extensions/truncations at either the N or C terminal end. Based on the alignment of several sequences of the peptides bound to MHC class II DR1 and DQ3.1, a putative motif can be derived (Chicz and Strominger, 1992; Sidney et al., 1994; Hammer et al., 1992). This motif contains up to four anchor positions, with 2–4 different amino acids tolerated at each position. The other positions in the peptide sequence appear to be very permissive and only certain residues such as proline and positively or negatively charged amino acids appear not to be well tolerated. Further, the fact that many different residues are accepted at various positions, combined with a substantial variation in peptide length, has hindered the identification of other anchor positions by simple sequence alignment. Recently, MHC Class II HLA-DR1 molecules have been crystallized with a single well defined peptide from influenza virus hemagglutinin (Stern et al., 1994). The structure shows that 12 hydrogen bonds between main-chain atoms along the peptide and HLA-DR1 provide the major component to the binding interaction. Five side chains of the peptide are accommodated by polymorphic pockets in the HLA-DR1 binding site. One of the side chains is buried in a large pocket which shows a strong preference for large hydrophobic residues and appears to be the major structural feature determining peptide binding to HLA-DR1. Four other peptide side chains can be accommodated by smaller and less hydrophobic pockets, which are more permissive in the side chains that they can accept. This permissiveness may explain the binding of a great number of different peptide sequences to HLA-DR1; most of the pockets can accommodate a wide variety of side chains, and the cooperative effect of hydrogen-bonding networks along the peptide back-bone provides a major part of the binding energy. Furthermore, the X-ray structure of the human Class II molecule HLA-DR1, readily explains the lack of a length constraint in Class II bound peptides; the peptide binding groove of Class II molecules is open at both ends (Brown and Wiley, 1993).

In summary, Class II MHC bind peptides of variable length in an extended conformation using a combination of multiple main chain H-bonds and class-selected binding pockets. In addition, polymorphism in residues lining the binding pockets explain the observed allelic selectivity for various peptide sequence motifs.

Conclusions and future directions

The findings reported have a number of implications for understanding T-cell immune responses. The allele-specific consensus motifs impose a selection of antigen presentation, resulting in a reduction of the number of peptides presented by a given MHC molecule. This limitation, however, is overcome by the expression of multiple Class I and Class II loci and of different alleles at a given locus.

The analysis of the three-dimensional structure of Class I and Class II MHC molecules bound to single peptides has provided the basic mechanisms of antigenic peptide binding to histocompatibility proteins. However, many aspects of the peptide-binding interactions remain to be analyzed. For example, certain Class I alleles can bind very long peptides and it is not known which mechanism of binding is used. The question of whether peptides induce conformational changes in the MHC molecule also remains open. The surface of bound peptide not in contact with MHC-molecules potentially is accessible for interactions with the T-cell receptor. However, the question of how the T-cell receptors discriminate among different bound peptide sequences has still to be answered. A three-dimensional structure of a complex between peptide-MHC and T-cell receptor will be essential to our understanding of this interaction.

Success in the identification of MHC-allotype-specific peptide-binding motifs offers the possibility of synthesizing putative epitopes for viral, bacterial and parasitic antigens. For this to happen, it is necessary to know all the potentially antigenic peptides from the various proteins and to be able to obtain more sequence information from peptides that are represented at subfemtomole concentrations. This level of sensitivity may be obtained by mass spectrometry.

Recent developments in the ionization processes of mass spectrometry have already extended the use of this technique for sequencing subpicomole quantities of proteins and peptides. Further improvements in the design and operation of the ion-trap mass spectrometer should allow sequencing at even femtomole and subfemtomole levels and accommodate molecules with weights of up to 30 000 (Cox et al., 1992). Another advantage of this sensitive technique is the ability to differentiate between isoleucine and leucine, which has not been possible on triple quadrupole mass spectrometer systems. The application of ion-trap technology to the structural analysis of naturally bound peptides found associated with MHC molecules in defined cell lines will undoubtedly uncover the mechanism by which MHC molecules recognize diverse peptide ligands.

The availability of specific T-cell lines will greatly aid in the identification of MHC-restricted antigenic peptides. T-cells are known to be able to identify as few as 60 MHC-peptide complexes on the surface of antigen-presenting cells. This number corresponds to a peptide concentration that is 100- to 1000-fold less than that required by the most sensitive sequencing procedures to date. However, technological advances in the ion-trap system should soon enable sequencing at the levels close to those detectable by T-cells. Systematic analysis of pathogen-derived and tumor-specific epitopes will be an area of intense investigation in the near future by mass spectrometry, especially in conjunction with X-ray crystallography. The application of these technologies toward the development of therapeutic agents will be useful for the treatment and prevention of human diseases.

Acknowledgments
This work was supported by grants from the Intramural AIDS Targeted Antiviral Program of the Office of the Director of the National Institutes of Health (to E. A.) and from the U.S. Public Health Service (GM 37537 and AI 33993 to D.F.H.).

References

Bernstein, F.C., Koetzle, T.F., William, G.J.B., Meyer, E.F., Jr., Brice, M.D., Rodgers, J.R., Kennard, O., Shimanouchi, T. and Tasumi, M. (1977) The Protein Data Bank. A computer-based archival file for macromolecular structures. *J. Mol. Biol.* 112: 535–542.

Bjorkman, P.J., Saper, M.A., Samraouri, B., Bennett, W.S., Strominger, J.L. and Wiley, D.C. (1987) Structure of the human class I histocompatibility antigen, HLA-A2. *Nature* 329: 506–512.

Brown, J.H., Jardetzky, T.S., Gorga, J.C., Stern, L.J., Urban, R.G., Strominger, J.L. and Wiley, D.C. (1993) Three dimensional structure of the human class II histocompatibility antigen HLA-DRI. *Nature* 364: 33–39.

Buus, S., Colon, S., Smith, C., Freed, J.H., Miles, C. and Grey, H.M. (1986) Interaction between a "processed" ovalbumin peptide and IA molecules. *Proc. Natl. Acad. Sci. USA* 83: 3968–3971.

Chen, Y., Sidney, J., Southwood, S., Cox, A., Sakaguchi, K., Henderson, R.A., Appella, E., Hunt, D.R., Sette, A. and Engelhard, V.H. (1994) Naturally processed peptides longer than 9 amino acid residues bind to the Class I MHC molecule HLA-A2.1 with high affinity and in different conformations. *J. Immunol.* 152: 2874–2881.

Chicz, R.M., Urban, R.G., Gorga, J.C., Lane, W.S., Stern, L.J., Vignali, D.A.A. and Strominger, J.L. (1992) Predominant naturally processed peptides bound to HLA-DR1 are derived from MHC-related molecules and are heterogenous in size. *Nature* 358: 764–768.

Chicz, R.M., Urban, R.G., Gorga, J.C., Vignali, D.A., Lane, W.S. and Strominger, J.L. (1993) Specificity and promiscuity among naturally processed peptides bound to HLA-DR alleles. *J. Exp. Med.* 178: 27–47.

Connolly M.L. (1983) Analytical molecular surface calculation. *J. Appl. Crystallogr.* 16: 548–558.

Corr, M., Boyd L.F., Frankel, S.R., Kozlowski, S., Padlan, E.A. and Margulies, D.H. (1992) Endogenous peptides of a soluble major histocompatibility Complex Class I molecule, H-2L^d_S: sequence motif, quantitative binding, and molecular modeling of the complex. *J. Exp. Med.* 176: 1681–1692.

Corr, M., Boyd, L.F., Padlan, E.A. and Margulies, D.H. (1993) H-2D^d exploits a four residue peptide binding motif. *J. Exp. Med.* 178: 1877–1892.

Cox, K.A., Williams, J.D., Cooks, R.G. and Kaiser, R.E., Jr. (1992) Quadrupole ion trap mass spectrometry: current applications and future directions for peptide analysis. *Biol. Mass. Spectrom.* 21: 226–230.

Demotz, S., Grey, H.M. and Sette, A. (1990) The minimal number of Class II MHC-Antigen complexes needed for T-cell activation. *Science* 249: 1028–1030.

Engelhard, V.H. (1994) Structure of peptides associated with Class I and Class II MHC molecules. *Ann. Rev. Imm.* 12: 181–208.

Falk, K., Rotzschke, O., Stevanovic, S., Jung, G. and Rammensee, H.G. (1991) Allele-specific motifs revealed by sequencing of self-peptides eluted from MHC molecules. *Nature* 351: 290–296.

Falk, K., Rotzschke, O., Grahovac, B., Schendel, D., Stevanovic, S., Gnau, V., Jung, G., Strominger, J.L. and Rammensee, H.G. (1993) Allele-specific peptide ligand motifs of HLA-C molecules. *Proc. Natl. Acad. Sci. USA* 90: 12005–12009.

Fremont, D.H., Matsumura, M., Stura, E.A., Peterson, P.A. and Wilson, I.A. (1992) Crystal structures of two viral peptides in complex with murine MHC Class I H2K^b. *Science* 257: 919–927.

Garrett, T.P.J., Saper, M.A. Bjorkman, P.J., Strominger, J.L. and Wiley, D.C. (1989) Specificity pockets for the side chains of peptide antigens in HLA-Aw68. *Nature* 342: 692–696.

Germain, R.N. and Rinker, A.G., Jr. (1993) Peptide binding inhibits protein aggregation of invariant-chain free class II dimers and promotes selective cell surface expression of occupied molecules. *Nature* 363: 725–728.

Guo, H.C., Jardetzky, T.S., Garrett, T.P.J., Lane, W.S., Strominger, J.L. and Wiley, D.C. (1992) Different length peptides bind to HLA-Aw68 similarly at their ends but bulge out in the middle. *Nature* 360: 364–367.

Hammer, J., Takacs, B. and Sinigaglia, F. (1992) Identification of a motif for HLA-DR1 binding peptides using M13 display libraries. *J. Exp. Med.* 176: 1007–1013.

Harding, C.V. and Unanue, E.R. (1990) Quantitation of antigen-presenting cell MHC Class II-peptide complexes necessary for T-cell stimulation. *Nature* 346: 574–576.

Huckzo, E.L., Bodnar, W.M., Benjamin, D., Sakaguchi, K., Zhu, N.Z., Shabanowitz, J., Henderson, R.A., Appella, E., Hunt, D.F. and Engelhard, V.H. (1993) Characteristics of endogenous peptides eluted from the Class I MHC molecule HLA-B7 determined by mass spectrometry and computer modeling. *J. Immunol.* 151: 2572–2588.

Hunt, D.F., Henderson, R.A., Shabanowitz, J., Sakaguchi, K., Michel, H., Sevilir, N., Cox, A., Appella, E. and Engelhard, V.H. (1992a) Characterization of peptides bound to the class I MHC molecule HLA-A2.1 by mass spectrometry. *Science* 255: 1261–1263.

Hunt, D.F., Michel, H., Dickinson, T.A., Shabanowitz, J., Cox, A., Sakaguchi, K., Appella, E., Grey, H.M. and Sette, A. (1992b) Peptides presented to the immune system by the murine class II major histocompatibility complex molecule I-A^d. *Science* 256: 1817–1820.

Jardetzky, T.S., Lane, W.S., Robinson, R.A., Madden, D.R. and Wiley, D.C. (1991) Identification of self peptides bound to purified HLA-B27. *Nature* 353: 326–329.

Kubo, R.T., Sette, A., Grey, H.M., Appella, E., Sakaguchi, K., Zhu, N.Z., Arnot, D., Sherman, N., Shabanowitz, J., Michel, H., Bodnar, W.M., Davis, T.A. and Hunt, D.F. (1994) Definition of specific peptide motifs for four major HLA-A alleles. *J. Immunol.* 152: 3913–3924.

Madden, D.R., Gorga, J.C., Strominger, J.L. and Wiley, D.C. (1991) The structure of HLA-B27 reveals nonamer self-peptides bound in an extended conformation. *Nature* 353: 321–325.

Madden, D.R., Gorga, J.C., Strominger, J.L. and Wiley, D.C. (1992) The three dimensional structure of HLA-B27 at 2.1 A° resolution suggests a general mechanism for tight peptide binding to MHC. *Cell* 70: 1035–1048.

Madden, D.R., Garboczi, D.N. and Wiley, D.C. (1993) The antigenic identity of peptide/MHC complexes: a comparison of the conformations of five viral peptides presented by HLA-A2. *Cell* 75: 693–708.

Padlan, E.A., (1977) Structural implications of sequence variability in immunoglobulins. *Proc. Natl. Acad. Sci.* USA 74: 2551–2555.

Roche, P.A. and Cresswell, P. (1990) High-affinity binding for an influenza hemagglutinin-derived peptide to purified HLA-DR. *J. Immunol.* 144: 1849–1856.

Rotzschke, O., Falk, K., Deres, K., Schild, H., Norda, M., Metzger, J., Jung, G. and Rammensee, H.G. (1990) Isolation and analysis of naturally processed viral peptides as recognized by cytotoxic T cells. *Nature* 348: 252–254

Rudensky, A.Y., Preston-Hurlburt, P., Hong, S.C., Barlow, A. and Janeway, C.A., Jr. (1991) Sequence analysis of peptides bound to MHC Class II molecules. *Nature* 353: 622–627.

Saper, M.A., Bjorkman, P.J. and Wiley, D.C. (1991) Refined structure of the human histocompatibility antigen HLA-A2 at 2.6 Å resolution. *J. Mol. Biol.* 219: 277–319.

Sidney, J., Oseroff, C., Del-Guercio, M.F., Southwood, S., Krieger, J.I., Ishioka, G.Y., Sakaguchi, K., Appella, E. and Sette, A. (1994) Definition of a DQ3.1 specific binding motif. *J. Immunol.* 152: 4516–4525.

Silver, M.L., Guo, H.C., Strominger, J.L. and Wiley, D.C. (1992) Atomic structure of a human MHC molecule presenting an influenza virus peptide. *Nature* 360: 367–369.

Stern, L.J. and Wiley, D.C. (1992) The human Class II MHC protein HLA-DR1 assembles as empty αβ heterodimers in the absence of antigenic peptide. *Cell* 68: 465–477.

Stern, L.J., Brown, J.H., Jardetzky, T.S., Gorga, J.C., Urban, R.G., Strominger, J.L. and Wiley, D.C. (1994) Crystal structure of the human class II MHC protein HLA-DR1 complexed with an influenza virus peptide. *Nature* 368: 215–221.

Townsend, A., Rothbard, J., Gotch, R., Bahadur, B., Wraith, D. and McMichael, A. (1986) The epitopes of influenza nucleoprotein recognized by cytotoxic T lymphocytes can be defined with short synthetic peptides. *Cell* 44: 959–968.

Townsend, A. and Bodmer, H. (1989) Antigen recognition by Class I-restricted T lymphocytes. *Ann. Rev. Immunol.* 7: 601–624.

Young, A.C.M., Zhang, W., Sacchettini, J.C. and Nathenson, S.G. (1994) The three-dimensional structure of H-2D^b at 2.4Å resolution: implications for antigen-determinant selection. *Cell* 76: 39–50.

Zhang, W., Young, A.C.M., Imarai, M., Nathenson, S.G. and Saccehettini, J.C. (1992) Crystal structure of the major histocompatibility complex Class I/H-2K^b molecule containing a single viral peptide: implications for peptide binding and T-cell receptor recognition. *Proc. Natl. Acad. Sci.* USA 89: 8403–8407.

Interface between Chemistry and Biochemistry
ed. by P. Jollès and H. Jörnvall
© 1995 Birkhäuser Verlag Basel/Switzerland

Catalytic antibodies: Evolution of protein function in real time

R.A. Lerner and K.D. Janda

Departments of Chemistry and Molecular Biology, The Scripps Research Institute, 10666 N. Torrey Pines Road, La Jolla, CA 92037, USA

Summary. Natural selection of enzyme function has evolved over millions to billons of years, whereas antibody induction operates over a period of weeks. If one considers how new protein functions are generated, one sees that both the immune system and natural selection have powerful methods for the generation of diversity. In this treatise we discuss antibody catalysis from an evolutionary standpoint as we consider how the elicitation of a catalytic antibody can be a useful tool for exploring the nature of biological catalysis.

Introduction

The essence of the field of antibody catalysis is the use of the programmable binding energy of the immune system to control the outcome of chemical transformations. A wide variety of catalysts has been generated; they often have large accelerations and control the regio- and enantioselectivity of reactions, even rerouting reaction pathways to give highly disfavored products (Lerner, Benkovic and Schultz, 1991; Lerner and Schultz, 1993). In some cases catalysts have been generated for reactions for which there are no known enzymes. These features of antibody catalysis have been the subject of other reviews (Lerner et al., 1991; Lerner and Schultz, 1993). In this review we cover some general principles but our focus is on the somewhat different but important issue of what antibody catalysis is teaching us about the evolution of natural enzymes. When one considers how new proteins functions are generated, one sees that both the immune system and natural selection have powerful methods for the generation of diversity (Fig.1). In both cases the process of mutation and selection uses time scales that are vastly different. Natural selection of enzyme function evolves over millions to billions of years whereas immune induction operates over a period of weeks. Thus, if one hopes to induce antibodies which are enzymes, an additional intervention is necessary and that is education. In effect, the induction of catalytic antibodies hinges on the ability to "teach" the antibody molecule to do chemistry. Thus, if we examine the simple diagram shown in Figure 2, we see the difference between laboratory science which attempts to mimic evolution and the field of catalytic antibodies. The highest art in laboratory systems which at-

Generations of New Functions in Biology

Protein evolution

Exxon Shuffling
Gene Duplication
Accumulation of point mutations
Natural selection
Timescale:
$10^2 - 19^9$ years

Immune response

VDJ rearrangement
Batteries of V, D and J gene elements
Somatic hypermutation
Clonal selection
Timescale
Weeks

Figure 1. Generation of diversity from a protein evolutionary standpoint and the immune system.

tempts to mimic evolution is the generation of a system which generates diversity and uses selection to yield new functions. Such is the case with RNA-based evolutionary systems where one starts at zero function and generates improved functions at each round of mutation and selection. In the field of antibody catalysis, on the other hand, the highest art form is to enter the game as nearly as possible to catalytic perfection. In practice, this is done by analyzing the nature of the rate-limiting transition state of a chemical reaction and then synthesizing an analogue of the transition state. Antibodies induced against the transition state analogue can be expected to lower the energy of the reaction and thereby increase its rate. This new ability to induce catalytic function in proteins in real time allows one to ask some important questions about

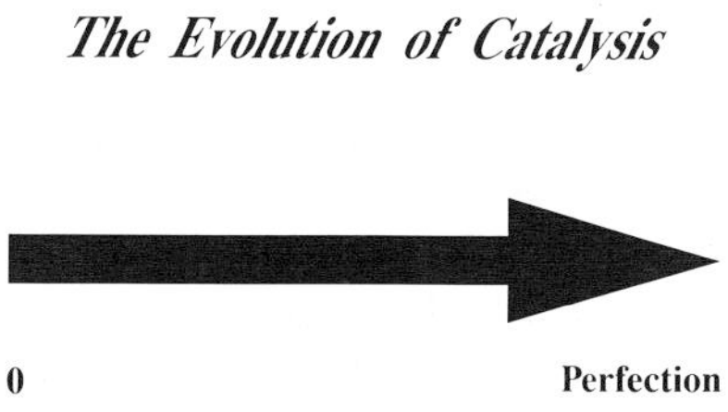

The Evolution of Catalysis

0   Perfection

Figure 2. Diagram depicting the difference between laboratory science, which attempts to mimic evolution, and the field of catalytic antibodies.

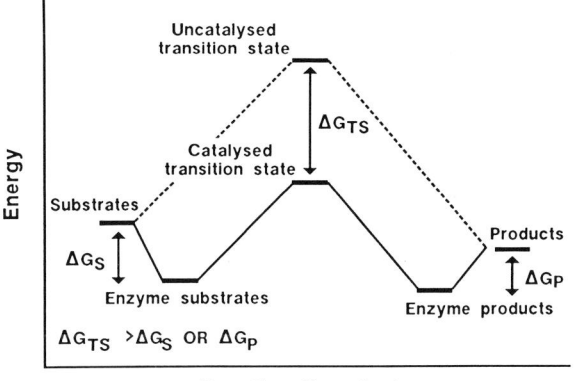

Figure 3. Generalized energy reaction coordinate diagram for conversion of substrate to product.

the evolution of catalytic activity in proteins. We would like to know: (1) how many ways can a protein catalyze a given chemical transformation?; (2) how difficult is it to evolve complex catalytic machinery?; and (3) when new catalysts are induced will they recapitulate the mechanisms of highly evolved proteins or will they show us new mechanisms long discarded by nature in favor of more efficient processes? This latter question is of extreme interest because it touches on the important concept of convergent evolution. Thus, if convergent evolution occurred in real time, one would gain powerful evidence that this process occurs because there are limited options for certain chemical transformations, especially if one demands some level of efficiency.

Some fundamentals

Before we begin to focus on the main theme of this review, which is what antibody catalysis can teach us about enzyme evolution, we need to review some fundamentals. All catalysts operate by lowering the energy barrier for a chemical transformation thereby increasing its rate. In Figure 3 we depict an idealized reaction coordinate scheme for conversion of a substrate to product. A catalyst can be expected to be most efficient if its complementary to the rate limiting transition state which in this example serves to lower its energy by the factor ΔG_{TS}. This principle can be examined by considering how one might make an enzyme for ester or amide hydrolysis. In Figure 4 we consider the hydrolysis of a simple ester. The carbonyl carbon of the planar substrate is attacked by a nucleophile (which in this case is the hydroxide ion) to yield the high energy tetrahedral intermediate which subse-

Transition State Stabilization

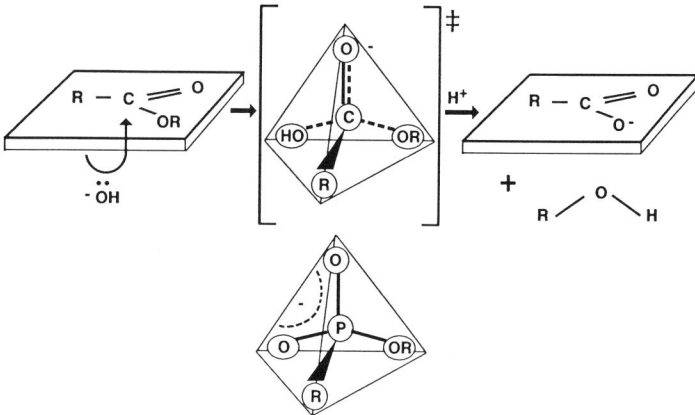

Figure 4. Hydrolysis of an ester functionality and the presumed transition-state which the reaction proceeds through.

quently collapses to the acid and alcohol products. A good catalyst should "anticipate" the critical features of the transition state which include the tetrahedral geometry, developing negative change on the carbonyl oxygen, and the changes in bond lengths in the transition state. The key idea in antibody catalysis is simply to induce an antibody

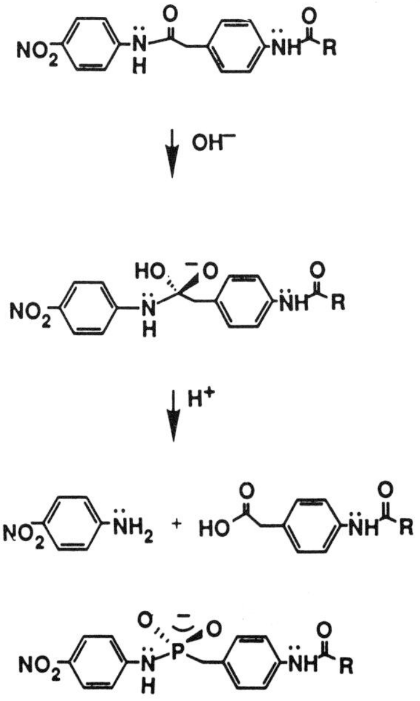

Figure 5. Antibody-catalyzed hydrolysis of an amide bond.

which is complementary to these key features of the transition state and, thus, uses its binding energy to lower its energy. Since the transition state is too unstable to use as an antigen, one must immunize with an analogue of the transition state. For example, the most widely used analogue for ester and amide hydrolysis is the phosphonate moiety shown in Figure 4. Its main features are that it is tetrahedral, negatively charged, and the phosphate-oxygen bond is some 20% longer than the carbon-oxygen bond, thereby mimicking the bond-breaking process which occurs in the transition state (Lerner et al., 1991; Lerner and Schultz, 1993). In practice, even difficult reactions such as that shown in Figure 5 can be catalyzed by antibodies to such transition state analogues. Here, antibodies generated to the phosphonamidate transition state analogue were used to catalyze the hydrolysis of the amide functionality of the nitro-anilide (Lerner et al., 1991). From a study of physical organic chemistry and reaction mechanisms, we know of the factors which constitute the energy barriers for reactions and which factors dominate particular reactions. To date, these factors have been utilized to generate many catalytic antibodies, but here we focus on only a few examples which are particularly instructive as to general principles. We have already mentioned lowering the energy of the transition state. But, another way to lower the energy barrier is by raising the energy of the substrate in a process termed ground state destabilization. Hilvert and his colleagues (Lewis et al., 1991) used this principle to decarboxylate the oxazole substrate shown in Figure 6. Here, the napthalene antigen was used to induce an antibody which would bring the substrate into an apolar environment where it would be of higher energy than in solvent, thereby facilitating the decarboxylation process. For many chemical transformations there is a large entropic component

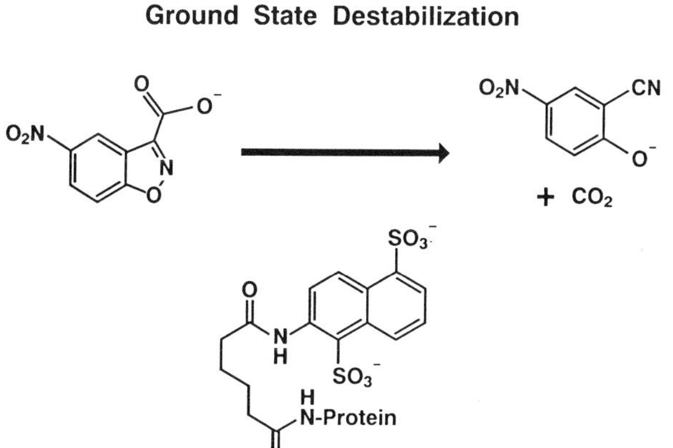

Ground State Destabilization

Figure 6. Antibody-catalyzed decarboxylation reaction and the hapten used to elicit antibodies which catalyzed the reaction shown.

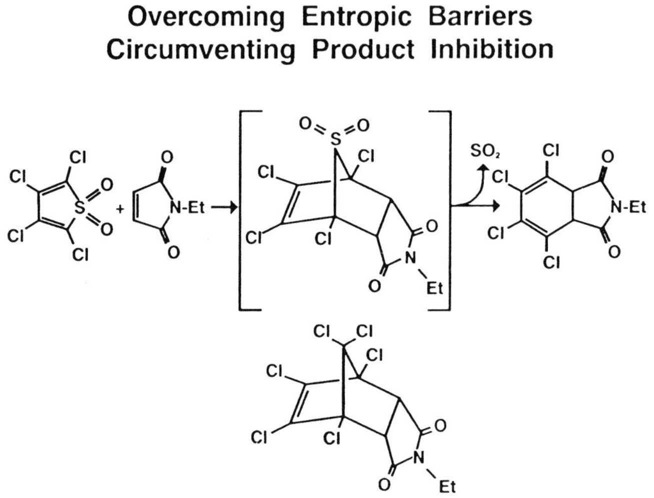

Overcoming Entropic Barriers
Circumventing Product Inhibition

Figure 7. Bimolecular Diels-Alder reaction catalyzed by an antibody. In parentheses is the presumed transition-state of the reaction and below this is the hapten/inhibitor for process.

to the energy barrier. This is especially severe for bimolecular reactions such as the Diels-Alder cycloaddition which has been catalyzed by Hilvert, Schultz, and ourselves (Hilvert et al., 1989; Braisted and Schultz, 1990; Gouverneur et al., 1993). The idea here is to use the binding energy of the antibody to facilitate the approach of the diene and dienophile in the transition state of this concerted process. In Figure 7 the Hilvert experiment is shown which involves the condensation of tetrachlorothiophene dioxide and N-ethyl maleimide followed by the loss of sulfur dioxide to give the rearranged phthalimide. The product like antigen anticipates the boat-shaped arrangement of the diene and dienophile as they approach each other. Since the antigen looks like the initial product, one might anticipate that the catalyst would be severely product inhibited. Hilvert and his colleagues neatly sidestepped this problem by utilizing the elimination of sulfur dioxide and subsequent rearrangement to give a product which no longer resembles the immunizing antigen. Finally, many catalysts require the presence of chemical functionalities in precise locations to effect catalysis. For example, all serine proteases require the presence of the catalytic triad of histidine, serine, and aspartic acid to allow the acylation and deacylation of serine in the process of covalent catalysis. In antibody catalysis functionalities are induced by placing complementary functionalities in the antigen in what is sometimes referred to as a "bait and switch strategy". This is illustrated in our experiment where we developed a catalyst to accomplish ester hydrolysis by a general base mechanism (Fig. 8; Janda et al., 1989). This reaction requires a general base in the appropriate location to remove the proton from water in the

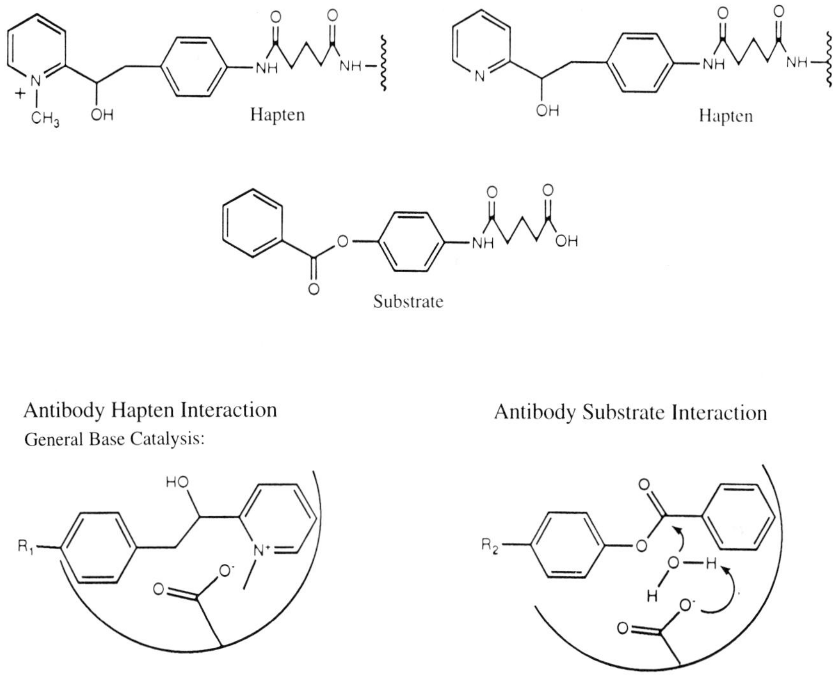

Figure 8. "Bait and Switch" strategy used to elicit catalytic residues within an antibodies combining site.

hydrolysis reaction. Thus the antigen includes a strategically placed tertiary amine which will be protonated at physiological pH. This positively charged amine functionality can be expected to induce a carboxylate in the antibody which can function as the required general base.

Putting the principles together to catalyze difficult chemistry

While each of these principles discussed above is by itself simple, when they are used in concert a great deal of chemical sophistication can be brought to bear on a given transformation. To conclude our brief discussion of the general principles of antibody catalysis and illustrate how several features can be used in concert to achieve difficult chemistry, we can consider the recent generation of a catalytic antibody which achieves a cation-olefin cyclization (Li et al., 1994).

One of the most important achievements of modern science has been the elucidation of the biological pathway by which cholesterol is synthesized from acetate (Olah and Schleyer, 1992; Johnson, 1968; Abe et al., 1993; Woodward and Bloch, 1953; Stork and Burgstahler, 1955; Es-

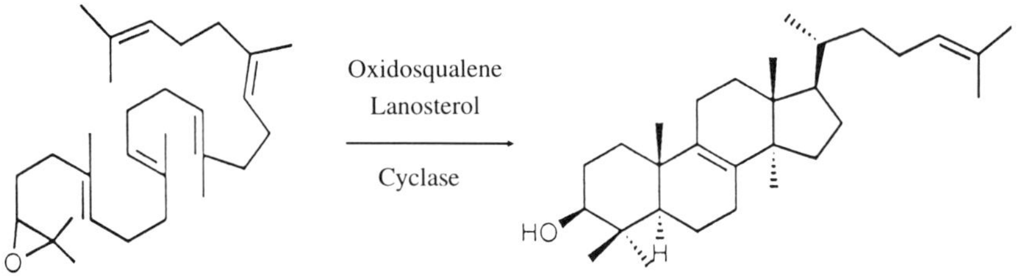

Figure 9. The cascade/rearrangement reaction of (3S)-2, 3-oxidosqualene to lanosterol which is catalyzed by oxidosqualene lanosterolcyclase.

chenmoser et al., 1955; Maudgal et al., 1958; Cornforth et al., 1965; Corey and Russey, 1966; Willett et al., 1967; Sharpless, 1968; Barton et al., 1975; van Tamelen, 1982; Schroepfer, 1982; Corey and Matsude, 1991; Kusano et al., 1991; Kelly et al., 1990; Buntel and Griffin, 1992; Bartlett, 1984; Johnson et al., 1987; Sutherland, 1991). Of all the stages involved in the biogenesis of cholesterol, the enzymatic transformation by which (3S)-2, 3-oxidosqualene is converted to lanosterol is of the greatest interest to the synthetic organic chemist because a molecule which has only one asymmetric center is converted into a single stereochemically pure molecule with six new asymmetric centers. This remarkable cascade of cyclizations and rearrangements is catalyzed by the enzyme oxidosqualene lanosterol-cyclase (EC 5.4.99.7) (Fig. 9). While generally viewed as a cationic cyclization process, the detailed understanding of the mechanism of this reaction has remained elusive in spite of much effort. This is in part true because the enzyme has proven difficult to purify, and only recently has progress been made in the cloning of the gene (Kelly et al., 1990; Buntel and Griffin, 1992).

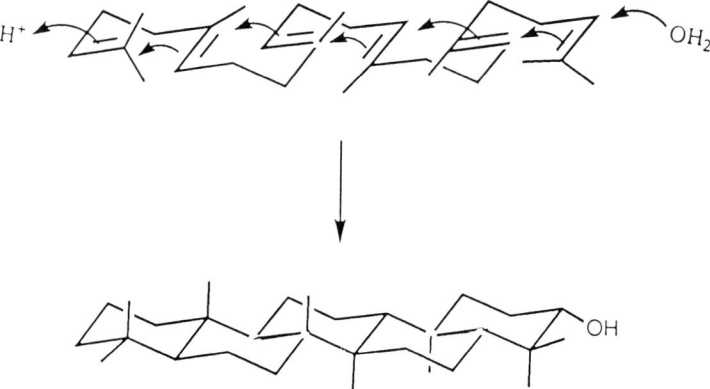

Figure 10. Cascade reaction forming five new fused ring systems which is catalyzed by the enzyme *tetrahymena pyriformis*.

In contrast to the as yet incomplete picture of the biological process, synthetic studies have produced a wealth of information on polyene cyclization reactions (Olah and Schleyer, 1992; Johnson, 1968; Abe et al., 1993; Woodward and Bloch, 1953; Stork and Burgstahler, 1955; Eschenmoser et al., 1955; Maudgal et al., 1958; Cornforth et al., 1965; Corey et al., 1966; Willett et al., 1967; Sharpless, 1968; Barton et al., 1975; van Tamelen, 1982; Schroepfer, 1982; Corey and Matsude, 1991; Kusano et al., 1991; Kelly et al., 1990; Buntel and Griffin, 1992; Bartlett, 1984; Johnson et al., 1987; Sutherland, 1991). Indeed, study of this reaction did much to foster the fruitful collision between conformational analysis and attempts to understand the stereochemical outcome of complex organic transformations. Given the complexity of the reaction, it would have been reasonable to believe that the conversion of an achiral substrate such as squalene to lanosterol, with absolute control of six sterogenic centers, was the sole purview of the enzymatic processes of nature. However, in a key insight to the stereoselectivity of this reaction, Stork and Eschenmoser and their colleagues independently hypothesized that the stereochemical outcome of the cyclization of all-trans polyolefins is intrinsic to the molecule and should yield the "natural" configuration (Fig. 10; Stork and Burgstahler, 1955; Eschenmoser et al., 1955). The major elements of their hypothesis are that cyclization occurs via chair-like conformations of the nascent rings and that addition to each double bond takes place in an antiparallel fashion. Thus, the all-trans double-bond stereochemistry of squalene is transformed into the trans, anti, trans stereochemistry of polycyclic terpenes. The proton-initiated cyclization leading to tetrahymenol is one striking example of this process. The perception of the Stork-Eschenmoser hypothesis led to a large number of synthetic experiments aimed at understanding the mechanisms by which polyolefins cyclize and the development of synthetic strategies based on this process. Arguably, the most complete set of studies on non-enzymatic cationic polyene cyclization reactions was carried out by Johnson and his colleagues (Johnson, 1968). Their work led to the recognition that the process must be initiated in a way which generates a cationic center on carbon without affecting the olefinic bonds. In spite of these advances and the now understood role of enzymes in the initiation of steroid biosynthesis, the construction, in the laboratory, of a catalyst for this reaction has not been achieved. This failure is no doubt due to the formidable problem of simultaneously controlling the initial generation and stabilization of the carbocation as well as the attendant entropic and stereoelectronic parameters intrinsic to the cyclization process.

Catalytic antibodies with their ability to simultaneously address multiple parameters which appear along a reaction pathway have already proven capable of controlling complex organic transformations including chemical disfavored reactions (Lerner et al., 1991; Lerner and

Schultz, 1993). They should, in principle, also be ideal catalysts for initiation and control of cationic cyclization where one needs to control conformational and stereoelectronic features of the transition state of the reaction.

The reaction

For the purposes of this review, we should extract the central point that the programmable binding energy of the antibody allows the experimenter to intervene in a highly detailed and precise way to control the route of a chemical reaction. The purpose of the illustration is also to further show how the individual features discussed above are brought together in a concerted process. Some of the earliest synthetic work on cationic cyclization reactions was on the formation of six-membered rings (Johnson, 1968). Given this rich history, we felt that a study of six-membered rings would be a good starting point for investigation of antibody catalyzed cationic cyclization. Typically, a cationic cyclization

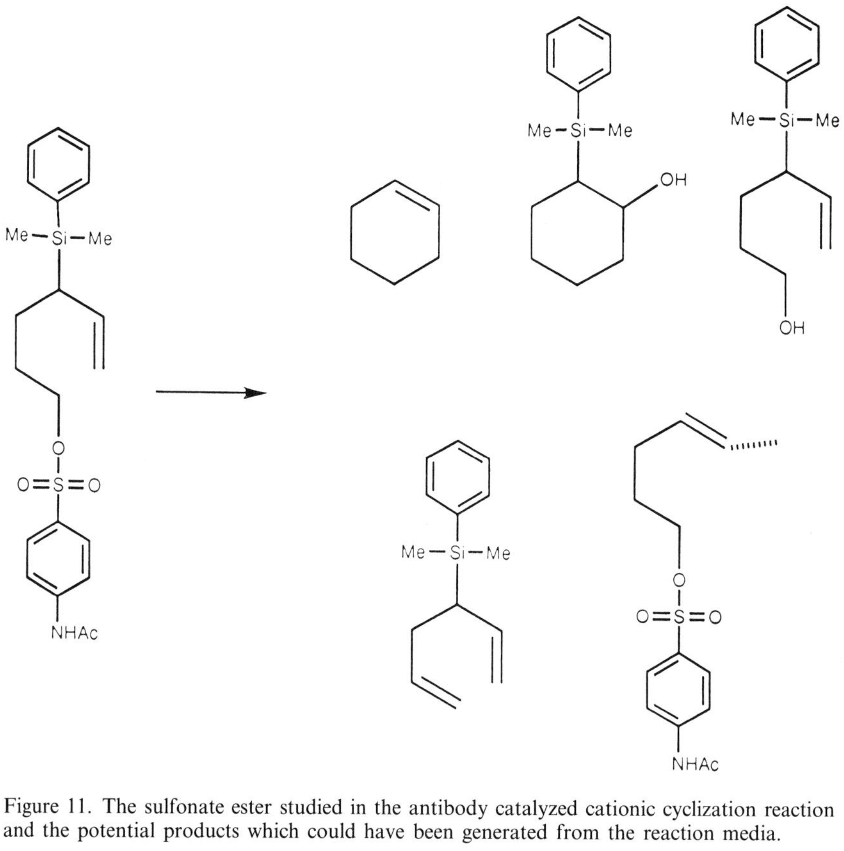

Figure 11. The sulfonate ester studied in the antibody catalyzed cationic cyclization reaction and the potential products which could have been generated from the reaction media.

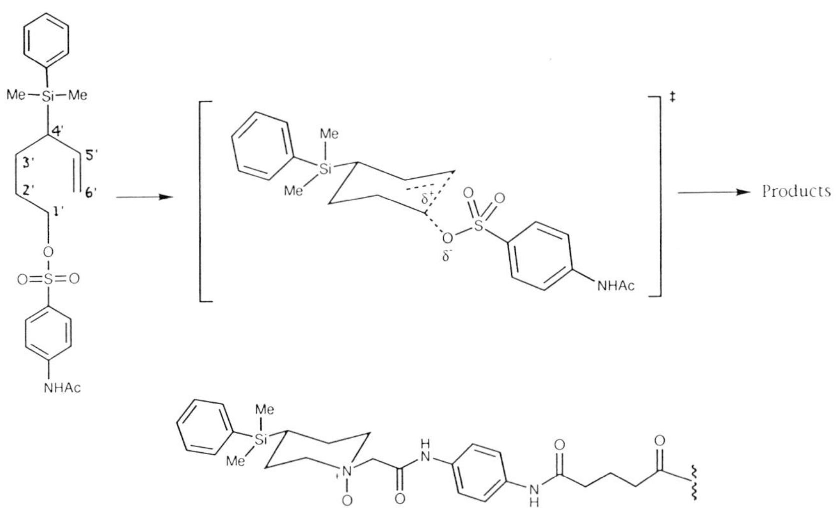

Figure 12. The hapten used to obtain catalytic antibodies for a cationic cyclization reaction and one possible transition-state for this type of process.

reaction is initiated by the formation of a carbocation, either by electrophilic addition to a double bond or by ionization at a sp3 hybridized carbon. Also, the reaction is thought to proceed via a quasi-chair like transition state which positions the olefin to capture the cationic center in a largely concerted process. The reaction then can terminate by a variety of mechanisms including capture of a cationic

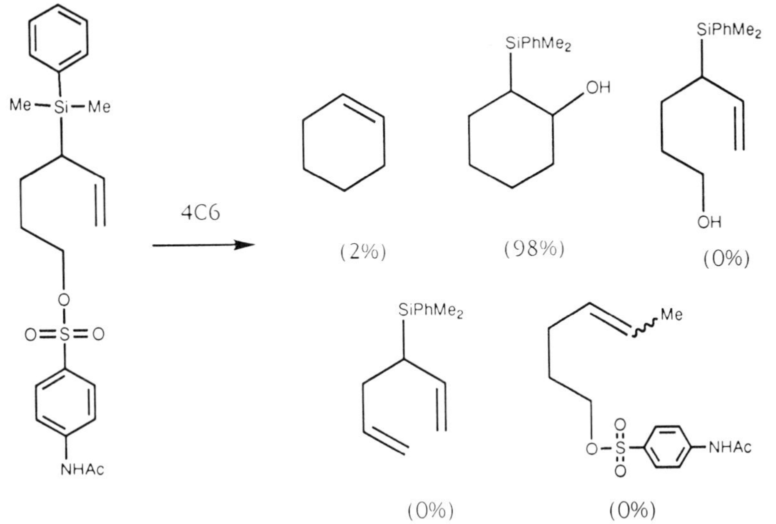

Figure 13. Antibody-catalyzed cationic cyclization reaction with the product distribution.

center by solvent, migration of a hydride or alkyl group, or loss of a proton to generate a new olefin. Thus, the programmable binding energy of the antibody molecule might be used to control the energy of the various carbocations and high energy intermediates which appear along the reaction coordinate, thereby altering the chemical outcome of the reaction and controlling its stereochemistry.

The reaction which we studied is the cyclization of the olefinic sulfonate shown in Figure 11 which, under solvolysis conditions, converts to many products. To catalyze this reaction we had to simultaneously address multiple features of this reaction if control of the reaction coordinate was to be achieved. It was necessary to initiate loss of the sulfonate, stabilize the developing cationic center, and achieve the chair-like geometry of the reactants which correctly position the olefin to participate in the reaction (anchimeric assistance) and capture the carbocation. The N-oxide shown in Figure 12 would seem to be an ideal hapten to accomplish this task in that it has a chair conformation and the cationic nitrogen center can be expected to induce a complementary anionic functionality in the antibody combining site which should stabilize the developing carbocation so that is is not prohibitively high in energy (Li et al., 1994). Thus, one has to overcome entropic barriers to organize the chair-like conformation of the reactants, induce functionalities in a regio-selective fashion to neutralize charges developing in the transition state, and favor the proper stereoloectronic features of the reaction such that the pi orbitals are properly aligned to participate in the process. The success of this approach is illustrated by the data in Figure 13 in which the product outcome of the antibody catalyzed reaction is given. Unlike background reactions which require harsh conditions (acid pH and high temperature) and give a complicated mixture of products, in the antibody catalyzed reaction only one significant product was formed in 98% yield (2-dimethylphenylsilyl cyclohexanol) (Fig. 13). This degree of control over an olefin-cation cyclization reaction is unprecedented and can be attributed to the ability of the antibody to rigidly enforce a reaction mechanism. In addition to the ability to develop new catalysts for the formation of carbon-carbon bonds, catalytic antibodies such as these may find practical use. In particular, since catalytic antibodies have already proven capable of facilitating disfavored chemical events, we can expect to reroute cyclization reactions by overcoming the unfavorable barriers intrinsic to Anti-Baldwin and Anti-Markovnikov processes. To obtain a single product in these more complicated processes we will also control the termination of the reaction. The design of the catalyst should allow one to perordain whether termination occurs by addition of an external nucleophile, migration of a hydride or an alkyl group, or by elimination. Such catalysts may allow formation of unique steroids as well as other organic compounds not easily obtainable by conventional organic synthesis.

Figure 14. Simple model of an acyl transfer reaction.

What catalytic antibodies are teaching us about the evolution of protein catalysts

There are not that many plays in the play book: To begin our examination of questions about the evolution of protein catalysis, we can first focus on the process of transesterification which is an important transformation in biology. This apparently simple reaction is depicted in Figure 14. In this reaction an ester reacts with an alcohol to yield a new ester and a new alcohol in an acyl transfer process. The operative phase is "apparently simple" but, in water, this reaction is far from simple. First, since the reaction is bimolecular, one must overcome a large entropic barrier. Second, whereas the alcohol is present in micromolar amounts, water, which is nearly as efficient a nucleophile, is present at 55 molar such that the conditions favor simple hydrolysis of the ester rather than condensation with the alcohol. Finally, the alcohol is not a very good nucleophile and must be deprotonated to the alkoxide. The difficulty of this process is underscored by the fact that without a catalyst this process does not occur in water even if a secondary alcohol and a reactive ester are allowed to react for weeks and all one sees is hydrolysis of the ester. Both the Schultz laboratory (Jacobsen et al., 1992) and our laboratory (Wirsching et al., 1991) generated efficient catalysts for this reaction. The rate enhancements for this reaction achieved by antibody catalysis were extremely large and approached the theoretical maximum that one would expect if one overcame all the loss of translational entropy. But, more importantly as the mechanisms of these antibody catalysts were examined, we began to see answers to some of the questions posed above. The transition state analogue and the reaction which we catalyzed are shown in Figure 15 and 16. From our discussion above, one might expect that the reaction would

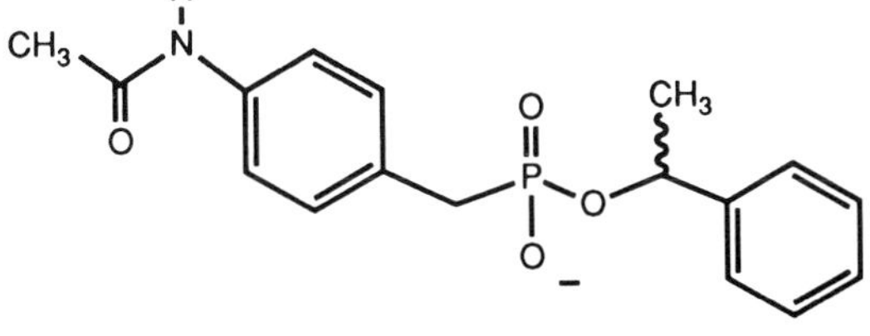

Figure 15. Hapten used to obtain catalytic antibodies that could kinetically resolve a mixture of alcohols and carry out a transesterification reaction.

be catalyzed in a simple way such that the phosphonate would induce a binding pocked to favor the charged tetrahedral intermediate and the antibody binding would overcome the entropic barrier by using its binding energy to approximate the secondary alcohol and the ester reactants. However, one of the catalysts was found to proceed by a much more complicated mechanism in that an acyl intermediate was formed and the catalysis proceeded by a ping-pong mechanism (Wirsching et al., 1991). Thus, remarkably, the antibody adopted a mechanism previously thought to be the sole purview of highly sophisticated catalysts honed through millions of years of evolution. Why should this be so? In retrospect, the answer seems fairly simple and revolves around the fact that even when the amino acid side chains act in concert there are, from the chemical point of view, not that many ways to effect difficult chemical transformation, especially under the mild conditions (water solution, physiological pH, and low temperature) which they must perform. In other words, there are just not that many plays in the book when only amino acid side chains can be used. Thus, when we find an efficient catalytic antibody it should not surprise us that it uses a mechanism which nature stumbled upon and improved. The only reason that we can evolve it in real time is that we give the system a head start via a chemical education. In this transacylation catalyst, we "taught" it about the entropic barrier and the shape and charge of the transition state and then left it to its own devices. In turn, the catalyst taught us that one of the few ways to be efficient is to proceed via a covalent intermediate and, thus, it called up that play. Thus, we have seen convergent evolution in real time where the driving force is the fact that there are only limited ways for proteins to accomplish difficult chemistry. When proteins accomplish chemistry beyond the ability of the simple amino acids, they use cofactors such as transition metals. Given, the rather limited range of chemistries available to proteins alone, it is perhaps no surprise that a full third of nature's protein enzymes use metal cofactors and prosthetic groups.

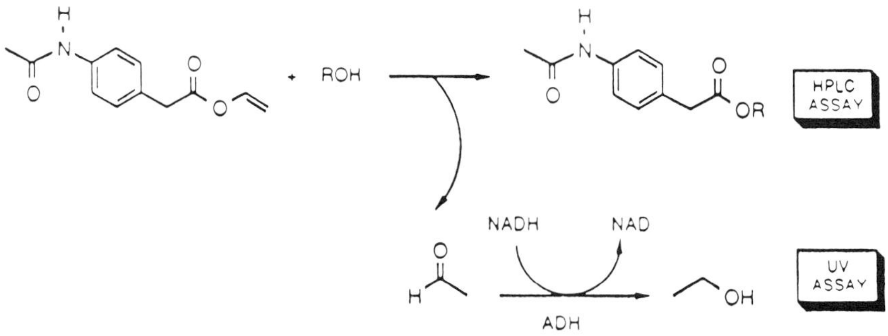

Figure 16. Transesterification reaction that was catalyzed by antibodies obtained to hapten shown in Figure 15. A two-assay format was used to characterize these antibodies which was also the first enzyme-catalytic antibody coupled assay.

Recently, we have seen a more remarkable example of convergent evolution in antibody catalysis. Scanlan, Fletterick and their colleagues have solved the crystal structure of a very efficient catalytic antibody which hydrolyzes unactivated amino acid esters (Zhou et al., 1994). One of the main purposes of generating this class of catalytic antibody is to allow further evolution by selection in auxotrophic bacteria which requires the amino acid that is the product of the catalyzed reaction. This part of the experiment has not yet been accomplished. However, along the way they found something quite remarkable. To determine why one of the antibodies was so efficient, they solved its crystal structure complexed to the inducing transition state analogue at an atomic resolution of 2.5 angstroms (Zhou et al., 1994). The antibody stabilized the oxyanion (oxyanion hole) using a lysine side chain and *place a histidine and serine in the precise position in which they occur in all serine proteases*! Thus, this young catalyst which evolved in the laboratory in real time adopted much of the catalytic machinery of enzymes evolved on an evolutionary time scale. The only selective pressure here was that the scientists only wanted to study antibodies which were efficient catalysts.

In another crystal structure, Wilson and his colleagues studied a catalytic antibody which has chorismate mutase activity and converts (−) chorismate to prephenate (Hilvert and Nared, 1988; Jackson et al., 1988; Haynes et al., 1994). The reaction and its transition state are shown in Figure 17. The reaction is thought to proceed through a transition state in which the lower energy extended conformation of the substrate is converted to a pseudo-diaxial geometry which is higher in energy but favorable for the stereoelectronic requirements of the reaction. While the reaction is concerted, it is somewhat asynchronous, thereby leading to some charge separation in the transitionstate (Fig. 17). Also, there is a stereogenic center in chorismate at the carbinol carbon and the crystal structure must explain the kinetic resolution of the enantiomers achieved by the catalytic antibody (Hilvert and Nared, 1988). In Figure 18, we present a view of the antibody binding pocket complexed to the oxabicyclic transition state analogue used to induce the catalyst. One sees from this view much of the basis of the catalysis. The binding pocket is compact, thus forcing the vinyl ether

Nature's Instructions versus Chemist's Instructions

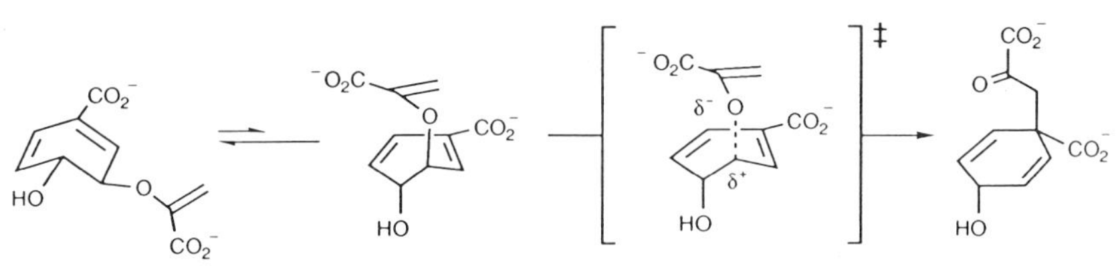

Figure 17. Antibody catalyzed reaction which converts chorismate to prephenate.

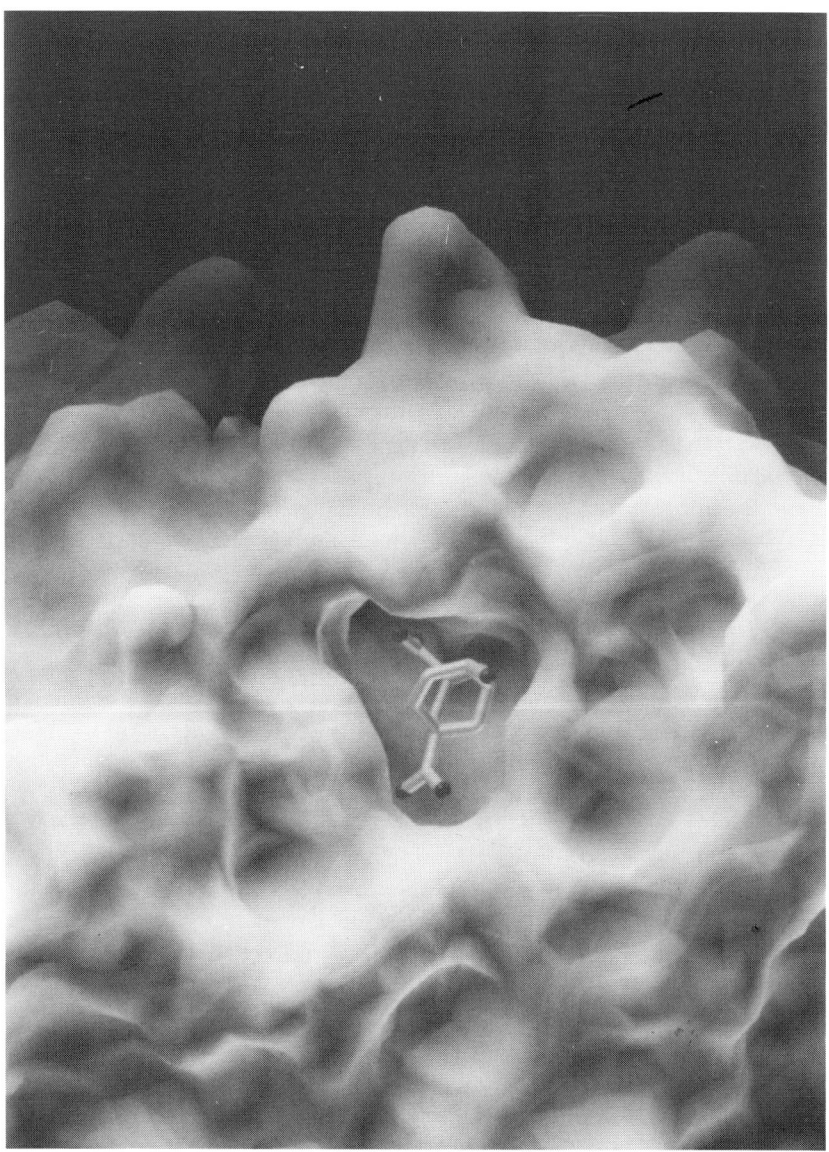

Figure 18. Crystal structure of antibody complexed to hapten which was used to elicit this antibody.

side chain of the substrate to adopt the higher energy conformation of the transition state. Thus, the antibody uses its binding energy to pay for adoption of the pseudodiaxial chair-like conformation. Also, the origins of the stereochemistry are evident in that the binding pocket is highly asymmetrical such that only one enantiomer of the substrate is allowed. Importantly, these authors were able to compare their structure to a natural chorismate mutase enzyme bound to the same transition state analogue recently solved by Chook et al., 1993). The overall conclusion of the comparison was that the difference in catalysis between the antibody and the natural enzyme was one of the degree rather than kind. One difference between the catalysts, however, concerned how the developing charge in the transition state was stabilized. The natural enzyme used some of its binding energy to favor the developing enolate character of the breaking bond in the transition state. This feature was not present in the catalytic antibody since it was not included in the design of the transition state analogue used to induce the antibody. Knowing this, one can now design better transition state analogues for this reaction. Interestingly, Schultz and his colleagues developed a catalytic antibody using the same transition state analogue (Jackson et al., 1988). The Schultz enzyme has a rate acceleration which is 100 times faster than the Hilvert enzyme and is only two orders of magnitude less than the natural enzyme. This structure is being solved by David Davies and his colleagues (personal communication). When this structure is complete, we will have a unique look at the evolution of enzymic function in that we will be able to compare three enzymes – one evolved on the evolutionary time scale and the other two in real time. Thus, for the first time, we can extend the study of protein catalysis to enzyme other than those offered by nature. We expect these "real time" catalysts to be both useful and to teach us much about their more sophisticated cousins.

References

Abe, I., Rohmer, M. and Prestwich, G.D. (1993) Enzymatic cyclization of Squalene and Oxisqualene to Sterols and Triterpenes. *Chem. Rev.* 93: 2189–2206.
Bartlett, P.A. (1984) Olefin cyclization processes that from carbon-heteroatom bonds. *In*: J.D. Morrison (ed.): '*Assymmetric Synthesis*', Vol. 3, Academic Press, New York, pp 341–409.
Barton, D.H.R., Jarman, T.R., Watson, K.C., Widdowson, D.A., Boar, R.B. and Damps, K. (1975) Bisosynthesis of 3beta-hydroxytriterpenoids and -steroids from (3S)-2,3-epoxy-2,3-dihydrosqualene. *J. Chem. Soc. Perkin Trans.* 1: 1134–1138.
Braisted, A. and Schultz, P.G. (1990) An antibody catalyzed bimolecular Diels-Alder reaction. *J. Am. Chem. Soc.* 112: 7430–7431.
Buntel, C.J. and Griffin, J.H. (1992) Nucleotide and deduced aminoacid sequences of the oxidosqualene cyclase from Candida albicans. *J. Am. Chem. Soc.* 114: 9711–9713.
Chook, Y.M., Ke, H. and Lipscomb, W.N. (1993) Crystal structures of the monofunctional chorismatic mutase from *Bacillus subtilis* and its complex with a transition state analogue. *Proc. Natl. Acad. Sci. USA* 90: 8600–8603.
Cornforth, J.W., Cornforth, R.H., Donninger, C., Popjak, G., Shimizu, Y., Ichii, S., Forchielle, E. and Caspi, E. (1965) The migration and elimination of hydrogen during biosynthesis of cholesterol and squalene. *J. Am. Chem. Soc.* 87: 3224–3228.

Corey, E.J., Russey, W.E. and Ortiz de Montellano, P.R. (1966) Metabolic fate of 10, 11-dihydrosqualene in sterol-producing rat liver homogenate. *J. Am. Chem. Soc.* 88: 4750–4752.

Corey, E.J. and Matsude, S.P.T. (1991) New mechanistic and stereochemical insights on the biosynthesis of sterols from 2,3-oxidosqualene. *J. Am. Chem. Soc.* 113: 8172–8174.

Eschenmoser, A., Ruzicka, L., Jeger, O. and Arigoni, D. (1955) Eine stereochemische Interpretation der biogenetischen Isoprenregel bei den Triterpenen. *Helv. Chim. Acta.* 38: 1890–1904.

Gouverneur, V.E., Houk, K.N., de Pascual-Teresa, D., Beno, B., Janda, K.D. and Lerner, R.A. (1993) Control of the exo and endo pathways of the Diels-Alder reaction by antibody catalysis. *Science* 262: 204–208.

Haynes, M.R., Stura, E.A., Hilvert, D. and Wilson, I.A. (1994) Routes to catalysis: Structure of a catalytic antibody and comparison with its natural counterpart. *Science* 263: 646–652.

Hilvert, D. and Nared, K.D. (1988) Stereospecific Claisen rearrangement catalyzed by an antibody. *J. Am. Chem. Soc.* 110: 5593–5594.

Hilvert, D., Hill, K.W., Nared, K.D. and Auditor, M.-T.M. (1989) Antibody catalysis of a Diels-Alder reaction. *J. Am. Chem. Soc.* 111: 9261–9262.

Jackson, D.Y., Jacobs, J.W., Sugasawara, R., Reich, S.H., Bartlett, P.A. and Schultz, P.G. (1988) An antibody-catalyzed Claisen rearrangement. *J. Am. Chem. Soc.* 110: 4841–4842.

Jacobsen, J.R., Prudent, J.R., Kovhersperger, L., Yonkovich, S. and Schultz, P.G. (1992) An efficient antibody-catalyzed aminoacylation reaction. *Science* 256: 365–367.

Janda, K.D., Weinhouse, M.I., Schloeder, D.M., Lerner, R.A. and Benkovic, S.J. (1989) Bait and switch strategy for obtaining catalytic antibodies with Acyle-Mansfer capabilities. *J. Am. Chem. Soc.* 112: 1274–1275.

Johnson, W.S. (1968) Nonenzymatic biogenetic-like olefin cyclization. *Acc. Chem. Res.* 1: 1–8.

Johnson, W.S., Lindell, S.D., Steele, J. (1987) *J. Am. Chem. Soc.* 109: 5852–5853.

Kelly, R., Miller, S.M., Lai, M.H., Kirsch, D.R. (1990) Cloning and characterization of the 2,4-oxidosqualene cyclase-coding gene of Candida albicans. *Gene* 87: 177–183.

Kusano, M., Abe, I., Sankana, U. and Ebizuka, Y. (1991) Purification and some properties of squalene-2,3-epoxide: Lanasterol cyclase from rat liver. *Chem. Pharm. Bull.* 39: 239–241.

Lerner, R.A., Benkovic, S.J. and Schultz, P.G. (1991) At the crossroads of chemistry and immunology: catalytic antibodies. *Science* 252: 659–667.

Lerner, R.A. and Schultz, P.G. (1993) Antibody catalysis of difficult chemical transformations. *Acc. Chem. Res.* 26: 391–395.

Lewis, C., Kramer, T., Robinson, S. and Hilvert, D. (1991) Medium effects in antibody-catalyzed reactions. *Science* 253: 1019–1022.

Li, T., Janda, K.D., Ashley, J.A. and Lerner, R.A. (1994) Antibody catalyzed cationic cyclization. *Science* 264: 1289–1293.

Maudgal, R.K., Tchen, T.T. and Bloch, K. (1958) B2-methyl shifts in the cyclization of squalene to lanosterol. *J. Am. Chem. Soc.* 80: 2589–2590.

Olah, G.A. and Schleyer, R. (1992) *Carbonium Ions.* Wiley Interscience, New York.

Schroepfer, G.J. (1982) Review of sterol biosynthesis. *Annu. Rev. Biochem.* 51: 555–585.

Sharpless, K.B. (1968) Ph.D. thesis, Stranford University.

Stork, G. and Burgstahler, A.W. (1955) The stereochemistry of polymere cyclization. *J. Am. Chem. Soc.* 77: 5068–5077.

Sutherland, J.K. (1991) Polymere cyclizations. *In*: B.M. Trost and I. Fleming (eds): *Comprehensive Organic Synthesis*, Vol. 3, Pergamon Press, Oxford, pp 341–377.

van Tamelen, E.E. (1982) Bioorganic characterization and mechanism of the 2,3-oxidosqualene to lanosterol conversion. *J. Am. Chem. Soc.* 104: 6480–6481.

Willett, J.D., Sharpless, K.B., Lord, K.E., van Tamelen, E.E. and Clayton, R.B. (1967) Squalene-2,3-oxide, an intermediate in the enzymatic conversion of squalene to lanasterol and cholesterol. *J. Biol. Chem.* 242: 4182–4192.

Wirsching, P., Ashley, J.A., Benkovic, S.J., Janda, K.D. and Lerner, R.A. (1991) An unexpectedly efficient catalytic antibody operating by ping-pong and induced fit mechanism. *Science* 252: 680–685.

Woodward, R.B. and Bloch, K. (1953) The cyclization of squalene in cholesterol synthesis. *J. Am. Chem. Soc.* 75: 2023–2024.

Zhou, G.W., Guo, J., Huang, W., Fletterick, R.J. and Scanlan, T.S. (1994) Crystal structure of a catalytic antibody with a serine protease active site. *Science* 265: 1059–1064.

Analysis of proteins and nucleic acids

Chemical techniques employed for the primary structural analysis of proteins and peptides

A.S. Inglis, G.E. Reid and R.J. Simpson

Joint Protein Structure Laboratory, Ludwig Institute for Cancer Research (Melbourne Branch) and The Walter and Eliza Hall Institute for Medical Research, Parkville, Victoria Australia

Summary. This chapter summarises modern microchemical approaches to the purification, identification and primary structure analysis of peptides and proteins. Discussion of high-sensitivity purification methods is restricted to two-dimensional polyacrylamide gel electrophoresis (2-DE) and microbore/capillary column reversed-phase high-performance liquid chromatography (RP-HPLC). Associated techniques are discussed, particularly with respect to analysis of the products with current automated amino acid sequencers and mass spectrometers.

Introduction

Dramatic advances in molecular genetics over the past decade have created the opportunity to isolate essentially any gene in an organism. These gene-cloning techniques, combined with our ability to express cloned genes in cells growing in culture now permit the production of large quantities of specific "recombinant" proteins for structure-function studies, vaccine production, therapeutics and clinical applications in general. To date, these techniques have been particularly valuable for studies on intrinsically scarce proteins such as hormones, enzymes, immunomodulators, serum proteins and viral antigens. The time taken from the initial description of a substance with biological activity to purification of suitable quantities for biochemical analysis, the cloning of the gene encoding the protein, large scale production of the recombinant protein, clinical trials and use of a therapeutic, is decreasing at an astounding rate. To understand this rapid turn of events it is important to understand where the significant breakthroughs have come from and what they will mean for modern structure-based drug discovery.

A representative structure-based drug discovery paradigm, indicating the syngergy of molecular biology, protein chemistry and physical chemistry is depicted in Figure 1. A typical initial step is the isolation of a protein from a natural source and the determination of sufficient amino acid sequence to enable the cloning and expression of the target protein. Subsequent structure-function studies on the recombinant target protein, using a number of approaches such as chemical modification, site-directed mutagenesis, inter-species chimeras, molecular modelling etc., may indicate a *"lead"* protein that exhibits some of the

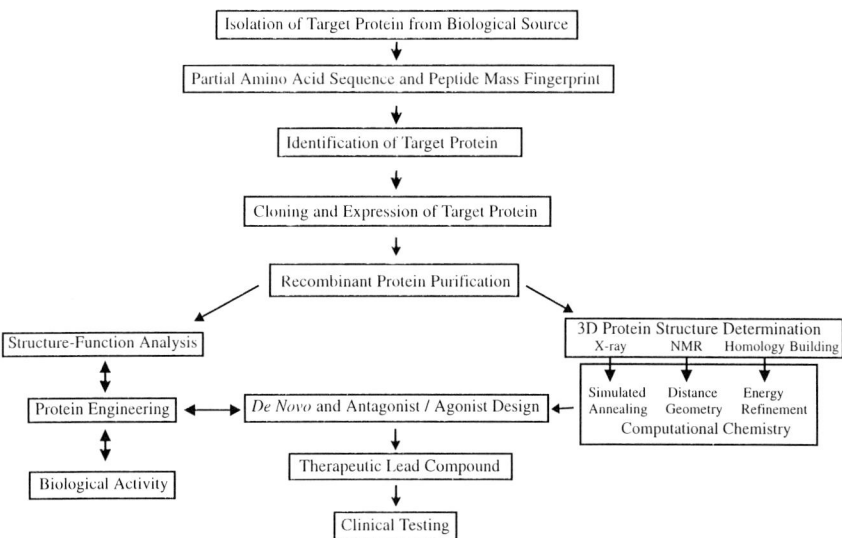

Figure 1. Paradigm for the generation of a "lead" protein for therapeutic purposes. Protein chemistry, along with molecular biology and physical chemistry plays a vital role in facilitating therapeutic design.

desired biological activity (e.g., agonist, antagonist). A powerful adjunct of any structure-function study is the three-dimensional structure determination, by X-ray crystallography or NMR, of the target protein. The raw data from such determinations are turned into a refined structure by the computational-chemistry applications of simulated annealing, distance geometry or molecular dynamics. The three-dimensional structure of a protein target can also be effectively constructed computationally by homology to a protein of known structure. In all of the above, computing plays a critical role. Since the results of computational experiments will only be as good as the framework within which the question is asked, modelling with the aid of supercomputers cannot *ipso facto* replace the laboratory. However, it can offer powerful insights that may suggest the direction of future experiments.

It is clear from the paradigm in Figure 1 that protein chemists and molecular biologists are dependent on each other for the characterisation of low-abundance proteins since many gene cloning strategies rely upon obtaining partial amino acid sequence data from the target protein. Such information is used to design oligonucleotide primers, suitable for use as probes for screening cDNA/genomic libraries or for gene cloning using polymerase chain reaction methodologies.

Protein structural analysis is complicated by the need to interface high-resolution protein and peptide isolation techniques with the latest generation amino acid sequence analysers and mass spectrometers.

These instruments impose a number of constraints on both the volume and composition of samples. Accordingly, one must consider these limitations during the design of protein and peptide purification strategies. Indeed, it is now generally recognised that the major obstacle in obtaining amino acid sequence data on subnanomole (low microgram) quantities of material is not the sensitivity of structural analysis *per se*, but the ability to isolate and micromanipulate samples at these low levels. Only too often do we hear that an important sample was inextricably lost between the isolation and sequence analysis steps. It is the intention of this chapter to highlight – with particular emphasis on current technologies used in the author's laboratory – micropreparative procedures employed for the high-sensitivity primary structural analysis of proteins and peptides.

Protein purification

High-performance liquid chromatography (HPLC) and polyacrylamide gel electrophoresis (PAGE) are the two high-resolution techniques which have found extensive use in both the isolation and characterization processes for trace-abundant proteins (Simpson et al., 1989b; Shively et al., 1989).

Although modern protein structural analysis techniques, whether by Edman degradation or mass spectrometry, permit sequence information to be obtained on less than 5 picomoles of starting material, the ability to purify and, if necessary, manipulate (e.g., buffer-exchange, concentrate, reduce and alkylate, fragment etc.) proteins and peptides at such low levels presents a considerable technical challenge. Most conventional protein purification procedures, when applied to low microgram amounts of material, usually result in severe loss of sample. This problem, however, can be circumvented by the use of two-dimensional gel electrophoresis (2-DE), microbore column (i.e., ≤ 2.1 mm internal diameter (ID) or capillary column (i.e., ≤ 0.5 mm ID) RP-HPLC and judicious attention to sample handling techniques. These issues will be discussed in turn.

Purification of proteins by acrylamide gel electrophoresis

PAGE is an extremely powerful tool for purifying and characterizing proteins and peptides (Laemmli, 1970; O'Farrell et al., 1975), particularly those are not amenable to separation by RP-HPLC (e.g., high molecular weight hydrophobic proteins). For example, 2-DE, where proteins are separated in the first dimension on the basis of their charge (isoelectric point) by isoelectricfocusing (IEF) and in the second dimension, on the basis of their size by SDS-PAGE, has been widely used for resolving extremely complex protein mixtures. This approach has been used extensively to study alterations in cellular protein expression patterns in response to various biological stimuli (Celis and Bravo, 1984). The first dimension IEF step is achieved using carrier ampholytes which form a pH gradient, typically, over the pH range 5–7 or 3.5–10; for extremely basic proteins non-equilibrium pH gradient electrophoresis (NEPHGE) at pH 7–9 or 9–11 (2% carrier ampholytes) is recommended (O'Farrell et al., 1977). Visualisation of proteins in the gel is typically achieved by Coomassie blue or silver staining or image analysis of intrinsically radiolabelled proteins.

A major problem with 2-DE is the irreproducibility of gel patterns, from one day to another as well as between laboratories. This problem, which stems from irreproducibility of the first dimensional IEF, has seriously hampered efforts to integrate existing 2-DE protein databases and to disseminate such information (Simpson et al., 1992). The major cause of this problem is that the pH gradient generated is not suitable for long term intervals, with the gradient drifting towards the cathodic end with prolonged focusing times. This results in poor resolution of basic proteins. The recent introduction of immobilised pH gradients IPGs (Görg et al., 1988) has overcome many of these problems. First, the IPG is not affected by the time of focusing and, thus, is a true equilibrium technique. Second, the IPG gel can be precast, and dried gel strips are now commercially available.

Although 2-DE has been widely used during the past 20 years, its usefulness has been limited by the problems associated with protein identification. However, in recent years tremendous advances have been made in technologies for direct N-terminal sequence analysis of picomole amounts of 2-DE resolved proteins (Fig. 2). Traditionally, electroelution and passive elution have been the methods of choice for recovering proteins from the gel matrix for structural analysis. Since samples recovered in this way are usually contaminated with high concentrations of SDS, buffer salts and acrylamide gel-related artifacts which are deleterious to protein sequence analysis, they need further manipulation, e.g., inverse gradient RP-HPLC (Simpson et al., 1987), organic solvent precipitation (Wessel and Flügge, 1984), entrapment of protein on a polyvinylidine difluoride (PVDF) filter, or polystyrene divinylbenzene based chromatography (Moyer and Burkhart, 1994) in order to obtain meaningful sequence data.

Alternatively, proteins can be electrotransferred from the acrylamide gel onto a chemically inert membrane such as polybase-treated siliconized glass, polyproplyene or PVDF (see Simpson et al., 1989b; Matsudaira, 1993; Patterson, 1994 and references therein) for direct sequencing. More recently, Teflon has been described as an alternative immobilising matrix (Burkhardt and Moyer, 1994).

Parameters affecting electroblotting yields include the composition and ionic strength of transfer buffers, use of semi-dry or tank-type

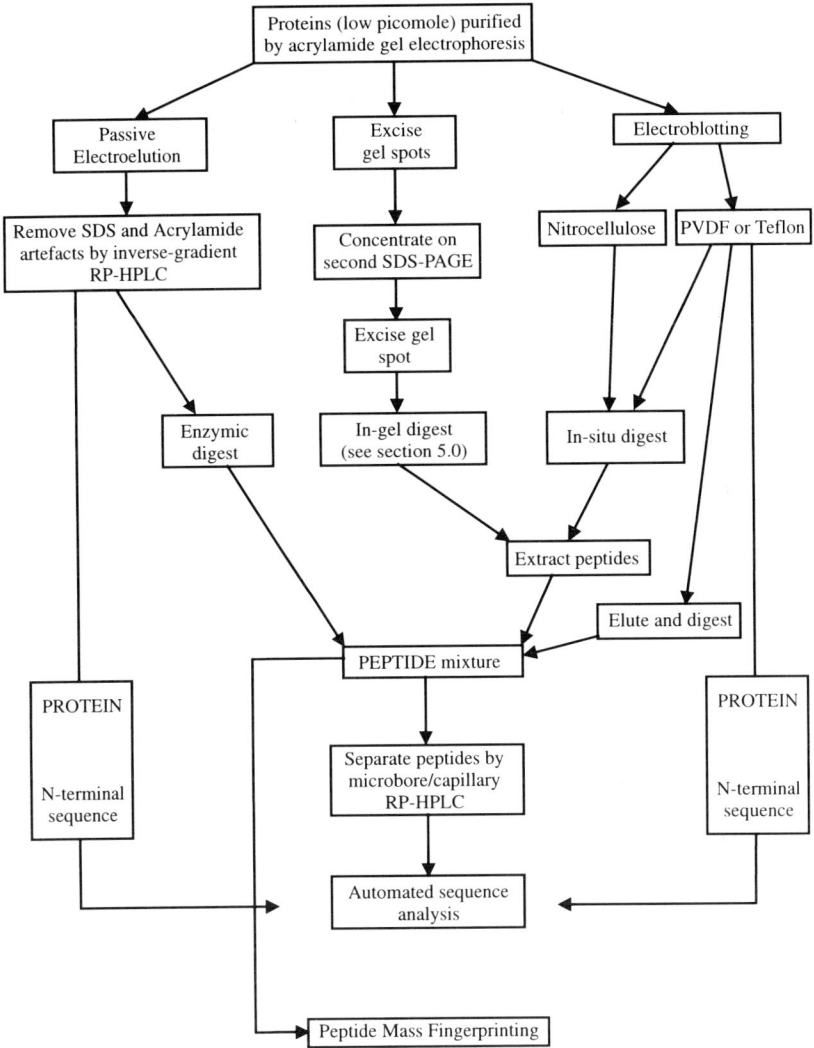

Figure 2. Low abundant proteins can be purified by SDS-PAGE or 2-DE gel electrophoresis and recovered by electroelution or electroblotting for N-terminal sequence analysis. For internal sequence information, the recovered protein, after concentrating on a second SDS-PAGE gel (Vanderckhove et al., 1993), is then digested and the generated peptides purified by microbore or capillary column RP-HPLC for structural analysis. Amino acid compositional data can be compared with protein databases for protein identification. Alternatively, the Coomassie blue staining protein can be digested *in situ* in the gel slice, a mass fingerprinting of the peptide obtained with the mass spectrometer and the protein identified using mass spectrometric methodologies.

electrotransfer units and transfer time (Xu and Shively, 1988; for reviews, see Simpson et al., 1989b; Matsudaira, 1993). SDS concentrations have been found to be critical. High concentrations assist migration from the gel but hinder protein binding to PVDF. Addition of methanol plus pre-soaking of the gel in the transfer buffer compensates for a deceased concentration of SDS in the gel and improves transfer recoveries. Higher N-terminal sequencing yields have been obtained by pre-electrophoresis with mercaptan and lower pH during electrophoresis (Moos et al., 1988). It is now generally recognised that substantial losses of protein are incurred during their electrotransfer from the gel matrix onto the immobilising membrane. Hence, for internal sequence analysis the *in-gel* proteolysis approach is generally preferred to the *in-situ* ("on membrane") proteolysis approach. The *in-gel* proteolysis approach, peptide mass fingerprinting, and protein identification by amino acid compositional analysis will all be discussed in the following sections.

Chromatographic methods for preparing proteins and peptides for structural analysis

With the steadily growing variety of reversed-phase, ion-exchange and affinity chromatography packings available for HPLC (see Simpson and Nice, 1989; Mant and Hodges, 1991), the task of the protein chemist in purifying trace-abundant proteins is becoming much easier. Capillary zone electrophoresis (CZE) is a relatively new technique, however with its short separation times, high resolution and sensitivity, it is a promising approach that is now providing proteins and peptides in quantities suitable for structural analysis (Bergman, 1993). For a review of this subject, the reader is referred to Monnig and Kennedy (1994).

Microbore and capillary column reversed-phase HPLC
For high-sensitivity (i.e., low microgram) protein and peptide isolation and micromanipulation, we, and others, have developed purification strategies that utilise short (30–100 mm length) microbore column RP-HPLC (Simpson et al., 1989a). More recently, we have extended the sensitivity of this chromatographic approach by utilising capillary columns (Moritz and Simpson, 1992). In addition to sensitivity, these chromatographic approaches are ideal for concentrating samples (trace enrichment) and removing buffer salts that otherwise interfere with sequence analysis using commercial instruments (with the exception of the Hewlett-Packard sequencer) and mass spectrometric analysis.

Compared with conventional column (4.6 mm ID) chromatography, microbore columns (1–2.1 mm ID) offer enhanced sensitivity and decreased peak volumes without loss in chromatographic performance. To

Table 1. Optimal protein load for common column diameters[1]

Column	Conventional	Microbore		Capillary	
Internal diameter (m/m)	4.6	2.1	1.0	0.5	0.05
Flow rate (ml/min)	1	0.2	0.04	0.01	0.0002
Protein load (μg)	100	20	2	0.2	0.001
Peak volume (μl)	500–2000	100–400	20–80	5–20	0.1–0.4

[1]Support: Brownlee RP-300. Protein: bovine α-lactalbumin. Gradient elution (0.15% v/v aq. TFA to 60% acetonitrile/40% aq. 0.12% TFA) over 60 min was used at equivalent linear flow velocities.

operate microbore columns at linear flow velocities equivalent to those used with conventional columns, flow rates must be decreased in a manner proportional to the square of the reduction in column diameter. For example, if the optimal flow rate for a 4.6-mm ID column is 1 ml/min, then a 1.0-mm ID column would be operated at ~50 μl/min to achieve the same linear flow velocity. Provided that the microbore

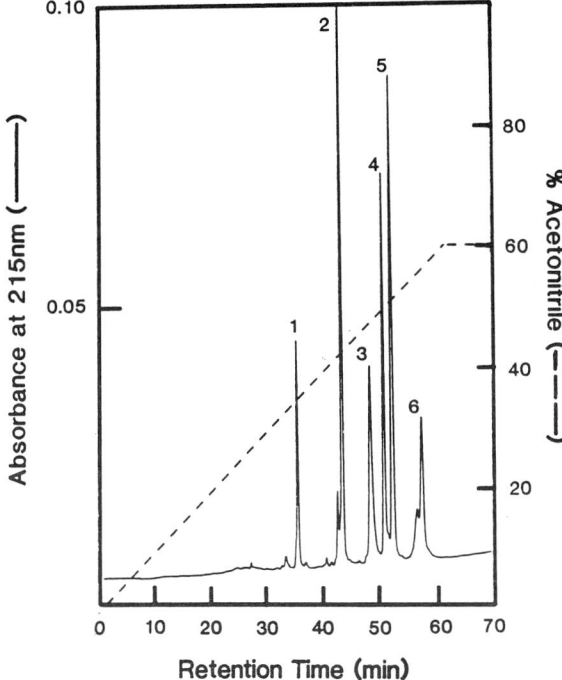

Figure 3. Separation of protein standards on a Brownlee RP-300 column (50 × 0.32 mm ID Chromatographic conditions: linear 60-min gradient from 0.10% (v/v) aq. TFA to 60% CH$_3$CN/40% water/0.10% (v/v) TFA. Column temp., 45°C. Flow rate, 3.6 μl/min. Protein standards: 1 = ribonuclease A; 2 = lysozyme; 2 = bovine serum albumin; 4 = carbonic anhydrase; 5 = myoglobin; 6 = ovalbumin. Sample load, 50 ng of each protein. (Reproduced from Moritz and Simpson (1992) with permission by *J. Chromatogr.*, Elsevier Publishers, Amsterdam.)

column is not overloaded, column efficiencies are maintained and samples will be recovered in smaller peak volumes (Simpson et al., 1989a). Further reduction in column internal diameter, i.e., capillary column chromatography (<0.5 mm ID), results in a subsequent increase in sample concentration. When a 0.32 mm ID capillary column was compared with 2.1 and 4.6 mm ID columns, a 25- and 125-fold increase in sensitivity of protein detection was achieved, respectively. Constant peak band widths, as a function of time, for the three columns indicated that comparable column efficiencies were achieved with capillary liquid chromatography. Peak recovery volumes for the 0.3-, 2.1- and 4.6-mm ID columns were 2.5, 133.5 and 660 µl, respectively (Moritz and Simpson, 1992). Table 1 lists some experimental data obtained for optimisation of the chromatography of α-lactalbumin using columns of different diameters. Detailed procedures for configuring conventional HPLC systems to provide accurate low flow rates (0.4–4 µl/min) and gradients necessary to operate capillary columns as well as the construction and slurry-packing of capillary columns have been described (Moritz and Simpson, 1992; Moritz et al., 1994).

Since chromatographic resolution is a function of both (i) the selectivity of the packing and (ii) the efficiency of the column, it necessarily follows that solute separation will be independent of column dimensions for a well-packed column operated under equivalent elution conditions. The potential of capillary column chromatography to resolve a mixture of protein standards at high sensitivity is shown in Figure 3. In this example, 50 ng amounts of five proteins were recovered in peak volumes (<5 µl) suitable for direct loading onto the sample disk of gas-phase sequencers without further manipulation. By careful attention to solvent preparation (i.e., the use of highly-polished water and the careful adjustment of the trifluoroacetic acid (TFA) concentration of both primary and secondary solvents in order to achieve a straight baseline) protein levels as low as 500 pg can be detected.

The advantage of capillary over wider-bore columns in RP-HPLC make it particularly attractive for coupling to detection devices such as electrospray mass spectrometers (Jardine, 1990; Ji et al., 1994; Moritz et al., 1994). The flow rates required for capillary chromatography are ideal for optimal mass spectrometric performance.

Primary structure analysis

Edman N-terminal degradation

In 1954, Sanger and his colleagues provided a land mark in protein chemistry when, after 10 years of intensive work on insulin at the University of Cambridge, they described the first chemical structure of a protein (for a review, see Thompson, 1955). Insulin, the pancreatic hormone that governs sugar metabolism in the body, is one of the smallest proteins, being just 51 residues long, but with the technology available then, and the added complication of the presence of both inter- and intra-chain disulfide bonds, this achievement was a watershed which thoroughly deserved the Nobel prize. The ability of the protein chemist to determine the covalent structure of the protein molecule has led to major advances in the biochemical area, and it has been seminal to both the development of three-dimensional pictures of the protein and an understanding of structure-function relationships.

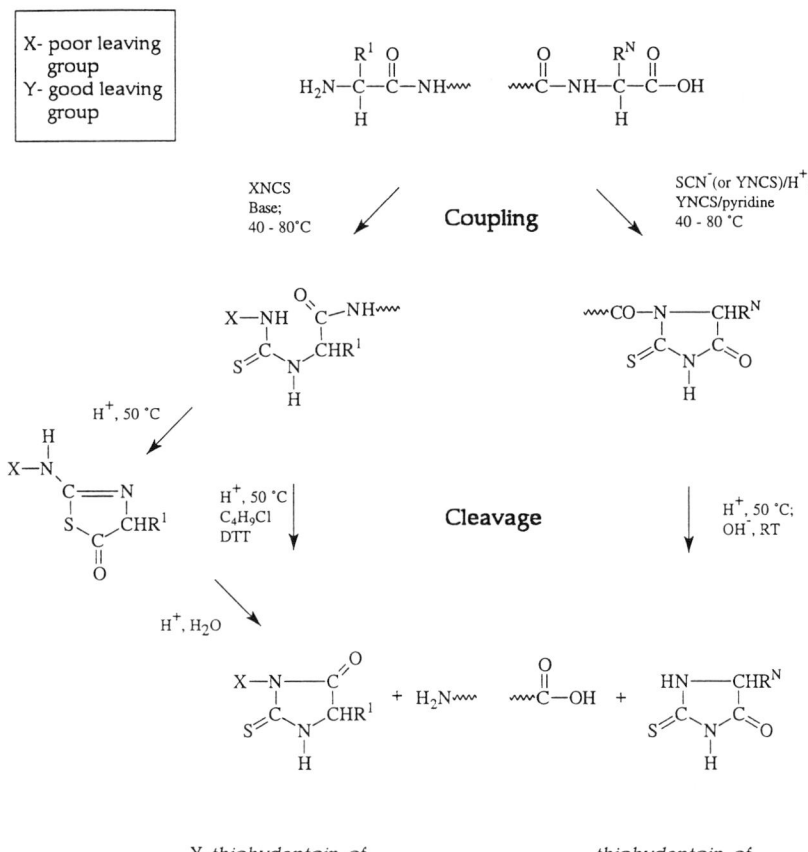

Figure 4. Procedures for removal of N- and C-terminal amino acids as thiohydantoins. Both derivatives may be detected in the UV because of their high absorbance at 268 nm (Emax ≅ 17 500). Classical Edman chemistry used phenyl isothiocyanate (x = phenyl) for N-terminal degradations.

All current N-terminal sequencing methods are based on the Edman procedure (1949) which consists essentially of: (i) a coupling reaction at pH 9 with phenylisothiocyanate (PITC) and the N-terminal amino group of a peptide or protein; (ii) a cleavage reaction with anhydrous acid – which is facilitated by the proximity of the nucleophilic sulfur atom of the derivatized N-terminus to the carbonyl carbon of the first peptide bond – to yield an amino acid thiazolinone and the peptide minus its N-terminal amino acid, and (iii) conversion of the unstable thiazolinone to a more stable phenylthiohydantoin (PTH) by treatment with aqueous acid. The procedure can be repeated on the shortened peptide to yield the second amino acid residue along the chain, and so on (see Fig. 4, x = phenyl). Cleavage and conversion can be accomplished with just one aqueous acid reaction but the anhydrous acid optimises the specific cleavage of the peptide bond of the N-terminal amino acid. Recent work (Farnsworth and Steinberg, 1993) has suggested that the thiohydantoin is actually formed preferentially when the cleavage is carried out in the presence of a thiol such as dithiothreitol (DTT).

Automation of the coupling and cleavage steps (Edman and Begg, 1967) enabled researchers to determine the sequence of the first 30–40 amino residues in a protein (0.3 µmole) routinely. The thiazolinones were manually converted in batches to the PTHs. Later workers were to add devices for automatic conversion of the separated thiazolinones to the PTH amino acids. These were followed by on-line detection of the PTHs using HPLC (see Tab. 2).

Automated gas-phase microsequencing

Automatic microsequencing came of age with the release of a commercial instrument which used miniaturized components and gaseous coupling and cleavage reactions of the protein immobilised on a glass fibre disc; washout was prevented by addition of a polyamide (Polybrene) to the glass support (Herwick et al., 1981). Conversion was performed automatically and addition of an on-line HPLC system allowed sequence analyses to be made on PTH-amino acids at the 10 pmole level. A

Table 2. Advances in amino acid sequence automation

Instrument		Sample required	Repetitive yield	Comments and drawbacks
Spinning cup	'70	0.3 µmole	92–95%	Manual conversion only, peptide washout
	'73			Auto conversion; improved values
	'78			Polybrene peptide anchor
	'78			HPLC detection of PTHs
Gas phase	'81	100 pmole	92–95%	Miniaturised; glass fibre support
ABI 477A	'85			On-line HPLC; 0.5 pmole detection
Hewlett-Packard	'92	3.5 pmole	97%	Protein on bi-phasic column

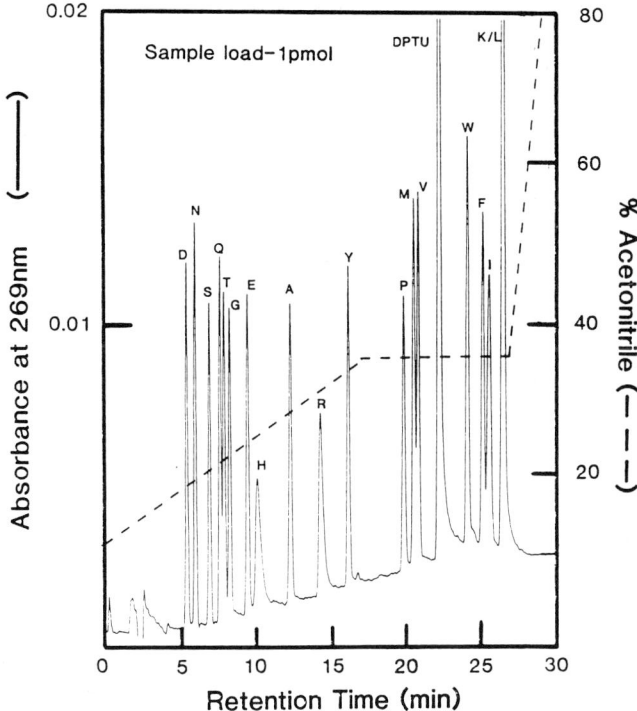

Figure 5. Separation of 1 pmol amounts of phenylthiohydantoin-amino acids by reversed-phase capillary liquid chromatography. Column, 150 × 0.32 mm I.D. Applied Biosystems PTH-C18 (packed by LC packings). Solvent A, 8.3 mM sodium acetate-5% (v/v) tetrahydrofuran (pH 4.1); solvent B, acetonitrile. Column temperature, 55°C. Flow rate, 5 μl/min. PTH-amino acid notation is shown using the one letter code for amino acids; DPTU denotes diphenylthiourea. (Reproduced from Moritz and Simpson (1992) with permission by *J. Chromatogr.*, Elsevier Publishers, Amsterdam.)

combination of the glass support system, highly reliable, inert valves, and especially purified chemicals contributed largely to a user-friendly and reliable instrument. In addition, the advent of HPLC and its application to the separation of Edman derivatives made a major contribution to the successful application of the Edman reaction because it provided not only a very sensitive method of detection, but also a quantitative one. With these two extremely important parameters, in one blow HPLC virtually removed the need for modified Edman reagents, because one could estimate yields of the released PTH-derivatives of the amino acid at each degradation cycle down to low picomole levels; capillary columns (0.32 mm ID) further extend the sensitivity to the low femtomole levels (Moritz and Simpson, 1992), see Figure 5.

Although analyses on the newer machines showed more consistent repetitive yields, they did not provide the high repetitive yields (<95%) essential for long sequencer runs. However, Hewlett-Packard found that addition of an alkylamine to the biphasic reactor column in their sequencer improves the repetitive yields (> 98%), presumably by both ensuring that the protein amino group is uncharged for the coupling reaction and providing a scavenger of potential blocking species in the reaction system.

Recent work (Bartlet-Jones, 1994) on sub-picomole amounts of peptides, using small reaction volumes containing a volatile isothiocyanate (trifluoroethylisothiocyanate, B.P. 93–94°C), has shown that it is not essential to preclude oxygen from the vial during coupling, at least for several consecutive residues. There was little evidence of desulfurization of the reagent or the coupled peptide. This raises the questions of whether oxidative side reactions are a major problem during coupling or whether the thiourea byproducts are involved in a protective role. In our experience, diphenylthiourea is destroyed at a faster rate than thiohydantoins in general, and its addition to very dilute solutions of thiohydantoins has a protective effect on them.

Solid-phase microsequencing

The solid phase sequencing methodology of Laursen (1971) closely followed the development of the spinning cup-liquid phase sequenator (Edman and Begg, 1967). In this approach the protein or peptide is *covalently* attached to an insoluble support. In theory, reagents may be removed and products extracted efficiently by thorough washing of the support. Covalent coupling to proteins is particularly advantageous for identifying very polar residues which are only poorly extracted from the sequencer with chlorobutane. Hence, cysteic acid, phosphoserine, glycosylated residues, etc. can be extracted with more polar solvents once the peptide or protein is attached chemically to the support. Aspects of solid phase sequencing have been incorporated into the design of the latest gas phase sequencers and, with growing interest in glycoproteins and phosphorylated proteins, and post-translational modifications in general, this methodology will undoubtedly be developed further.

Alternative N-terminal reaction methodologies

Several variations on the Edman reagent, phenylisothiocyanate, have been proposed, primarily to increase the analytical sensitivity of the method. Both coloured (Chang and Creaser, 1976) and fluorescent isothiocyanates have been used for specific purposes, but none has superseded the Edman reagent for routine sequencing purposes, primarily because they have not given quantitative coupling, presumably due to either the increased size of the reagent or deactivation of the isothiocyanate group.

Inman and Appella (1977) first demonstrated that anilinothiazolinones (ATZs) (see Fig. 4) could be reacted with nucleophiles, particularly primary alkylamines, to form a phenylthiocarbamyl (PTC)-amino acid derivative in high yields. Tsugita et al. (1989) subsequently proposed that the sensitivity of the sequencing procedure could be increased significantly if the ATZs were derivatised with a fluorescent reagent such as 4-aminofluorescein. The problem with the general approach, however, is that the thiazolinone is not in fact formed exclusively upon cleavage and many amino acids form a PTH derivative. By opening the ring in aqueous base in the presence of a reducing agent, Farnworth and Steinberg (1993) showed that the PTC-amino acid could be formed and converted to the ATZ in high yield using boron trifluoride etherate in trifluoroacetic acid. Sequence analysis of 500 femtomole of α-lactalbumin with fluorescent detection of the amino fluorescein was then achievable using an automated procedure.

While excellent results on 2–10 pmoles of peptide have been obtained by upgrading commercial instruments in terms of sensitivity (Erdjument-Bromage et al., 1993) theoretically, further miniaturisation of the instrument would be advantageous: preliminary concentration would not be necessary for application of the sample to capillary HPLC or capillary electrophoresis instruments, background and cost would be decreased because miniaturisation would result in a corresponding decrease in the volumes of reagents required. However, reduction in size alone does not necessarily lead to any great improvement in performance; our approach has been to use both a modified chemistry and a new simplified design. Internal volumes have been reduced by using fused silica capillary tubing throughout, and employing a microcell with only 100 nanolitre internal volume. The system lends itself to low femtomole analysis by capillary HPLC.

C-terminal degradation

Although there were active efforts to develop chemistry for sequential removal of C-terminal amino acids from a protein some 25 years prior to Edman's first report on N-terminal sequencing, advances have lagged badly behind N-terminal methodology. Despite a plethora of potentially useful chemical modifications of the C-terminal amino acid, which include conversions of the carboxyl group to thiohydantoins, carboxylic acid esters, alcohols, acylureas, isothio-ureas, azides and hydrazides, the published procedures have given difficulties when tried by others (for review see Inglis, 1991). Consequently, enzymatic methods using carboxypeptidases have provided most of the C-terminal sequence data obtained over the years. An alternative solution has been to isolate the C-terminal peptide from a digest of the protein and analyse it using either the N-terminal degradation procedure or mass spectrometry.

Chemistry

The most widely studied method for C-terminal sequencing has been one in which the thiohydantoin of the C-terminal amino acid is first formed, then the adjacent peptide bond is cleaved in either acidic or basic medium. This yields an amino acid thiohydantoin (TH) and a shortened protein which may be subjected to a further degradation cycle. The C-terminal thiohydantoin procedure can be used repetitively and there are no major interfering side reactions as evidenced by nine manual degradations of Leu-enkephalin (Inglis et al., 1989). This approach has similarities to the Edman reaction for N-terminal sequencing, as shown in Figure 4. While the latter employs initial coupling with "stable" isothiocyanates such as phenylisothiocyanate, the C-terminal procedure uses either thiocyanate salts or organic isothiocyanates which readily disproportionate releasing the reactive isothiocyanate ion.

While the reaction mechanisms may differ – the amino group is best coupled at pH 9 (to remove H^+) in a polar environment, whereas the carboxyl group has been found to couple under anhydrous conditions over a wider range of pHs (from slightly basic to acidic conditions) – both coupling reactions cause a weakening of the adjacent peptide bond which is subsequently cleaved to release relatively stable thiohydantoin derivatives of the terminal amino acids. The common PTHs (x = phenyl in Fig. 4) are less polar than the THs but they have similar properties, the UV spectra especially being important because of the high extinction coefficient (E max $\sim$ 17 500) at around 269 nm which may be used for identification purposes.

Coupling

The temperature of the coupling reaction has usually been kept at around 50°, in keeping with temperatures used for N-terminal sequencing. However, there was no apparent drawback when the procedure was used at 80°. A peptide containing aspartic acid in the penultimate position was also analysed successfully, thereby ruling out doubts concerning interactions of the side chain carboxyl groups (Inglis et al., 1989). A short pre-treatment ("activation") with acetic acid-acetic anhydride (1:4) was given, then followed by addition of 1.4 M thiocyanic acid (HSCN) in acetone (1 part HSCN: 6 parts activation mixture). Subsequently this pre-treatment was deleted without adverse results (Inglis et al., 1992).

When the organic isothiocyanates, trimethylsilylisothiocyanate (TMS-ITC) and diphenlyphosphorocyanatidate (DPP-ITC) were tried as reagents, it was found that yields were better in solutions close to neutrality, or slightly basic, and the preferred reagent – DPP-ITC – in the presence of pyridine was found to give complete reaction of most amino acids in under 1 hour at 50° (Bailey et al., 1992). However, proline does not react at all under these conditions. Additions of TFA to an ammonium thiocyanate solution in acetic acid-acetic anhydride (Inglis et al., 1992) gave a facile proline coupling; this has now been confirmed by others (Bailey et al., 1994). At 40° a significant yield of a clean product was obtained which suggests that the optimum coupling temperature for this procedure may be well below 80°C.

Cleavage

A range of bases has been tried for cleavage of the peptidylthiohydantoins (Inglis et al., 1989). Cleavage is very fast in 0.5 M KOH in 33% methanol (3 min at 25°C) and slows as the amount of organic solvent increases. Experiments with a number of potential basic reagents confirmed that addition of a reductant (such as DTT) to the cleavage mixture was desirable to preserve the integrity of the product. After trying both acidic and basic reagents, Bailey et al. (1992) concluded that sodium trimethylsilanolate ($Na^+(CH_3)_3SiO^-$) was the reagent of choice.

An interesting extension of the thiohydantoin chemistry has been made by Boyd et al. (1992) who found that cleavage occurs under mild conditions when the sulfur of the peptidyl-thiohydantoin is alkylated. By cleaving with trifluoroacetic acid in the presence of trimethylsilylisothiocyanate, both cleavage and thiohydantoin formation on the new C-terminal amino acid may be effected. Promising results have been obtained but the procedure is so far not successful with threonine and serine, in particular. It has been suggested that proline could also cause difficulties at this cleavage step. Unlike other amino acids, the proline nitrogen is already part of a pyrrolidine ring and so becomes quaternary on cyclization. To preserve this less stable thiohydantoin structure, cleavage with acid was proposed (Inglis and De Luca, 1993) to eliminate the possibility that base would remove the charge on this nitrogen, and promote ring opening rather than cleavage of the peptide bond. This approach has been used successfully for the analysis of proline in the Hewlett-Packard C-terminal sequencer.

Future perspectives

Advances in the chemistry and instrumentation over the past 6 years have confirmed that the C-terminal thiohydantoin methodologies have potential to rival the N-terminal sequencing methods. The chemistry can be carried out in commercial N-terminal sequencers with appropriate modification for reagent and temperature requirements. The initial sequencer runs were made on either peptides attached to glass beads or proteins (nanomole quantities) attached to PVDF-DITC membranes (Inglis, 1991). Recently, it was shown that proteins can be sequenced when immobilised on Zitex membranes (Bailey et al., 1992). Only short sequence runs were made, but the level of sequencing is being lowered gradually and is now subnanomolar. However, repetitive yields and background still fall far short of those obtained by N-terminal sequencing.

It now remains for parameters that are important for high repetitive yields to be catalogued and incorporated in automatic methodology. The availability of commercial instruments should accelerate the development.

Mass spectrometry

Together with fast atom bombardment (FAB) ionisation (Barber et al., 1981), the development in 1988 of two mass spectrometric ionisation techniques, namely electrospray ionisation (ESI) (Fenn et al., 1988) and matrix assisted laser desorption ionisation (MALDI) (Karas et al., 1988), have changed the way in which the structural analysis of biomolecules is now approached. These novel techniques have rapidly become indispensable tools for the biological laboratory. A variety of techniques has been developed for the structural analysis of peptides using mass spectrometry. These range from the use of mass spectrometry for sequence analysis using ion fragmentation techniques, to the development of novel Edman-like reagents that rely on the mass spectrometer as a detector for derivatised amino acids and peptide ladder sequencing. For an extensive review of mass spectrometry see Burlingame et al., 1994.

Mass spectrometric sequence analysis
Several methods exist for the structural analysis of peptides using tandem mass spectrometry. The electrospray ionisation process is an efficient method to transfer peptides and proteins from solutions into the gas phase. Singly or multiply protonated ions are formed, generating ions of a mass/charge ratio having the general formula $(M + nH)^{n+}$ where n equals the number of protons or charges associated with the ion. As this is a "soft" ionisation technique, fragmentation of the molecule to yield sequence information does not occur. In order to obtain comprehensive amino acid sequence information, the triple quadrupole mass spectrometer may be used. From the mixture of all ions present at a given time in a quadrupole mass analyser Q1, a parent ion of selected mass/charge ratio is chosen and allowed to undergo characteristic low energy (<300 eV) collision-induced dissociations (CID) with a collision gas such as argon in mass analyser Q2 to yield fragment daughter ions. Manipulation of the kinetic energy of the ion and the concentration of gas inside Q2 determines the degree of fragmentation. These product daughter ions are then separated by mass analyser Q3. These low-energy collisions, resulting in random fragmentation of the peptide backbone, generate a laddered series of ions from

both the N- and C-termini. By fitting together the "jigsaw" of fragment ion masses, a complete structure can be constructed (Cox et al., 1994). Low energy CID however, does not allow the determination of the isobaric amino acids isoleucine or leucine as they share the same mass.

High-energy decomposition of ions (>300 eV), possible in a four-sector or hybrid mass spectrometer, results in cleavage of both peptide backbone and amino acid side chains, thus allowing the determination of Ile and Leu residues (Papov et al., 1994).

More recently, sequence information has been produced from MALDI-time of flight (MALDI-TOF) instruments equipped with linear/reflectron capabilities. A phenomenon known as metastable decay (or post source decay) has been observed to occur in the field-free region of the mass spectrometer following desorption. These metastable ions, similar to those observed in high energy collisions, may be analysed by a dual stage reflectron operated in stepped potential mode (Kaufmann et al., 1993).

Protein ladder sequencing
A novel approach to the Edman sequencing strategy, suitable for analysis by mass spectrometry, was described by Chait et al. (1993). A small amount of an N-terminal amino acid blocking reagent (phenylisocyanate) was incorporated in the coupling solution, generating a small amount of blocked peptide chain at each cycle. A "ladder" of masses corresponding to the terminated peptides was generated and analysed by MALDI-TOF mass spectrometry, the mass difference between them identifying the amino acid removed by each cycle of Edman degradation. This is akin to earlier "subtractive Edman" procedures that met with only limited success because of the lack of sensitivity and speed of the analytical methods used. An alternative approach has been proposed (Bartlet-Jones et al., 1994) which does not require the use of a blocking reagent but instead generates a "laddered" set of peptides simply by adding an aliquot of starting peptide at each cycle of degradation; a volatile isothiocyanate reagent (trifluoroethyl isothiocyanate) was used to eliminate solvent extractions and loss of peptides. Analysis by MALDI-TOF mass spectrometry provided sub-picomole (700 fmol) sensitivity. The great sensitivity of the mass spectrometer, the powerful data readout facilities and the speed of analysis all combine to give an attractive procedure. Degradation cycles are very fast (35–40 min) and multiple samples can be processed simultaneously using this approach; peptides of up to 2000 Da have been successfully analysed.

Alternate Edman derivatives suitable for mass spectrometric analysis
Aebersold et al. (1992) have recently reported the use of a new isothiocyanate derivative which incorporates a trimethylammonium group. This reagent, 3-[4′(ethylene-N,N,N-trimethylamino)phenyl]-2 isothio-

cyanate (PETA-PITC), yields derivatives that are rapidly detected at high sensitivity by ion-spray/electrospray mass spectrometry. 3-Pyridyl isothiocyanate and 3-pyridylmethyl isothiocyanate have also been designed for sensitive detection by ESI/MS and have been shown to give satisfactory coupling and cleavage reactions (Dharmasiri et al., 1994).

C-terminal sequence analysis by mass spectrometry
Various researchers have examined the use of mass spectrometry for the analysis of ragged-end C-termini generated by carboxypeptidase digestion or limited C-terminal hydrolysis. Acid cleavage has been used to yield C-terminal sequence information when used in conjunction with mass spectrometric analysis (Tsugita et al., 1992). High concentrations of organic acids, for example, 90% pentafluoropropionic acid, result in successive cleavage of the peptide chain from the C-terminus. Cleavage also occurs on the carboxyl-side of aspartic acid and the amino-side of serine residues. Prolyl bonds were resistant to this type of cleavage. While this approach is still in its infancy it is promising and indicates the power of modern mass spectrometric techniques in protein chemistry.

Peptide mapping and internal sequence analysis strategies

Obviously, if one had a procedure that would remove amino acids sequentially from one end of the protein chain with 100% efficiency, the primary structure determination would be straightforward. Such efficiencies are rare in chemical reactions and this goal cannot yet be reached, although continuous sequences of up to 100 residues have been obtained with automatic sequencers. To obtain the complete sequence, the chain must be cleaved specifically at positions adjacent to a particular amino acid using a range of enzymatic or chemical techniques (see Tab. 3) thus generating a set of peptide fragments ending (or beginning) with this amino acid; the cleavage process can then be repeated at a different amino acid to yield another set of characteristic peptide fragments. Each peptide is then purified to homogeneity prior to sequence analysis. A knowledge of the amino acid composition, mass and sequence of a sufficient number of peptides indicates unequivocally the order of the peptides along the chain, and hence the amino acid sequence of the protein. The complexity of the problem thus increases as the protein size increases. For trace-abundant proteins these approaches are not amenable due to insufficient material. For this reason, as well as speed, the DNA sequencing approach of sequencing cDNAs and genomic fragments encoding the target protein is preferred.

Additionally, analysis of the peptides ("fingerprints" or "maps") generated by the specific fragmentation of a protein enables one to obtain sequence data from N-terminally blocked proteins as well as to

Table 3. Cleavage reagents for protein structure

Cleavage reagent	Site	Source	Comments
Trypsin	Arg-x, Lys-x	Bovine pancreas Pig pancreas	Specific, does not cleave Arg-Pro and Lys-Pro bonds Aminoethylcysteine provides an additional site
Endoproteinase, Lys-C	Lys-x	Lysobacter enzymogenes	High specific activity; very effective for *in-gel* digests
Achromobacter protease, Lys-C	Lys-x	Achromobacter lyticus	High specific activity; very effective for *in-gel* digests
Clostripain,	Arg-x	Clostridium histolyticum	Yields may be lower than with trypsin; cleaves Arg-Pro
V8 protease	Glu-x, Asp-x Cysteic acid-x	Staphylococcus aureus	Specificity influenced by pH and buffer salt, cleaves cysteic acids also
Endoproteinase, Asp-N	x-Asp	Pseudomonas fragi	Very specific
Asparaginylendopeptidase	Asn-X	Jack bean	pH 5.5–6.5; stable to denaturants
Chymotrypsin	Trp-x, Tyr-x, Phe-x, Leu-x, Met-x	Bovine pancreas	Useful for limited digestion applications
Thermolysin	x-Leu, x-Val, x-Ile, x-Ala	Bacillus thermoproteolyticus	Broad specificity may be used at 60°
Pepsin	diverse	Porcine gastric mucosa	Relatively non-specific: prefers aromatics or Leu; low pH optimum; no disulfide interchange occurs
Cyanogen bromide	Met-x		Very specific, converted to homoserine lactone; yields > 70%, Ser, Thr bonds difficult; Trp-X can cleave
Dilute acid (pH 2)	Asp		Use formic acid for insoluble proteins and dilute; some amide cleavage; yields > 69%
Formic acid (80%)	Asp-Pro		Yields of 40% with high specificity
HCl 6M	Multiple		For random generation of small peptides
Hydroxylamine	Asn-Gly		Also breaks imide link to Gly; yields approx. 50%
DMSO/HCl/HBr; Iodosobenzoic acid; BNPS-skatole; N-Chloro succinimide	Trp-x		*Several alternative reagents available*; oxidation and modification of amino acids likely

[1] The peptide bond cleaved lies between the designated amino acid and residue 'x'.

(i) localise disulfide bonds, (ii) identify sites of post-translational modifications such as deamidation, oxidation, glycosylation and phosphorylation, (iii) characterise other structural modifications such as truncation of either the N- or C-terminal ends, (iv) identify amino acid substitutions and (v) determine the authenticity of genetically engineered protein products. Multiple wave length detection systems are useful adjuncts to RP-HPLC peptide mapping approaches since they identify aromatic amino acid-containing peptides (Fell, 1983; Simpson et al., 1989b). These residues, particularly tryptophan (with its unique nucleotide codon) are extremely useful in gene cloning strategies. Peptide mapping by RP-HPLC is particularly powerful when interfaced with mass spectrometry (see the section below, peptide-mass fingerprinting).

Application of peptide mapping to the complete sequence determination of proteins

Microbore RP-HPLC peptide purification procedures, when combined with Edman degradation, are particularly useful for the primary structural analysis of proteins. Typically, peptide mixtures are obtained by enzymatic or chemical digestion and fractionated on a reversed-phase 300 Å, C8 column using a linear gradient of acetonitrile in 0.1% aqueous TFA.

In order to obtain peptides in a pure state suitable for sequence analysis (Simpson et al., 1989b) variations in chromatographic conditions are usually required. Chromatographic selectivity can be varied by changing mobile phase parameters such as pH and buffer salt (e.g., phosphate, TFA, ammonium bicarbonate, unbuffered sodium chloride) or the nature of the chromatographic packing (see Fig. 6). For short highly charged peptides that are not retained on conventional reversed-phase columns the addition of an ion-pairing reagent such as sodium hexylsulfonate to the mobile phase will increase their hydrophobicity and hence, their retention time (Simpson et al., 1987).

Using this approach, we have determined the complete amino acid sequence of murine interleukin-6 (IL-6) (Simpson et al., 1988) and interleukin-9 (Simpson et al., 1989b) using less than 2 nmole of starting material and in a time comparable to that required with recombinant DNA methodologies. The peptide purification strategy for murine IL-6 is shown in Figure 6. Peptides generated by tryptic digestion of S-carboxymethyl-IL-6 were initially fractionated on a C8 reversed-phase column using a linear gradient of acetonitrile in aqueus 0.1%.

Upon initiation of a protein sequence analysis, if after several cycles of Edman degradation it is apparent that the protein is N-terminally blocked, useful internal sequence information can be obtained by first halting the sequence run, performing *in situ* cyanogen bromide cleavage

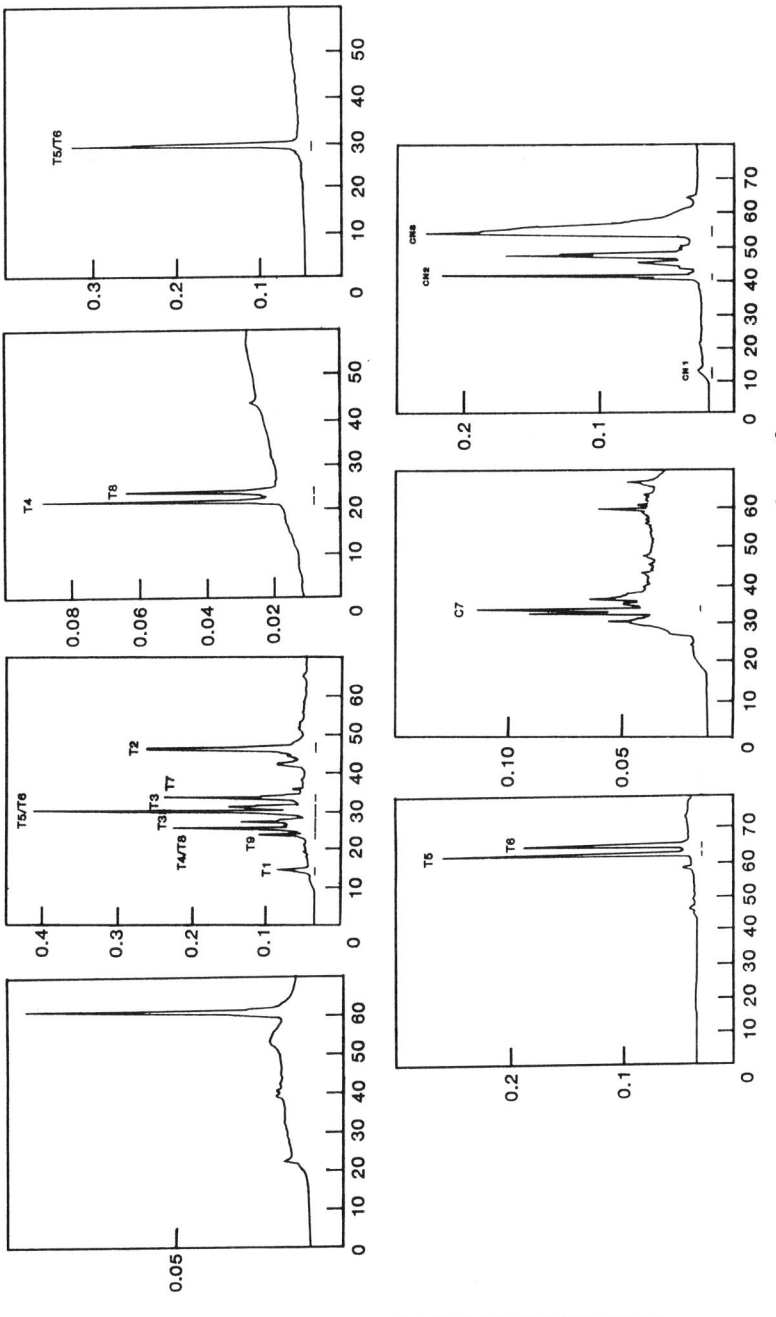

Figure 6. Microbore RP-HPLC peptide mapping of murine IL-6. (A) Desalting of S-carboxymethyl-IL-6 on Brownlee RP-300 (50 × 1 mm I.D.). (B) Separation of tryptic peptides of SCM-IL-6 on Brownlee RP-300 (30 × 2.1 mm I.D.). Linear 60 min gradient from 0.1% TFA-60% CH_3CN. Flow rate, 100 μl/min. (C) Rechromatography of peptide T4/T8. Same conditions as in B but TFA replaced with 1% NaCl. (D, E) Purification of tryptic peptides T5 and T6. T5/T6 were not resolved using conditions described in C (see D) but were separated on 5-μm ODS-Hypersil (50 × 1 mm I.D.). (F) Ion-pair chromatography. Peptides not retained using conditions in Figure 1B were purified on ODS-Hypersil (50 × 1 mm H_3PO_4/30 mM sodium hexylsulfonate (pH 2.5) to 50% CH_3CN. (G) Separation of CNBr fragments. Conditions: same as Figure B. Sample (in 85% aq. TFA) was diluted 100-fold with H_2O and loaded in 10 μl at 2 ml/min.

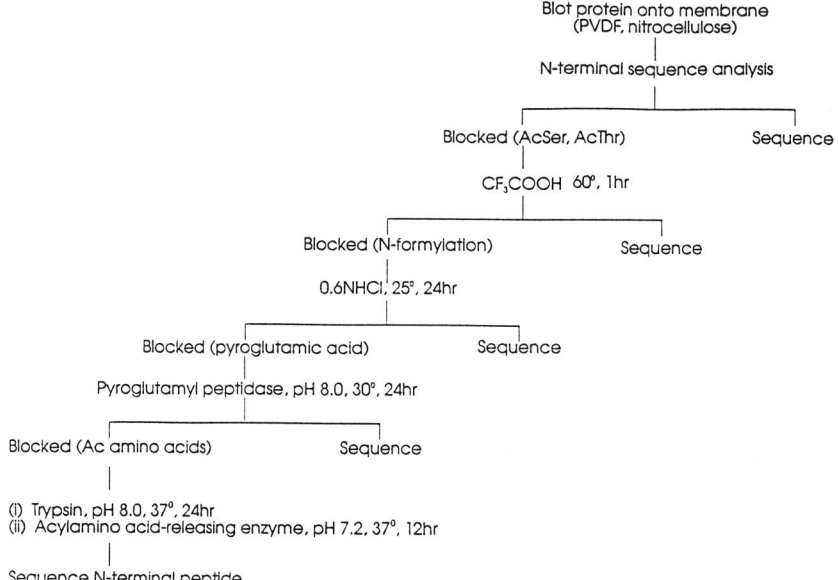

Figure 7. Deblocking strategy for N-terminally blocked proteins. Chemical and enzymatic procedures were used in turn to remove either the blocking group or the modified N-terminal amino acid. The acylamino acid-releasing enzyme does not digest protein with high molecular masses.

(Simpson and Nice, 1984) and then recommencing the analysis. This was particularly useful for the N-terminally blocked IL-9 (Simpson et al., 1989b). Alternatively, one can use the deblocking strategy of Tsunasawa and Hirano (1993) depicted in Figure 7.

In-gel peptide mapping of a 2-DE gel resolved protein

Figure 8 shows the capillary RP-HPLC/ESI-MS analysis of peptides isolated from a 2-DE resolved protein spot. Details of the *in-gel* digestion and extraction of peptides are given in Moritz et al. (1994). (A useful comparison of various in-gel digestion strategies is also given in the Association of Biomolecular Resource Facilities (ABRF) News, issue 3, (ABRF News, 1992).) Peptides were separated on a capillary column (0.2 mm ID) RP-HPLC and the eluted peptides introduced directly into the mass spectrometer to yield the total ion current (TIC) profile shown in Figure 8. Although UV detection greatly assists in the interpretation of the TIC profile, provides an invaluable guide to sample quality/quantity and attests to the performance of the systems, the use of TFA containing buffers is not recommended due to signal suppression effects during mass spectrometric analysis. In this instance, acetic acid-containing buffers were used to obtain optimal sensitivity.

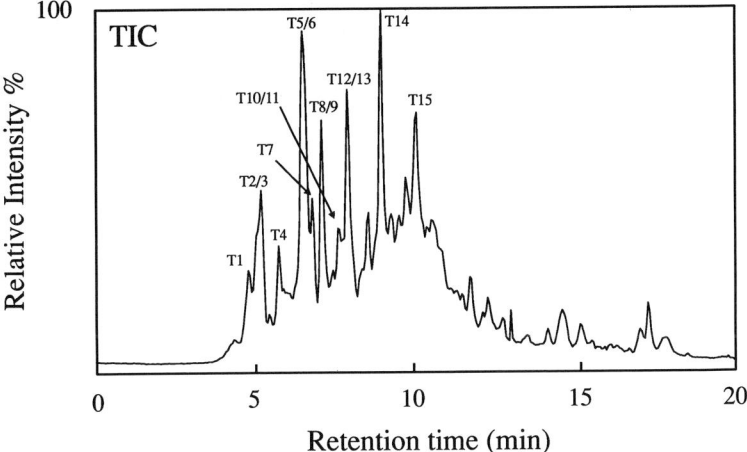

Figure 8. Rapid capillary RP-HPLC/ESI-MS analysis of an in-gel tryptic digest of protein #4 from LIM 1215 cells. Sample load: 10 μl, 10% of total digest (40 moderately Coomassie blue-stained protein spots from identical 2D gels); 500–2000 fmol. Column 150 × 0.2 mm ID. Brownlee RP-300; Flow rate: 2.0 μl/min. Gradient: linear 15 min from 0–80% where solvent A was 0.1 M aq. acetic acid; solvent B was acetonitrile. Shown is the total ion current profile of the chromatographic eluent.

Recent developments in high sensitivity protein identification

Peptide-mass fingerprinting

Digestion of a protein with specific proteases generates unique sets of peptides which can be used to characterize the protein. It is common to use the HPLC chromatogram of the enzyme digest as a "fingerprint" for this purpose. Alternatively, the output from a HPLC column may be interfaced with a mass spectrometer for the provision of obtaining a peptide molecular weight fingerprint. This, of course, is useful in determining features of protein structure such as assignment of disulfide bonds, post-translational modifications, mutations etc.

Recently, several groups have developed algorithms using existing protein sequence databases that facilitate the rapid identification of the protein of interest (for a recent review see Patterson, 1994) based solely on the masses of peptides generated upon digestion with a specific enzyme. The MOWSE fragment database (Pappin et al., 1993), which is derived from over 50 000 proteins, generates a theoretical list of peptide digest fragments from all of the protein sequences listed in the databases and uses search and scoring algorithms based on the observed distribution frequency of peptides in the database compared with experimentally observed values. It is often possible to identify a protein from just four to six peptide masses from a digest.

Table 4. Identification of 2-DE resolved protein spot #4 from LIM 1215 cells by peptide-mass fingerprinting (MOWSE algorithm), employing electrospray mass spectrometry

i) Input of raw peptide mass data.

```
Begin
Reagent       Trypsin
Tolerance     1
SeqMW         60 000
Filter        50
Datastart
785  843  844  869  901  912  960  1045  1206  1215  1233  1345  1919  2210  2561
Dataend
```

ii) Ranked protein output from the MOWSE database.

No. of hits = 30 (MAX. ALLOWED)
No. of database entries scanned = 72 018

1. P60_HUMAN   MITOCHONDRIAL MATRIX PROTEIN P1 PRECURSOR (P60 LYMPHOCYTE PROTEIN
2. P60_CRIGR   MITOCHONDRIAL MATRIX PROTEIN P1 PRECURSOR (P60 LYMPHOCYTE PROTEIN
3. P60_MOUSE   MITOCHONDRIAL MATRIX PROTEIN P1 PRECURSOR (P60 LYMPHOCYTE PROTEIN
4. P60_RAT   MITOCHONDRIAL MATRIX PROTEIN P1 PRECURSOR (P60 LYMPHOCYTE PROTEIN
5. A41931   heat shock protein hsp60, hsp60 = chaperonin-mouse
6. MMHSP60A   MMHSP60A HSP60 protein: (555 AA).-*Mus musculus*
7. S35637   high mobility group 1 protein homolog-rat
8. CSMCPN601   CSMCPN601 LOCUS CSMCPN601 2023 bp RNA PLN 04-FEB-1993-Mitoch
9. S16750   probable transposase-Calothrix sp.
10. S16892   probable transposase (insertion sequence IS702)-Calothrix sp.
11. ALRPP60A   ALRPP60A LOCUS ALRPP60A 2241 bp ds-DNA VRL 01-OCT-1992-Rous
12. KSRC_RSVPA   SRC TYROSINE KINASE TRANSFORMING PROTEIN (EC 2.7.1.112) (PP60-
13. CGL_HUMAN   CYSTATHIONINE GAMMA-LYASE (EC.4.4.1.1).-HOMO SAPIENS (HUMAN)
14. S52784   S52784 cystathionase; For the protein sequence (NCBI gibbsq 12)
15. B27671   spectrin alpha chain, nonerythroid-human (fragment)

iii) MOWSE peptide fragment database result for protein spot #4, identified as mitochondrial matrix protein P1 precursor (heat shock protein 60).

1. P60_HUMAN 2.400034e+06 61054 0.733
MITOCHONDRIAL MATRIX PROTEIN P1 PRECURSOR (P60 LYMPHOCYTE PROTEIN) (CHAPERONIN)

MW	START	END	SEQ
2561	97	121	LVQDVANNTNEEAGDGTTTATVLAR
1919	251	268	ISSIQSIVPALEIANAHR
1345	61	72	TVIIEQSWGSPK
1233	406	417	VGGTSDVEVNEK
1215	482	493	NAGVEGSLIVEK
960	421	429	VTDALNATR
912	293	301	VGLQVVAVK
901	397	405	LSDGVAVLK
844	474	481	IPAMTIAK
844	474	481	IPAMTIAK
785	463	469	IGIEIIK

NO MATCH 2210 1206 1045 869

The molecular weight of the intact protein is not essential and mass errors of several daltons can be tolerated. Thus, inexpensive time of flight mass spectrometers can be used. ESI tandem mass spectrometers and MALD ionisation instruments have provided appropriate mass data from low femtomole amounts of peptide, which is an order of sensitivity greater than that currently obtainable with commercial protein sequencers. The peptide-mass fingerprinting approach has been used to identify proteins expressed by normal and colon carinoma human cell lines, resolved using 2-DE (Ji et al., 1994). Peptides derived from the *in-gel* tryptic digest of a 2-DE resolved protein spot (protein #4 from the human colon carcinoma cell line LIM 1215; Reid et al., 1995) shown in Figure 8 were separated by capillary HPLC and directly analysed in the mass spectrometer using electrospray ionisation. Molecular masses of peptides obtained in this way were screened against the MOWSE database (Tab. 4). The search parameters had a mass tolerance of ± 1 dalton and a nominal molecular mass of 60 000 Da with a $\pm 50\%$ mass tolerance. Of the 15 peptide masses used in the search, 11 directly matched the molecular masses of predicted tryptic peptides from heat shock protein 60 (also known as mitochondrial matrix protein P1 precursor, or P60 lymphocyte protein; see Tab. 4). Three of these matches were confirmed by tandem mass spectrometric sequencing. Of the four masses constituting "misses" in the search, one of these was subsequently found to be derived from residues 98–107 of porcine trypsin.

Identification of proteins by amino acid compositional analysis

Classically, amino acid analyses were made on a protein hydrolysate using ion exchange chromatography to separate amino acids which were quantified after derivatisation with ninhydrin and by analysis of the coloured product at 570 nm (Spackman et al., 1958). The instrumentation has been refined over the years and is still capable of competing favourably with more modern alternatives at the 1–2 µg level (~ 50 picomole) of sample peptide.

Because of the need for analysis of picomole amounts of sample to match the sensitivity of purification and amino acid sequencing methods, a variety of reagents has been used to enhance the detection levels of amino acid derivatives. These include o-phthaldialdehyde (OPA), which has been used for both pre- and post-column derivatisation, phenylisothiocyanate (PTC), DABSYL chloride and FMOC chloride, all of which are used to derivatise the amino acids prior to HPLC chromatography using reversed-phase columns. These derivatives may all be detected at or below the 1-picomole level using spectrophotometric or fluorescence detectors (Haynes et al., 1991).

Based on a recent study of the Association of Biomolecular Resource Facilities (ABRF) (Yuksel et al., 1994), in which 50 participants carried out amino acid analyses of their choice on a 30-residue phosphopeptide (20 µg), pre-column analyses are about twice as popular as the post-column ninhydrin method, the PTC pre-column derivatisation being the method of choice of 39% of the participants. Only three participants obtained correct analyses with less than 5% error, two using PTC methodology, one the post-column ninhydrin method.

Attomole and subattomole quantities of derivatised amino acids may be detected using capillary zone electrophoresis and fluorescence detection (Yu and Dovichi, 1989). However, a major problem with analysis of amino acids at the picomole level and below is the relatively large background of amino acids that can be present. Glycine and serine are the most common contaminants but other amino acids may also be present. Accordingly, it is essential to use scrupulously cleaned containers; hydrolysis tubes, in particular, should be heated at 550°C for 4 h after washing. Microcapillary tubes have been used successfully (Liu and Boykins, 1989). Hydrolysis times of 24 h or more at 110°C with 6N HCl are still commonly used, but good results may be obtained for shorter periods at higher temperatures, for example, 4 h, 150°C. Short times of hydrolysis (2 h, 110°C) give the best yields for phosphorylated amino acids; 48 to 72 h may be required for cleavage of Ile-Ile bonds.

Vapour phase hydrolysis overcomes the problem of background from amino acids in the hydrolysis mixture but does not guard against traces of oxidant on the protein or oxygen in the hydrolysis flask. Tryptophan, especially, is easily destroyed under such conditions. In the past it has been analysed best when hydrolysed under vacuum in 4 M methanesulfonic acid containing tryptamine (Simpson et al., 1976). Recently it has been shown that vapour phase hydrolysis with 3 M mercaptoethanesulfonic acid, which is not only a suitable hydrolysing acid but also an antioxidant, gives excellent results when applied to proteins on PVDF membranes (Nakazawa and Manabe, 1992).

In recent times amino acid analysis has tended to become a secondary rather than a primary technique in many protein research laboratories. This is not so much because its role has been superseded, but rather that developments have not kept pace with other techniques such as those of protein/peptide purification and sequencing. While the latter may be applied reliably to low picomole amounts of material, 50 picomoles or more are generally required for an accurate amino acid analysis. Hence, researchers with scarce amounts of material have been forced to rely on criteria other than amino acid composition to establish the purity or nature of their preparations. Amino acid analysis, nonetheless, can be invaluable in establishing the purity and authenticity of a protein and confirming amino acid sequence data of peptides. Indeed, with currently available search algorithms (Shaw, 1993; Hobohm et al., 1994) it is now possible to identify proteins by simply comparing their compositions with those of known proteins in the available protein and DNA database. PROP-SEARCH, a search algorithm recently described by Hobohm et al. (1994), determines the percentage composition of each amino acid in every protein in the database as well as the molecular mass and isoelectric point. Only the 16 commonly analysed amino acids are considered; Glu and Gln, Asp and Asn are counted as Glx and Asx, respectively, while Trp and Cys values are excluded. Matching of the unknown proteins' compositional data with those in the database is made and a score indicative of the closeness of the match is calculated.

Best results were obtained with 25 nmole or more, protein hydrolysed and analysed, although amounts of 10 nmole were satisfactory.

With the rapid expansion of the protein (and putative protein) databases – as a consequence of the genomic sequence projects – the above methodology together with that of peptide mass fingerprinting will certainly improve our understanding of the underlying mechanisms of gene expression and regulation.

References

Aebersold, R., Bures, E.J., Namchuk, M., Goghari, M.H., Shushan, B. and Covey, T.C. (1992) Design, synthesis, and characterization of a protein sequencing reagent yielding amino acid derivatives with enhanced detectability by mass spectrometry. *Protein Science* 1: 494–503.

Bailey, J.M., Nikfarjan, F., Shenoy, N.R. and Shively, J.E. (1992) Automated carboxy-terminal sequence analysis of peptides and proteins using diphenylphosphoroisothiocyanatidate. *Protein Science* 1: 1622–1633.

Bailey, J.M., Tu, O., Issai, G., Ha, A. and Shively, J.E. (1994) *C-Terminal sequence analysis of polypeptides containing C-terminal proline*. Protein Society Meeting, San Diego. Poster No 151-S.

Barber, M., Bordoli, R.S., Sedgwick, R.D. and Tyler, A.N. (1981) Fast atom bombardment of solids as an ion source in mass spectrometry. *J. Chem. Soc. Chem. Commun.* 293: 270–275.

Bartlet-Jones, M., Jeffery, W.A., Hansen, H.F. and Pappin, D.J.C. (1994) Peptide ladder sequencing by mass spectrometry using a novel volatile degradation reagent. *Rapid Commun. Mass Spectrom.* 8: 737–742.

Bergman, T. (1993) Capillary electrophoresis in structural characterization of polypeptides. *In*: K. Imahori and F. Sakiyama (eds): *Methods in Protein Sequence Analysis*. Plenum Press, New York, pp 21–28.

Boyd, V.L., Bozzini, M., Zon, G., Noble, R.L. and Mattaliano, R.J. (1992) Sequencing of peptides and proteins from the carboxy terminus. *Anal. Biochem.* 206: 344–352.

Burkhart, W.A. and Moyer, M.B. (1994) *Protein Science* 3, (Suppl.) 1: 80.

Burlingame, A.L., Boyd, R.K. and Gaskell, S.J. (1994) Mass spectrometry. *Anal. Chem.* 66: 634R–683R.

Celis, J.E. and Bravo, R. (1984) *Two Dimensional Gel Electrophoresis of Proteins: Methods and Applications*. Academic Press, New York.

Chait, J.E., Wang, R., Beavis, R.C. and Kent, S.B.H. (1993) Protein ladder sequencing. *Science* 262: 89–92.

Chang, J.Y. and Creaser, E.H. (1976) A novel method for protein sequence analysis. *Biochem. J.* 157: 77–85.

Cox, A.L., Skipper, J., Chen, Y., Henderson, R.A., Darrow, T.L., Shabanowitz, J., Engelhard, V.H., Hunt, D.F. and Slingluff, Jr., C.L. (1994) Identification of a peptide recognised by five melanoma specific human cytotoxic cell lines. *Science* 264: 716–719.

Dharmasiri, K.A.N. and Watson, J.T. (1994) *New Edman reagents for peptide sequencing and detection by electrospray mass spectrometry*. Proceedings of the American Society for Mass Spectrometry, Poster No. 59 (Abstract).

Edman, P. (1949) A method for the determination of the amino acid sequence in peptides. *Arch. Biochem. Biophys.* 22: 475–476.

Edman, P. and Begg, G.S. (1967) A protein sequenator. *Eur. J. Biochem.* 1: 80–91.

Erdjument-Bromage, H., Geromanos, S., Chodera, A. and Tempst, P. (1993) Successful peptide sequencing with femtomole level PTH-analysis: a commentary. *In*: R.H. Angeletti (ed.): *Techniques in Protein Chemistry IV*. Academic Press, San Diego, pp 419–426.

Farnsworth, V. and Steinberg, K. (1993) Automated subpicomole protein sequencing using an alternative postcleavage conversion chemistry. *Anal. Biochem.* 215: 200–210.

Fell, A.F. (1983) Biomedical applications of derivative spectroscopy. *Trends Anal. Chem.* 2: 63–66.

Fenn, J.B., Mann, M., Meng, C.K., Wong, S.F. and Whitehouse, C.M. (1989) Electrospray ionization for mass spectrometry of large biomolecules. *Science* 246: 64–71.

Görg, A., Postel, W. and Gunther, S. (1988) The current state of the two-dimensional electrophoresis with immobilized pH gradients. *Electrophoresis* 9: 531–546.

Haynes, P.A., Sheumack, D., Greig, L.G., Kibby, J. and Redmond, J.W. (1991) Applications of automated amino acid analysis using 9-fluorenylmethyl chloroformate. *J. Chromatogr.* 588: 107–114.

Hewick, R.M., Hunkapiller, M.W., Hood, L.E. and Dreyer, W.J. (1981) A gas-liquid solid phase peptide and protein sequenator. *J. Biol. Chem.* 256: 7990.

Hobohm, U., Houthaeve, T. and Sander, C. (1994) Amino acid analysis and protein database compositional search as a rapid and inexpensive method to identify proteins. *Analytical Biochemistry* 222: 202–209.

Inglis, A.S., Wilshire, J.F.K., Casagranda, F. and Laslett, R.L. (1989) C-terminal sequencing: a new look at the Schlack-Kumpf thiocyanate degradation procedure. *In*: B. Wittmann-Liebold (ed.): *Methods in Protein Sequence Analysis*. Springer-Verlag, Berlin, pp 137–144.

Inglis, A.S. (1991) Chemical procedures for C-terminal sequencing of peptides and proteins. *Anal. Biochem.* 195: 183–196.

Inglis, A.S., Duncan, M.W., Adams, P. and Tseng, A. (1992) Formation of proline thiohydantoin with ammonium thiocyanate: progress towards a viable C-terminal amino-acid sequencing procedure. *J. Biochem. Biophys. Methods* 25: 163–171.

Inglis, A.S. and DeLuca, C. (1993) A new chemical approach to C-terminal microsequence analysis via the thiohydantoin. *In*: K. Imahori and F. Sakiyama (eds): *Methods in Protein Sequence Analysis*. Plenum Press, New York, pp 71–78.

Inman, J.K. and Appella, E. (1977) Identification of anilinothiazolinones after rapid conversion to Na-phenylthiocarbamyl-amino acid methylamides. *In*: C.H.W. Hirs and S.N. Timasheff (eds): *Methods in Enzymology*. Academic Press, San Diego, pp 374–385.

Jardine, I. (1990) Electrospray ionization mass spectrometry of biomolecules. *Nature* 345: 747–748.

Ji, H., Whitehead, R.H., Reid, G.E., Moritz, R.L., Ward, L.D. and Simpson, R.J. (1994) Two-dimensional electrophoretic analysis of proteins expressed by normal and cancerous human crypts: application of mass spectrometry to peptide-mass fingerprinting. *Electrophoresis* 15: 391–405.

Karas, M. and Hillenkamp, F. (1988) Laser desorption ionization of proteins with molecular masses exceeding 10 000 daltons. *Anal. Chem.* 60: 2299–2301.

Kaufmann, R., Spengler, B. and Lutzenkirchen, F. (1993) Mass spectrometric sequencing of linear peptides by product-ion analysis in a reflection time-of-flight mass spectrometer using matrix assisted laser desorption ionization. *Rapid Commun. Mass Spectrom.* 7: 902–910.

Laemmli, U.K. (1970) Cleavage of structural proteins during the assembly of the head of bacteriophage T4. *Nature* 227: 680–685.

Laursen, R.A. (1971) Solid-phase Edman degradation: an automatic peptide sequencer. *Eur. J. Biochem.* 20: 89–102.

Liu, T.Y. and Boykins, R.A. (1989) Hydrolysis of proteins and peptides in a hermetically sealed microcapillary tube: high recovery of labile amino acids. *Anal. Biochem.* 182: 383–387.

Mant, C.T. and Hodges, R.S. (1991) *HPLC of Peptides and Proteins: Separation, Analysis and Conformation*. CRC Press, Boca Raton.

Matsudaera, P. *A Practical Guide to Protein and Peptide Purification of Microsequencing*, Second Edition. Academic Press, San Diego.

Monnig, C.A. and Kennedy, R.T. (1994) Capillary electrophoresis. *Anal. Chem.* 66: 280R–314R.

Moos, M., Nguyen, N.Y. and Liu, T-Y. (1988) Reproducible high yield sequencing of proteins electrophoretically separated and transferred to an inert support. *J. Biol. Chem.* 263: 6005–6008.

Moritz, R.L., Ward, L.D. and Simpson, R.J. (1992) High-sensitivity peptide mapping utilising reversed-phase microbore and microcolumn liquid chromatography. In: R.H. Angeletti (ed.): *Techniques in Protein Chemistry III*. Academic Press, San Diego, pp 97–106.

Moritz, R.L. and Simpson, R.J. (1992) Application of capillary reversed-phase high-performance liquid chromatography to high-sensitivity protein sequence analysis. *J. Chromatogr.* 599: 119–130.

Moritz, R.L., Reid, G.E., Ward, L.D. and Simpson, R.J. (1994) Capillary HPLC: A method for protein isolation and peptide mapping. In: *Methods: A Companion to Methods in Enzymology* 6: 213–226.

Mortiz, R.L., Eddes, J., Ji, H., Reid, G.E. and Simpson, R.J. (1994) Rapid separation of proteins and peptides using conventional silica-based supports: Identification of 2-D gel proteins following in-gel proteolysis. In: *Techniques in Protein Chemistry VI*. Academic Press, pp 311–319.

Moyer, M.B. and Burkhart, W.A. (1994) Alternative reversed-phase supports for direct electroelution and *in situ* digestion on the Hewlett-Packard sequencing column. *J. Protein. Chem.* 13: 534.

Nakazawa, M. and Manabe, K. (1992) The direct hydrolysis of proteins containing tryptophan on polyvinylidene difluoride membranes by mercaptoethansulfonic acid in the vapor phase. *Anal. Biochem.* 206: 105–108.

O'Farrell, P.H. (1975) High resolution two-dimensional electrophoresis of proteins. *J. Biol. Chem.* 250: 4007–4021.

O'Farrell, P.Z., Goodman, H.M. and O'Farrell, P.H. (1977) High resolution two-dimensional electrophoresis of basic as well as acidic proteins. *Cell* 12: 1133–1141.

Pappin, D.J.C., Hojrup, P. and Bleasby, A.J. (1993) Rapid identification of proteins by peptide-mass fingerprinting. *Curr. Biol.* 3: 327–332.

Papov, V.V., Gravina, S.A., Mieyal, J.J. and Bieman, K. (1994) The primary structure and properties of thioltransferase (glutaredoxin) from human red blood cells. *Protein Science* 3: 428–434.

Patterson, S.D. (1994) From electrophoretically separated protein to identification: Strategies for sequence and mass analysis. *Anal. Biochem.* 221: 1–15.

Reid, G.E., Ji, H., Eddes, J.S., Moritz, R.L. and Simpson, R.J. (1995) Non-reducing two-dimensional polyacrylamide gel electrophoretic analysis of human colonic proteins. *Electrophoresis*; in press.

Shaw, G. (1993) Rapid identification of proteins. *Proc. Natl. Acad. Sci. USA* 90: 5138–5142.

Shively, J.E., Paxton, R.J. and Lee, T.D. (1989) Highlights of protein structural analysis. *Tibs* 14: 246–252.

Simpson, R.J., Neuberger, M.R. and Liu, T. (1976) Complete amino acid analysis of proteins from a single hydrolysate. *J. Biol. Chem.* 251: 1936–1940.

Simpson, R.J., Moritz, R.L., Nice, E.C. and Grego, B. (1987) A high performance liquid chromatography procedure for recovering subnanomole amounts of protein from SDS-gel electroeluates for gas-phase sequence analysis. *Eur. J. Biochem* 165: 21–29.

Simpson, R.J., Moritz, R.L., Lloyd, C.J., Fabri, L.J., Nice, E.C., Rubira, M.R. and Burgess, A.W. (1987) Primary structure of ovine pituitary basic fibroblast growth factor. *FEBS Lett.* 224: 128–132.

Simpson, R.J., Moritz, R.L., Rubira, M.R. and Van Snick, J. (1988) Murine hybridoma/plasmacytoma growth factor: Complete amino acid sequence and relation to human interleukin-6. *Eur. J. Biochem.* 176: 187–197.

Simpson, R.J. and Nice, E.C. (1989) Strategies for the purification of subnanomole amounts of proteins and polypeptides for microsequence analysis. In: A.R. Kerlavage (ed.):*The Use of HPLC in Receptor Purification and Characterization*, in series *Receptor Biochemistry and Methodology*. A.R. Liss, Inc., New York, pp 201–244.

Simpson, R.J., Moritz, R.L., Begg, G.S., Rubira, M.R. and Nice, E.C. (1989a) Micropreparative procedure for high sensitivity sequencing of peptides and proteins. *Anal. Biochem.* 177: 221–236.

Simpson, R.J., Moritz, R.L., Rubira, M.R., Gorman, J.J. and Van Snick, J. (1989b) Complete amino acid sequence of a new murine T-cell growth factor (P40). *Eur. J. Biochem.* 183: 715–722.

Simpson, R.J., Tsugita, A., Celis, J.E., Garrels, J.I. and Mewes, H.W. (1992) Workshop on two-dimensional gel protein databases. *Electrophoresis* 13: 1055–1061.

Spackman, D.H., Stein, W.H. and Moore, S. (1958) An automated amino acid analyser. *Anal. Chem.* 30: 1190.

Thompson, E.O.P. (1955) The insulin molecule. *Scientific American* 192: 36–41.

Tsugita, A., Kamo, M., Jones, C.S. and Shikama, N. (1989) Sensitization of Edman amino acid derivatives using the fluorescent reagent, 4-aminofluorescein. *J. Biochem.* 106: 60–65.

Tsugita, A., Takamoto, K., Kamo, M. and Iwadate, H. (1992) C-terminal sequencing of protein: A novel partial acid hydrolysis and analysis by mass spectrometry. *Eur. J. Biochem.* 206: 691–696.

Tsunasawa, S. and Hirano, H. (1993) Deblocking and subsequent microsequence analysis of N-terminally blocked proteins immobilized on PVDF membranes. *In*: K. Imahori and F. Sakiyama (eds): *Methods in Protein Sequence Analysis*. Plenum Press, New York, pp 45–53.

Vandekerckhove, J., Rider, M., Rasmussen, H.-H., De Boeck, S., Puype, M., Van Damme, J., Gesser, B. and Celis, J. (1993) Routine amino acid sequencing on 2D-Gel separated proteins: A protein elution and concentration gel system. *In*: K. Imahori and F. Sakiyama (eds): *Methods in Protein Sequence Analysis*. Plenum Press, New York, pp 11–20.

Vanfleteren, J.R., Raymackers, J.G., Van Bun, S.M. and Meheus, L.A. (1992) Peptide mapping and microsequencing of proteins separated by SDS-PAGE after limited in situ acid hydrolysis. *Biotechniques* 12: 550–557.

Wessel, D. and Flugge, U.I. (1984) A method for the quantitative recovery of protein in dilute solution in the presence of detergents and lipids. *Analytical Biocehmistry* 138: 141–143.

Xu, Q. and Shively, J.E. (1988) Microsequence analysis of peptides and proteins. VIII. Improved electroblotting of proteins onto membranes derivatized glass-fiber sheets. *Anal. Biochem.* 170: 19–30.

Yu, M. and Dovichi, N.J. (1989) Attomole amino acid determination by capillary zone electrophoresis with thermooptical absorbance detection. *Anal. Chem.* 61: 37–40.

Yuksel, K.U., Andersen, T.T., Apostol, I., Fox, J.W., Crabb, J.W., Paxton, R.J. and Strydom, D.J. (1994) Amino acid analysis of phospho-peptides: ABRF-93AAA. *In*: J.W. Crabb (ed.): *Techniques in Protein Chemistry V*. Academic Press, San Diego.

UV and nucleic acids

T. Douki and J. Cadet

CEA/Département de Recherche Fondamentale sur la Matière Condensée-SESAM/LAN, F-38054 Grenoble Cedex 9, France

Introduction

Nucleic acids and particularly DNA are the main cellular targets of UVC (220–290 nm) and UVB (290–320 nm) radiations, which are partly responsible for the mutagenic and skin carcinogenic effects of solar light (Taylor, 1990). However, the stratospheric ozone layer, whose decrease may have some deleterious consequences on earth life (Henriksen et al., 1990), prevents significant interaction of solar UVC with DNA. On the other hand, UVB, which overlaps with the upper end of the DNA absorption spectrum, is responsible for most of the induction of solar DNA damage through direct excitation (Moan and Peak, 1989; Cadet et al., 1992). In this survey, emphasis has been placed on recently available information concerning the characterization of the major DNA UV lesions. In addition, the main chemical and biochemical assays which have been recently developed for monitoring DNA photoproducts are critically reviewed. Two fast expanding topics in the field of DNA photobiology, whose studies also require the use of accurate chemical approaches, are surveyed. The first one deals with the use of specifically modified deoxyribooligonucleotides which are introduced in shuttle vectors and subsequently transfected in cells to study the mutagenic properties of individual DNA photodamage. The second example is the investigation of the mechanism of action of bacterial photolyases, a class of enzymes which specifically repairs UV-induced pyrimidine cyclobutane dimers upon subsequent exposure of cells to visible light.

UVC-induced pyrimidine dimerization

In the last three decades, extensive studies of the UVC photoreactions of DNA components have led to the identification of the main lesions (for a comprehensive review, see Cadet and Vigny, 1990). Most of the investigations have been carried out on model compounds since the level of modifications induced in DNA is usually too low to obtain enough material for the accurate determination of the structure and chemical properties of the photoproducts. Earlier studies have mostly involved monomeric models, including pyrimidine bases, nucleosides and nucleotides (Varghese, 1972; Wang, 1976). More recently, dinucleoside monophosphates have been widely used since they better take into account the steric constraints of the DNA structure and also favor intramolecular reactions between the nucleobase moieties. Extensive spectroscopic studies, including UV, NMR and mass spectrometry analysis, have allowed the determination of the chemical structure of the main photoproducts and their secondary decomposition products (Cadet and Vigny, 1990). Most of the UVC DNA photoproducts have a dipyrimidine structure. In addition, a few monomeric lesions are also generated, in a relatively lower yield, at least, in DNA.

Cyclobutane type dimers

The major class of UVC-induced DNA photoproducts is represented by the cyclobutane type dimers. Cyclobutadipyrimidines are generated by reversible [2 + 2] photoaddition of the C5–C6 double bonds of two adjacent pyrimidines (Fig. 1). Different diastereoisomers can be produced through the photoreaction. A first isomery involves the position of the two pyrimidine rings with respect to the plane of the cyclobutane ring. If the two bases are located on the same side of the cyclobutane ring, the photoproduct exhibits a *cis* configuration. On the other hand,

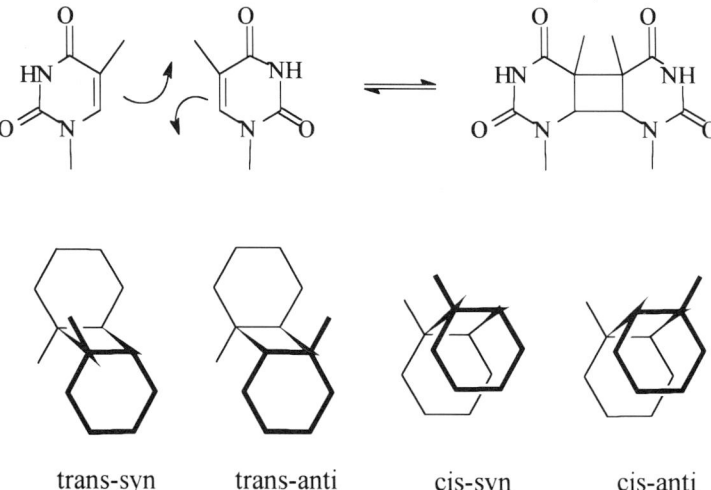

Figure 1. Formation and structure of the four diastereoisomers of the thymine cyclobutane dimers.

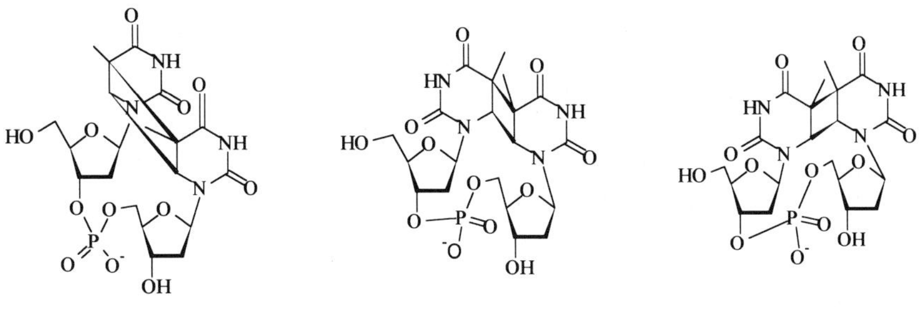

cis-syn   trans-syn I   trans-syn II

Figure 2. Diastereroisomeric cyclobutane dimers isolated from far-UV irradiated TpT.

the photoproduct has a *trans* stereochemistry if the pyrimidine moieties are disposed on the two opposite sides of the cyclobutane plane. In addition, if the C5 atoms of the two pyrimidine bases are linked to each other, the cyclobutane dimer has a *syn* configuration, whereas it is *anti* if the C5 atom of a pyrimidine base is covalently bound to the C6 atom of the adjacent base.

The relevant DNA photoproduct first characterized is the *cis-syn* cyclobutadithymine, followed by several base and nucleosides cyclobutane type dimers (for early reviews, see Varghese, and Wang, 1976). An extensive Fourier transform mass spectrometry analysis of a series of uracil and thymine cyclobutane dimers has been reported recently (Hettich et al., 1990).

In the last 15 years, most of the structural assignment of cyclobutane dimers has been carried out on dinucleoside monophosphates, including the thymine-cytosine (TpdC) (Liu and Yang, 1978) and thymine-thymine (TpT) (Kan et al., 1988) sequences. Two main cyclobutadipyrimidines exhibiting a *cis-syn* and *trans-syn* (*trans-syn* I) configuration (Fig. 2) have been obtained by UVC photolysis of each of the dinucleodise monophosphates. The latter work has been recently completed by the isolation and characterization of a second minor TpT *trans-syn* dimer (*trans-syn* II) (Kao et al., 1993). In addition, the photolysis of a dinucleoside monophosphate with a cytosine-thymine sequence (dCpT) has allowed the isolation and characterization of two cyclobutane type dimers exhibiting a *cis-syn* and a *trans-syn* configuration, respectively (Douki and Cadet, 1992). As for TpT and TpdC, the *cis-syn* cyclobutadipyrimidine is the major isomer, as would be expected in a right-handed B-DNA. Because of the steric constraints, only *syn* cylcobutane dimers are produced within double-stranded DNA. The only exception known concerns non adjacent *cis-anti* cyclobutadipyrimidines which are generated in single-stranded alternating copolymers exhibiting loops (Nguyen and Minton, 1988).

The conformational investigations on cyclobutane dimers of dinucleodise monophosphates are facilitated by their rigid structure. For this purpose, ^{1}H NMR studies, particularly nuclear Overhauser effect measurements, have been widely applied to cyclobutadipyrimidines. A first series of results has dealt with the determination of the orientation of the base moieties with respect to the deoxyribose rings, which can be either *SYN* (C-2 carbonyl directed above the sugar moiety) or *ANTI* (C-6 directed toward the deoxyribose ring). It was shown on a series of thymine and uracil containing dinucleoside monophosphates that the conformation of the 5' and 3'-end nucleosidic moieties of the *cis-syn* diastereoisomers are preferentially *ANTI* and *SYN*, respectively (Koning et al., 1991). Similar approaches have been used to determine the conformational features of the *trans-syn* cyclobutane dimers I and II of both TpT and TpdU. The absolute stereoconfiguration of the cyclobutane rings has a major influence on the conformational features of the two nucleosidic residues. The major diastereoisomer (*trans-syn* I) exhibits a *SYN/ANTI* conformation, whereas the *trans-syn* II diastereoisomer has an *ANTI/SYN* structure (Koning et al., 1991; Tabaczynski et al., 1993). Substitution of the C5 atom of the pyrimidine moieties by a methyl group greatly influences the puckering mode (Kim, J.K. et al., 1993). The NMR data, including chemical shifts, coupling constants, and NOE measurements, have been used to determine the complete geometrical structure of the *cis-syn* dUpT cyclobutane dimer by using molecular mechanics calculations (Blommers et al., 1992). These experiments illustrate the utility of isolating photoproducts from model compounds since the structural information provided cannot be obtained from irradiated DNA. The conformational features inferred from NMR studies, together with X-ray diffraction data on the S(p) diastereoisomer of the cyanoethyl ester of the *cis-syn* cyclobutane dimer of TpT (Hruska et al., 1986), can be used for theoretical study of the conformational changes associated with the presence of a *cis-syn* cyclobutadithymine within a DNA helix. Theoretical simulations have been recently performed on a d(CGCGAATTCGCG)$_2$ in which a *cis-syn* cyclobutadithymine had been incoporated (Rao and Kollman, 1993). The presence of the thymine dimer has been found to induce significant kinks and bends in the vicinity of the photodamage.

The pyrimidine (6-4) pyrimidone photoproducts

Pyrimidine (6-4) pyrimidone adducts, which constitute a second major class of DNA photoproducts, were discovered more recently than the cyclobutane dimers. Their formation involves a [2 + 2] addition of the C5–C6 bond of the 5' pyrimidine to either the carbonyl function or the enamine group at the C4 atom of the 3' end base (Fig. 3). The oxetane

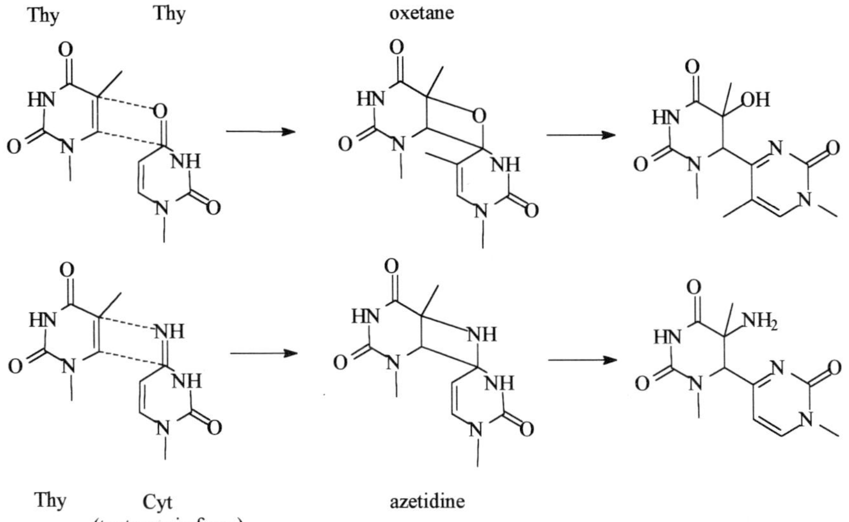

Figure 3. Mechanism of formation of pyrimdidine (6-4) pyrimidone photoproducts.

or azetidine intermediates produced are unstable molecules, which are further hydrolyzed into compounds exhibiting either a 5-hydroxy-5,6-dihydro- or a 5-amino-5,6-dihydro-pyrimidine structure on the 5′ end, and a pyrimidone ring on the 3′ end. Interesting spectroscopic features of the (6-4) adducts are due to the presence of the pyrimidone moiety which leads to a redshift of the maximum in the UV absorption spectrum. The maximum is centered around 315 nm for dinucleoside monophosphates when the 3′-end base is an unsubstituted pyrimidone (corresponding to a cytosine residue in the starting compound). A 10 nm bathochrome shift is observed when the 3′ base is a 5-methylpyrimidone, arising from the photoconversion of a thymine residue. The (6-4) photoproducts are highly fluorescent (Blais et al., 1994), exhibiting a maximum in the excitation and emission wavelengths around 320 nm and 380 nm, respectively. The fluorescence quantum yield is dependent on the sequence, with values of 0.03, 0.007, and 0.015 for the (6-4) adducts of TpT, TpdC, and dCpT, respectively. The hydrolysis of the phosphodiester bond increases by three- to seven-fold the fluorescence quantum yield of the resulting base (6-4) photoadducts.

As for the cyclobutane dimers, early structural work on (6-4) photoproducts was carried out by using mostly monomeric model compounds (Varghese, 1972; Wang, 1976). However, in the last decade, emphasis was placed on the characterization of the (6-4) adducts of dinucleoside monophosphates, including those of TpT (Rycyna and Alderfer, 1985) and TpdC (Franklin et al., 1985). Recently, the photolysis of a third dinucleoside monophosphate, dCpT, was investigated (Douki et al.,

1991). The related pyrimidine (6-4) pyrimidone photoproduct exhibits characteristic UVB absorption and fluorescence properties. In addition, the absolute configuration of the C5–C6 atoms of the 5' end base, as inferred from NOE measurements, is identical to those of the TpT and TpdC adducts (Douki and Cadet, 1992). It should be noted that the attempts to generate (6-4) photoproduct upon far-UV exposure of dCpdC have remained unsuccessful. On the other hand, far-UV photolysis of thymidylyl-(3'→5')-2'-deoxy-4-thiouridine has led to the characterization of the corresponding (6-4) adduct (Fourrey et al., 1989) together with a transient thietane (Clivio et al., 1991), the likely precursor of the latter photoproduct. This represents the first experimental evidence of the mechanism of formation of the (6-4) photoproducts.

It should be added that UVC photolysis of 5-methylcytosine containing dinucleoside monophosphates, including m^5dCpT, m^5dCpdC and Tpdm^{5}dC, leads to the formation of (6-4) adducts as efficiently as in the corresponding thymine compounds (Douki and Cadet, 1994). This is a major observation since 5-methylation of cytosine has been previously reported to prevent the formation of (6-4) photoproducts. However, additional work is required to further investigate the reactivity of 5-methylcytosine when incorporated in double-stranded DNA.

Secondary reactions of pyrimidine photoproducts

Photoisomerization of the (6-4) adducts
The pyrimidine (6-4) pyrimidone adduct of TpT has been shown to undergo a secondary transformation upon UVB photolysis (Johns et al., 1964). However, the structure of the resulting compound has been identified only recently as the Dewar valence isomer of the pyrimidone ring (Taylor and Cohrs, 1987). The formation of the Dewar photoproduct may be explained in terms of [2 + 2] intramolecular cycloaddition of the pyrimidone ring (Fig. 4) via a singlet excited state (Lin and Taylor, 1989). The photoisomerization also applies to other (6-4) photoproducts since the Dewar valence isomers of TpdC (Taylor et al., 1990) and dCpT (Douki et al., 1991) have been isolated and character-

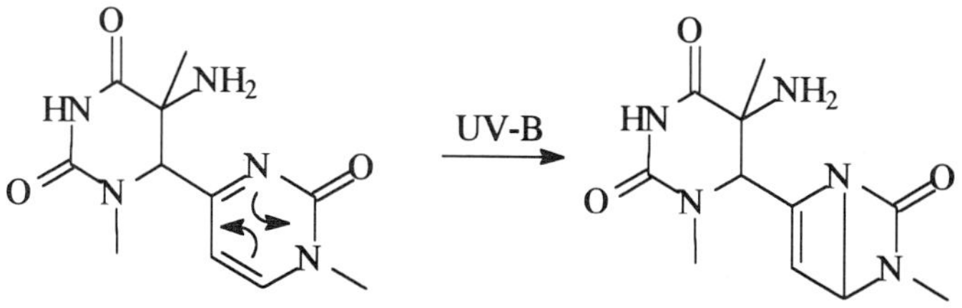

Figure 4. Formation of the Dewar valence isomer of the T-C (6-4) photoproduct.

ized. The quantum yield for the photoconversion of the (6-4) photoproducts of TpT and TpdC into the corresponding Dewar valence isomers have been determined as 0.020 and 0.018, respectively (Lemaire and Ruzsicska, 1993). In addition, it has been established that the Dewar valence isomer could be directly generated by UVC light irradiation of dCpT (Douki et al., 1991), TpT (Kan et al., 1992) and TpdC (Douki and Cadet, unpublished data).

Deamination
Deamination is a well known property of cytosine and 5-methylcytosine bases within DNA. It corresponds to the substitution of the 4-amino group of the base ring by a hydroxyl group, leading to the formation of uracil and thymine, respectively. The rate of deamination of cytosine is very low ($k = 10^{-7}$ min^{-1} at 90°C). In contrast, the deamination rate constant is much higher for the 5,6-saturated cytosine derivatives ($k = 10^{-3}$ min^{-1} for 5,6-dihydrocytosine at 25°C). This also applies to photoproducts containing 5,6-dihydrocytosine residues. However, precise information on the quantitative aspects of the deamination of the cyclobutane type dimers, the (6-4) adducts and their related Dewar isomers of dCpT and TpdC has been obtained only recently (Fig. 5).

The deamination rate of the photoproducts of dCpT strongly depends on the structure of the dinucleoside monophosphate since the (6-4) adduct and its Dewar valence isomer are 100 times more stable than the *cis-syn* cyclobutane dimer (Douki and Cadet, 1992). The deamination of the cyclobutane type dimers depends on their configuration, with a reaction rate being 50 fold lower for the *trans-syn* I photodimer than for the *cis-syn* diastereoisomer. Similar observations have been made for the TpdC cyclobutane dimers (Lemaire and Ruzsicska, 1993). The latter study showed that the rate-limiting step in the deamination reaction of the cyclobutane dimers is the nucleophilic attack on the 5,6-saturated cytosine residue by an hydroxide anion, thus providing important information on the deamination mechanism. These results are of biological importance since deamination of cytosine-containing cyclobutadipyrimidines has been shown to also occur in cellular

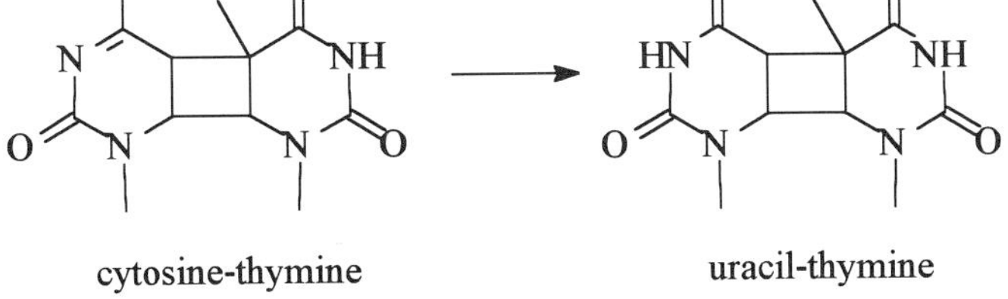

Figure 5. Deamination of a C-T cyclobutane dimer.

DNA (Rubio and Bockrath, 1989). In particular, deamination is likely to be involved in mutagenesis since repair enzymes, such as photolyase (*vide infra*), may reverse the T-U cyclobutane dimer (deaminated T-C photoproduct) to thymine and uracil, the latter base constituting a mutagenic hot spot (Bridges, 1992; Tessman et al., 1994).

Interestingly, 5-methylcytosine photoproducts deaminate much more slowly than the related cytosine derivatives (Douki and Cadet, 1994). However, deaminated photoproducts are generated upon irradiation of 5-methylcytosine containing dinucleoside monophosphates, suggesting that deamination occurs in the excited state of the molecule.

Measurements of dipyrimidine photoproducts

Measurement of lesions in cellular DNA is a requisite step for the assessment of their biological role. In this respect, several sensitive and specific assays, based on chemical or biochemical approaches, have been developed (for a recent review see Cadet and Weinfeld, 1993). Information provided by model studies is very useful since isolated photoproducts can be used, for example, for either the optimization of the chromatographic analysis or the characterization of antibodies. In addition, data on the stability of the photoproducts towards chemical hydrolysis or enzymatic digestion are very useful for the development of indirect assays for DNA damage, as described below.

Immunological techniques

One of the major improvements in the measurement of DNA photodamage is the use of antibodies directed towards dipyrimidine photoproducts. Monoclonal antibodies have been raised against the three major classes of adducts, including the cyclobutane dimers (Roza et al., 1988; Strickland et al., 1992), the (6-4) photoproducts (Mori et al., 1988; Mori et al., 1991; Mizuno et al., 1991) and their Dewar valence isomers (Matsunaga et al., 1993). However, the specificity of the assay is limited, as illustrated by the only five-fold lower affinity of the 64M-1 antibody for the TpT (6-4) photoproducts with respect to the corresponding TpdC photoadduct (Matsunaga et al., 1990). This implies that, in the present state, the immunological approach cannot provide qualitative and quantitative information on the formation of an individual photoproduct. Another limitation of the immunological assays devoted to the measurement of the (6-4) photoproducts is the lack of the calibration of the assay. However, this could be achieved by using an independent technique such as the HPLC-fluorescence detection technique (*vide infra*).

Antibodies can be used for the measurement of photoproducts in association with other techniques. Radioimmunoassay (Clarkson et al., 1983) or immunoprecipitation (Mitchell et al., 1990a) allows the determination of the averaged amount of a defined class of photoproduct in DNA on the basis of radioactivity measurements. The use of radioimmunoassay directed against pyrimidine (6-4) pyrimidone photoproducts and cyclobutane type dimers has recently found an interesting application. The relative yield of cyclobutadipyrimidines formed in UVC irradiated human cell chromosomes is similar in core and linker DNA. In contrast, the photo-induced formation of (6-4) photoadducts is six-fold greater in linker than in core nucleosome DNA (Mitchell et al., 1990c). An alternative approach involves the use of antibodies coupled to enzymes and further detected by a colorimetric technique (ELISA) (Olsen et al., 1989; Matsunaga et al., 1991), or a fluorescent probe (Berg et al., 1989). An interesting application of the antibodies raised against UVC photoproducts is the detection of DNA photolesions in cells, using immunohistochemistry (Olsen et al., 1989), laser cytometry (Mori et al., 1990; Berg et al., 1989) and flow cytometry (Berg et al., 1993). These techniques provide interesting information on the formation of dipyrimidine photoproducts in tissues and single cells.

Alkali-liability of the (6-4) adducts

In the last decade, an indirect method for the measurement of the (6-4) photoproducts was developed (Franklin et al., 1982). This was based on a gel-sequencing electrophoresis of UVC irradiated DNA of defined sequence, subsequent to alkaline treatment. The latter step was supposed to specifically cleave DNA strand at sites of (6-4) photoproducts (Lippke et al., 1981). Quantitation of the DNA strand breaks by either radioactive (Franklin et al., 1982) or fluorescence labeling (Koechler et al., 1991) together with sequence analysis allows the determination of the yield of formation of the different (6-4) photoproducts.

However, the quantitative aspect of the assay has been questioned recently, on the basis of a study involving an immunological assay (Mitchell, 1988). It was found that the (6-4) adducts are only labile under drastic alkaline conditions, under which the cyclobutane dimers are also decomposed. In contrast, it appeared that the Dewar valence isomers are much more sensitive to alkaline treatment than their (6-4) precursors. This indirect observation was further confirmed by studies on the photoproducts of dCpT and TpT (Douki and Cadet, 1992; Kan et al., 1992) showing that, upon alkaline treatment, the base moiety of the (6-4) adduct is modified whereas the dinucleoside monophosphate structure is maintained. In contrast, alkaline treatment of the corresponding Dewar valence isomers leads to their quantitative decomposition with cleavage of the phosphodiester bond. Consequently, it has been proposed to use the alkali-lability assay for the measurement of the (6-4) photoproducts after their preliminary UVB photoisomerization (Mitchell et al., 1990b). Using this approach, the ratio between the amount of cyclobutane dimers and (6-4) photoproducts has been shown to be 3:1, whereas it was previously estimated to be 10:1. However, the possibility of direct formation of Dewar valence isomer by 254 nm photolysis, observed upon irradiation of TpT, dCpT and TpdC, must be taken into account when measuring the (6-4) adducts by using this technique.

Use of enzymes for measurements in whole DNA

Base excision repair enzymes can be used for the indirect measurement of photolesions in DNA. As for the alkali assay directed against the (6-4) photoadducts, quantitation of DNA nicks provides information on the importance and the location of the damage if a gel sequencing method is used. This approach has been widely applied to the measurement of the cyclobutane type dimers since *M. luteus* endonuclease and T4 endonuclease V, are able to cleave the N-glycosidic bond of cyclobutadipyrimidines at the 5' end (Haseltine et al., 1980; Sage, 1993). An interesting appliction of the T4 endonuclease/gel electrophoresis sequencing assay has allowed the determination of the sequence specificity of the formation of cyclobutadipyrimidines in DNA following UVB irradiation (Mitchell et al., 1992). It was found that the induction of cyclobutane type dimers is greater in the presence of 5' flanking pyrimidines than purines. Measurement of the overall UV photoproducts in DNA can also be achieved by using the (A)BC excinuclease, a non-specific repair enzyme. It catalyses the excision of the phosphodiester backbone upstream and downstream from the dimeric lesion, which may be either a cyclobutane type dimer or a (6-4) photoproduct (Thomas et al., 1989).

A more sophisticated detection of the strand breaks induced by cleavage at a site of a photoproduct is based on the amplification of the damaged gene by polymerase chain reaction if a 5' phosphate group is available. This sensitive approach thus provided interesting information on the rate of formation of lesions in different genes at the nucleotide level. The assay has been applied to the detection of UVC induced alkali-labile sites within DNA, which were suggested to be (6-4) photoproducts (Pfeifer et al., 1991). However, there is doubt about the quantitative aspect of the measurement since the necessary UVB photoisomerization of the (6-4) photoproducts into Dewar valence isomers had not been performed prior to the alkaline treatment. A difference in the rate of induction has been observed between active and inactive human X chromosomes. The method has been extended to the detection of cyclobutane dimers, by incubating irradiated DNA with endonuclease V (Pfeifer et al., 1993). Another indirect assay for UVC photoproducts in DNA is based on the inhibition of restriction enzyme activity when a dipyrimidine lesion is located at a restriction site (Kovács et al., 1993).

Chromatographic assays involving hydrolysis of DNA

The bulk of assays described above are based on the indirect measurement of the photolesions within DNA. Another approach involves the release of the lesion from DNA by chemical or enzymatic hydrolysis. This is followed by the specific detection of the released base or nucleoside after its separation from the other DNA components by gas or liquid chromatographic methods. One of the advantages of this procedure is the individual measurement of each of the DNA photolesions. Photoproducts derived from model compounds are usually used for the optimization of the measurement. Use of model photoproducts also allows the determination of the stability of the photodamage during the hydrolysis step.

A well established assay involves the release of thymine and cytosine cyclobutane dimers upon treatment of cellular DNA, preliminarily radiolabeled at either thymine or cytosine residues. This involves formic acid hydrolysis of extracted DNA, followed by separation of the bases by high-performance liquid chromatography and subsequent detection of the photoproducts by radioactive measurement (Niggli, 1990, 1992). An alternative approach, which does not require radiolabeling of DNA, is based on the use of gas chromatography (CG) separation coupled with electron capture detection for the measurement of cyclobutadipyrimidines (Ramsey and Ho, 1989). However, derivatization of the cyclobutadipyrimidines is required prior to the GC/MS analysis. Acidic hydrolysis of pyrimidine (6-4) adducts from DNA has been attempted. However, usual acidic reagents, including formic, trifluoroacetic or hydrochloric acids lead to a significant decomposition of the (6-4) adducts. On the other hand, hydrogen fluoride stabilized in pyridine, a milder hydrolysis reagent, has been found to allow the quantitative release of the (6-4) photoproducts of thymine-thymine (TT), thymine-cytosine (TC) and cytosine-thymine (CT), and their subsequent measurement by HPLC coupled with a sensitive fluorescence detector (Douki et al., 1994).

Use of modified oligonucleotides

Modified oligonucleotides of more than 4-mers length are relevant DNA models for conformational investigations. In addition, these compounds can be used *in vitro* to investigate the role of a specific modification on the action of critical cellular enzymes such as DNA polymerase. Another use of oligonucleotides containing defined photoproducts is their incorporation in plasmids, used as shuttle vectors for the transfection of bacterial and mammalian cells. This approach provides information on the nature of the mutations induced by lesions located at specific sequences. Another interesting application of modified oligonucleotides is provided by the determination of the molecular mechanism of action of DNA repair enzymes (*vide infra*).

Synthesis

Preparation of oligonucleotides containing a photoproduct located at a specific position can be achieved in two ways. A first approach is the synthesis of a modified dinucleoside monophosphate, which is subsequently incorporated in an oligonucleotide after appropriate protection of the reactive functions (Taylor et al., 1987; Taylor and Brockie, 1988; Taylor and Nadji, 1991). This has been successfully applied to the thymine *cis-syn* and *trans-syn* cyclobutane type dimers. The corresponding TpT photoproducts have been first prepared by photosensitization and isolated by HPLC. In a subsequent step, the cyclobutane dimers have been functionalized by addition, in particular, of a phosphoramidite group to their 3'-OH end. Then, the resulting building block obtained has been incorporated into a 22-mer, by using DNA synthesis automated systems. A similar approach has also been used for the synthesis of *ras-proto* oncogenes containing either a *cis-syn* or a *trans-syn* cyclobutadithymine (Kamiya et al., 1993). However, this synthetic pathway is only possible for photoproducts which, like cyclobutane dimers, are stable under the conditions required for DNA synthesis. In particular, this strategy cannot apply to the preparation of (6-4) adducts and Dewar isomers containing oligonucleotides since these photoproducts are sensitive to both acidic and alkaline conditions. A milder alternative involves the irradiation of a short oligonucleotide, which is further enzymatically elongated after purification. For instance, irradiation of d(AATTAA) hexamer has allowed the isolation of the (6-4) photoproduct and its related Dewar valence isomer. The modified oligomers have been then converted into 49-mers by enzymatic DNA synthesis (Smith and Taylor, 1993).

Replication studies using oligonucleotides

The induction of a mutation by a DNA lesion is primarily associated with a mistranslation of the modified residue(s) during DNA replication. Therefore, relevant information on the mutation potentiality of a defined photoproduct can be inferred from DNA replication studies. Under these conditions, *Pol I*, a polymerization enzyme of *E. coli*, is not completely blocked, *in vitro*, by the presence of a *cis-syn* cyclobutane type dimer (Taylor and O'Day, 1990). However, the bypass yield is low and termination of the replication opposite the 3′ end thymine residue of the cyclobutane dimer is the major event. The termination site depends on the concentration of enzyme, since with a higher substrate/polymerase ratio, the replication ends mainly one nucleotide prior to the 3′ end of the dimeric lesion. In contrast, both *cis-syn* and *trans-syn* I thymine cyclobutane dimers are absolute blocks for the calf thymus polymerase δ, illustrating a difference of behavior for similar enzymes in prokaryotic and eukaryotic cells (O'Day et al., 1992).

Altogether, these experiments show that the presence of a damaged base may dramatically decrease the replication rate. Consequently, an activation of the replication enzymes is required to increase the translocation of the lesions. This has been observed *in vitro* for the calf thymus *pol δ* which is able to extend the DNA strand past the *cis-syn* and *trans-syn* cyclobutadithymines when activated by proliferating nuclear antigen (O'Day et al., 1992). The bypass rate of the lesions increases with higher concentration of nucleotide triphosphates in the medium. The latter parameter also influences the termination site of the replication which is one nucleotide prior to the lesion in low concentrated solution. In contrast, the termination is opposite to the 3′ or the 5′ end of the photoproduct for increasing concentrations of nucleotides.

The bypass of a lesion present in a shuttle vector after transfection into a host cell also requires an induction step, referred to, in prokaryotic cells, as the SOS response. The SOS response consists of a series of events, mediated by the synthesis of specific proteins, leading to the bypass of the lesions. This allows the DNA replication, since blockage of the function would lead to death of the cell. This has been observed for the cyclobutane type dimers of thymine (Banerjee et al., 1988; Gibbs and Lawrence, 1993) and uracil (Gibbs and Lawrence, 1993). The *trans-syn* cyclobutadithymine is poorly bypassed in uninduced *E. coli* cells. However, the bypass occurs with a much higher accuracy (96%) than in SOS-induced cells (Banerjee et al., 1990). Requirement of the SOS response for the bypass of the lesions has also been observed for the TT and TC (6-4) adducts (LeClerc et al., 1991; Horsfall and Lawrence, 1993), as for the related Dewar valence isomers.

The bypass frequency may vary from one lesion to the other in SOS-response-induced cells. The *trans-syn* I thymine cyclobutane type dimer more efficiently blocks the replication machinery than the *cis-syn* diastereoisomer (Gibbs et al., 1993a). Similarly, both the TT (LeClerc et al., 1991) and TC (6-4) adduct (Horsfall and Lawrence, 1993) are more easily bypassed than their Dewar valence isomers.

Mutational properties of DNA photolesions

Incorporation of a plasmid with known modifications in host cells is a very powerful approach for the evaluation of the mutagenic properties of DNA damage. For instance, the contribution of (6-4) photoadducts to the overall mutagenicity of UV light has been assessed by transfecting cells with UVC irradiated plasmids used as shuttle vectors. The plasmids had been treated by photolyase prior to their transfection, in order to remove cyclobutane type dimers (Bourre et al., 1989). Other approaches, involving either reversed transcription PCR (Dumaz et al., 1993) or direct DNA sequencing of both oligonucleotide strands of purified PCR products (Ziegler et al., 1993) have been used to determine the spectrum of specific UV mutation in the p53 tumor suppression gene. Seven of the mutation hotspots observed in the p53 gene of basal cells carcinoma, which are primarily CC→TT or C→T changes at dipyrimidine sites, are specific to skin cancers (Ziegler et al., 1993). All the mutations observed in the p53 gene of skin tumors from DNA-repair-deficient Xeroderma pigmentosum patients are located at CC sequences, leading, in most cases, to tandem CC→TT mutations. Preferential repair of both cyclobutadipyrimidine and (6-4) photoadducts occurs on the transcribed DNA strand in human tissues (Dumaz et al., 1993). However, these approaches cannot allow the identification of the contribution of each individual photoproduct. This can be achieved by transfecting the cells, primarily *E. coli* but also yeast cells, with plasmids carrying a photoproduct at a specific site. Interesting information has been obtained on the biological role of UVC-induced lesions, regarding (i) the bypass frequency of the lesions during replication, (ii) the mutation rate when this occurs, (iii) the base introduced opposite to the lesion (Gibbs et al., 1993b).

The first parameter studied has been the role of the chemical nature of the lesion on mutagenesis. It has been observed that the Dewar valence isomer of TpdC leads to more errors when bypassed than its (6-4) precursor (Horsfall and Lawrence, 1993). Opposite results have been obtained for the related TT photoproducts (LeClerc et al., 1991). The *cis-syn* cyclobutane dimers are less mutagenic than the *trans-syn* diastereoisomers at both TT (Gibbs and Lawrence, 1993; Gibbs et al., 1993a) and uracil-uracil (UU) sites (Gibbs and Lawrence, 1993). These results have allowed a better understanding of the molecular basis of the mutagenicity of far-UV photoproducts.

Until recently, mutagenesis of dipyrimidine photoproducts was mainly rationalized in terms of preferential insertion of adenine opposite to the lesion which behaves like an abasic site for the replication enzymes ("A rule"). However, recent data obtained from shuttle vector experiments have questioned this assessment (Gibbs et al., 1993b). First, the type of mutation induced is different from one photoproduct to the other. This is not expected if only the "A rule" is involved. In addition, in a series of shuttle vectors with the same sequences but bearing a different damage, an abasic site induces a very different mutation pattern than the TT and TC photoproducts (Horsfall and Lawrence, 1993). These observations provide evidence that the modified pyrimidine moieties are miscoding rather than noncoding. For instance, 85% of the bypassed TT (6-4) adducts lead to a 3' end T→C transition (LeClerc et al., 1991). This result can be explained by the ability of the pyrimidone ring to pair with guanine through hydrogen bonds involving the N-3 and O-2 atoms. This hypothesis is confirmed by the observation of a lower mutagenicity of the TC (6-4) photoproduct since the 3' end insertion of guanine would not induce transversion. For the cyclobutane dimers, the stereochemistry of the lesions appears to dramatically modify the mutational spectrum, as the result of differences in hydrogen bonding ability. T→A and T→C are the major mutations opposite both the TT and UU cyclobutadipyrimidines. They occur at the 5' end with the *trans-syn* isomer and at the end 3' end with the *cis-syn* cyclobutane type dimer (Gibbs and Lawrence, 1993; Kamiya et al., 1993). Cyclobutadithymines induce a greater extent of adenine insertion than the UU cyclobutane type dimers. This observation has been rationalized in terms of influence of the C5 substitution. Transversions and transitions are not the only mutagenic events induced by UVC photoproducts which may also lead to frameshift mutations. The induction of this type of mutation in thymine tracks occurs at the second thymine residue and involved a series of misalignment-realignment reactions of the DNA strand, as shown by the study of the replication of 59-mers oligonucleotides containing a *cis-syn* cyclobutadithymine at various positions in a $(T)_6$ sequence (Wang and Taylor, 1992).

  E. coli and *S. cerevisiae* cells exhibit different mutation spectra after transfection by the same TT *cis-syn* cyclobutane dimer containing shuttle vectors. *E. coli* cells allow a bypass of the lesions only in 16% of the cases with a 7.6% error, whereas yeast cells have a higher bypass rate (80%), but a much higher accuracy in the base incorporation opposite to the photodamage (Gibbs and Lawrence, 1993). The presence of the *trans-syn* I cyclobutadithymine leads to the same consequences in both cell lines. Altogether, these results showed that both the structure of the photoproducts and the nature of the cells play a major role in the mutation induced by a single photolesion (Lawrence et al., 1993). Deamination has been proposed to be involved in the differences between species in response to cyclobutane type dimers (Jiang and Taylor, 1993). In fast replicating cell lines, the mutations are induced by the miscoding properties of the cyclobutadipyrimidines that are not repaired. On the other hand, in slowly replicating cells, the photolyases have enough time to reverse the cyclobutane type dimer. If deamination also occurs at a cytosine site in the cyclobutane dimer prior to photorepair, it is likely that the mutations are induced by the presence of the uracil residue.

Enzymatic photoreversion of pyrimidine dimers

The formation of a photoproduct in DNA is a major deleterious event for the cell since it can lead to death or the appearance of mutations. This may be prevented by repair enzymes which are able to restore the integrity of the genome. As already mentioned, *M. luteus* endonuclease and T4 endonuclease V are able to cleave the 5'-end N-glycosidic bond of cyclobutane type dimers. The resulting abasic sites are then removed by the same enzymes, which also exhibit AP endonuclease properties. The X-ray structure and the mechanism of action of T4 endonuclease V has recently been determined (Morikawa et al., 1992; Doi et al., 1992). The binding of the enzyme to an oligonucleotide containing a TT *cis-syn* cyclobutane type dimer at a specific site was investigated by NMR (Lee et al., 1994). In addition, photo-CIDNP experiments have provided evidence that the aromatic segment at the C-terminus of the enzyme is involved in the binding reaction. The photoproducts may also be excized as short oligonucleotides by the (A)BC excinuclease (Sancar and Tang, 1993). This much less specific repair system is able to repair bulky adducts, including cyclobutane type dimers and (6-4) photoproducts (Svoboda et al., 1993).

Another possibility for the repair of a dimeric photoproduct involves the reversion of a lesion to the starting bases. A photoreactivating enzyme directed towards the TT and TC (6-4) photoproducts has been recently isolated (Todo et al., 1993). However, its exact repair mechanism remains to be established. The best known of the DNA photorepair enzymes are the photolyases, which split cyclobutane type dimers upon visible light irradiation (Kim and Sancar, 1993; Sancar, 1994). Photolyases have been identified in several prokaryotic species, but appear to be absent in human cells (Feng et al., 1993; Chao, 1993). The X-ray structure of photolyases of *E. coli* (Park et al., 1993) and *A. midulans* (Miki et al., 1993) have been determined. Much attention has been paid recently to the determination of the mechanism of action of photolyases, by using chemically synthesized models and various physical techniques.

Specificity

A first series of investigations has dealt with the binding properties of the enzyme. As for the mutagenesis studies, oligonucleotides containing a lesion at a specific site constitute major tools. It was shown that not only the thymine cyclobutane dimer (Kemmink et al., 1988) but also the thymine-uracil (TU) and the UU *cis-syn* cyclobutane dimers are substrates for the photolyases (Kim and Sancar, 1991). In addition, the *trans-syn* cyclobutadithymine is reversed by the photolyase upon illumination with visible light (Kim et al., 1993b). However, the *cis-syn* cytosine cyclobutane dimer is much less efficiently repaired than the corresponding thymine photoproduct. In addition, even though the binding of the enzyme was 10^4 times lower with RNA than with DNA, the efficiency of the splitting of cyclobutadipyrimidine is the same, once the enzyme is bound to the modified nucleic acid (Kim and Sancar, 1991). Further information on the binding properties of the enzyme has been provided by studying the repair of *cis-syn* cyclobutane type dimer containing polythymidine short oligonucleotides. It was observed that four nucleotides are necessary for a complete binding of the enzyme (Jordan et al., 1989). Further precision on the binding site of the photolyase was provided by using 49-mers containing two cyclobutadithymines located on opposite strands, in either a 3′ or a 5′ staggered configuration. The enzyme binds to a four-nucleobase-long fragment, including the phosphate upstream and the three phosphates downstream to the dimer (Svoboda et al., 1993).

Photochemical mechanism

Extensive studies have been devoted to the determination of the photochemical and photophysical processes involved in the splitting of pyrimidine dimers by photolyases (Fig. 6) (Kim and Sancar, 1993; Sancar, 1994). It is now well established that the bulk of the photolyases contains two chromophores, characterized as a reduced flavin adenine dinucleotide associated with either a methenyl tetrahydrofolate or a 8-hydroxydeazaflavin residue. This "second chromophore" is a light harvesting molecule transferring its energy in a singlet excited state to the flavin. However, its presence is not necessary for the efficiency of the enzyme (Lipman and Johns, 1992; Kim et al., 1992b). Then, the excited flavin residue is able to induce the formation of a pyrimidine dimer radical, observed by picosecond laser photolysis (Okamura et al., 1991) and time-resolved EPR (Kim et al., 1992a).

Model compounds
The nature of the pyrimidine cyclobutane dimer radical produced, either an anion or a cation, has been the matter of major debates. With model compounds, both species can be produced depending on the photosensitizer. Using a series of synthetic molecules with either an indole (Kim et al., 1990; Kim and Rose, 1990) or an arylamine (Kim and Rose, 1992; Groy et al., 1991) moiety covalently linked to a pyrimidine dimer, it has

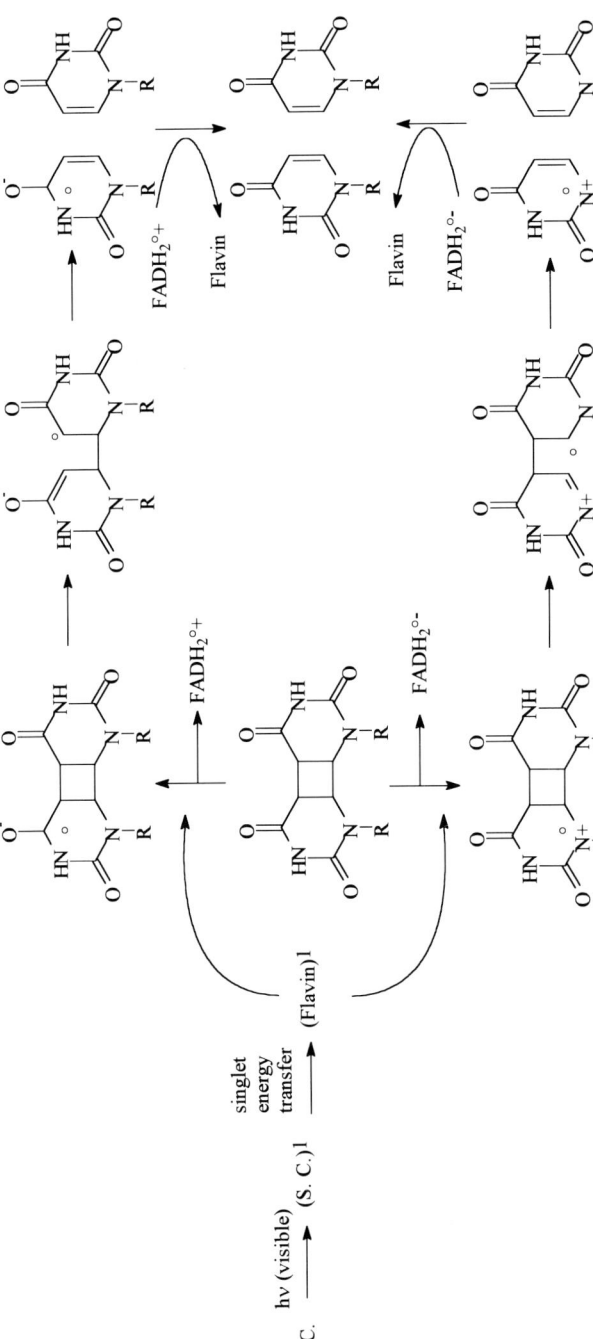

Figure 6. Possible mechanisms of action of photolyases (S.C.: second chromophore).

been shown that an efficient splitting of the cyclobutane ring occurs upon exposure to visible light, associated with the formation of a radical anion. In the latter systems, the photoreversion of cyclobutadipyrimidines is highly solvent dependent, with a higher efficiency in low polarity media. This observation is consistent with an electron transfer mechanism (Kim et al., 1990; Hartzfeld and Rose, 1992). On the other hand, a radical cation is formed when the photosplitting is achieved with a quinone derivative (Young et al., 1990).

The splitting mechanism has been investigated by taking advantage of the possibility to induce both types of dimer ion radicals, depending on the photosensitizer used. The formation of both the dimer radical anion and radical cation lead to the reversal of cyclobutadipyrimidines (Burdi and Begley, 1991; Yang and Begley, 1993). The mechanism of the reaction is supposed to be sequential, with an initial cleavage occurring at the 5,5 bond for the radical cation. On the other hand, the initial cleavage of the 6,6 cyclobutane bond is involved in the splitting of the anion radical (McMordie and Begley, 1992). It should be added that the possibility of a concerted reaction for the splitting of the cyclobutane ring, subsequent to the formation of the pyrimidine radical anion, has also been postulated (Mordie et al., 1993).

Flavin and enzyme
Studies of the splitting of pyrimidine cyclobutane dimer by flavin-mediated photosensitization, which is the photosensitizer present in the native enzyme, have not allowed an unambiguous conclusion as to whether an anion or a cation radical is involved. Observation of secondary deuterium isotope effects cannot be rationalized in terms of cleavage of only the 5,5 or 6,6 bond (Witmer et al., 1989). Moreover, the splitting occurs through a cation radical when the oxidized protonated flavin is used (Hartman and Rose, 1992a; Heelis et al., 1993) whereas an anion is involved with the deprotonated reduced flavin (Hartman and Rose, 1992b). However, most of the available data are in favor of the anion mechanism, even though the cation radical pathway cannot be totally ruled out. First, flavin is in a reduced deprotonated state in the native enzyme (Kim et al., 1993a). A singlet excited state of the reduced deprotonated flavin has been detected by laser flash photolysis. In addition, thermodynamic considerations have shown that, in solution, the cyclobutane dimer anion pathway, occurring through single electron transfer from $FADH^-$ to cyclobutadipyrimidine, is favored with reduced deprotonated flavin (Yeh and Falvey, 1992). Dynamic nuclear polarization detection also provided experimental evidence for the implication of the dimer anion radical in solution (Hartman et al., 1992).

References

Banerjee, S.K., Christensen, R.B., Lawrence, C.W. and LeClerc, J.E. (1988) Frequency and spectrum of mutations produced by a single *cis-syn* thymine-thymine dimer in a single-stranded vector. *Proc. Natl. Acad. Sci. USA* 85: 8141–8145.

Banerjee, S.K., Borden, A., Christensen, R.B., LeClerc, J.E. and Lawrence, C.W. (1990) SOS-dependent replication past a single *trans-syn* T-T cyclobutane dimer gives a different mutation spectrum and increased error rate compared with replication past this lesion in uninduced cells. *J. Bacteriol.* 172: 2105–2112.

Berg, R.J.W., de Gruijl, F.R., Roza, L. and van der Leun, J.C. (1993) Flow cytometry immunofluorescence assay for quantification of cyclobutyldithymine dimers in separate phases of the cell cycle. *Carcinogenesis* 14: 103–106.

Blais, J., Douki, T., Vigny, P. and Cadet, J. (1994) Fluorescence quantum yield determination of pyrimidine (6-4) pyrimidone photoadducts. *Photochem. Photobiol.* 59: 402–404.

Blommers, M.J.J., Lucasius, C.B., Kateman, G. and Kaptein, R. (1992) Conformational analysis of a dinucleotide photodimer with the aid of the genetic algorithm. *Biopolymers* 32: 45–52.

Bourre, F., Benoit, A. and Sarasin, A. (1989) Respective roles of pyrimidine dimers and pyrimidine (6-4) pyrimidone photoproducts in UV mutagenesis of simian virus 40 DNA in mammalian cells. *J. Virol.* 63: 4520–4524.

Burdi, D and Bergley, T.P. (1991) Mechanistic studies on DNA photolyase. 3. The trapping of the one-bond-cleaved intermediate from a photodimer radical cation model system. *J. Am. Chem. Soc.* 113: 7768–7770.

Bridges, B.A. (1992) Mutagenesis after exposure of bacteria to ultraviolet light and delayed photoreversal. *Mol. Gen. Genet.* 233: 331–336.

Cadet, J., and Vigny, P. (1990) Photochemistry of nucleic acids. In: H. Morisson (ed.): *Bioorganic Photochemistry*, Vol. 1, Wiley, New York, pp 1–272.

Cadet, J., Anselmino, C., Douki, T. and Voituriez, L. (1992) Photochemistry of nucleic acids in cells. *J. Photochem. Photobiol. B: Biol.* 15: 277–298.

Cadet, J. and Weinfeld, M. (1993) Detecting DNA damage. *Anal. Chem.* 65: 675–682.

Chao, C. C.-K. (1993) Lack of DNA enzymatic photoreactivation in HeLa cell-free extracts. *FEBS Lett.* 336: 411–416.

Clarkson, J.M., Mitchell, D.L. and Adair, G.M. (1983) The use of an immunological probe to measure kinetics of DNA repair in normal and UV-sensitive mammalian cell lines. *Mutat. Res.* 112: 287–299.

Clivio, P., Fourrey, J.-L. and Gasche, J. (1991) DNA photodamage mechanistic studies: characterization of a thietane intermediate in a model reaction relevant to the "6-4 lesions". *J. Am. Chem. Soc.* 113: 4581–4583.

Doi, T., Recktenwald, A., Karaki, Y., Kikuchi M., Morikawa, K., Ikehara, M., Inokowa, T., Hori, N. and Ohtsuka, E. (1992) Role of the basic amino acid cluster and Glu-23 in pyrimidine dimer glycosylase acitivity of T4 endonuclease V. *Proc. Natl. Acad. Sci. USA* 89: 9420–9424.

Douki, T., Voituriez, L. and Cadet, J. (1991) Characterization of the (6-4) photoproduct of 2′-deoxycytidylyl-(3′-5′)-thymidine and of its Dewar valence isomer. *Photochem. Photobiol.* 53: 293–297.

Douki, T. and Cadet, J. (1992) Far-UV photochemistry and photosensitization of 2′-deoxycytidylyl-(3′-5′)-thymidine: isolation and characterization of the main photoproducts. *Photochem. Photobiol.* 15: 199–213.

Douki, T. and Cadet, J. (1994) Formation of cyclobutane dimers and (6-4) photoproducts upon far-UV photolysis of 5-methylcytosine-containing dinucleoside monophosphates. *Biochemistry* 33: 11942–11950.

Douki, T., Voituriez, L. and Cadet, J. (1995) Measurement of pyrimidine (6-4) pyrimidone photoproducts in DNA by a mild acidic hydrolysis-HPLC fluorescence detection assay. *Chem. Res. Toxicol.* 8: 244–253.

Dumaz, N., Drougard, C., Sarasin, A. and Daya-Grosjean, L. (1993) Specific UV-induced mutation spectrum in the p53 gene of skin tumors from DNA-repair deficient Xeroderma pigmentosum patients. *Proc. Natl. Acad. Sci. USA* 90: 10529–10533.

Feng, Y., Kim, S.-T. and Sancar, A. (1993) Evidence for lack of DNA photoreactivating enzyme in humans. *Proc. Natl. Acad. Sci. USA* 90: 4389–4393.

Fourrey, J.-L., Gasche, J., Fountain, C., Guittet, E. and Favre, A. (1989) Sequence dependent photochemistry of di(deoxynucleoside) phosphates containing 4-thiouracil. *J. Chem. Soc., Chem. Commun.* 1334–1336.

Franklin, W.A., Lo, K.M. and Haseltine, W.A. (1982) Alkaline lability of fluorescent photoproducts produced in ultraviolet-irradiated DNA. *J. Biol. Chem.* 257: 13535–13543.

Franklin, W.A., Doetsch, P.W. and Haseltine, W.A. (1985) Structural determination of the ultra-violet light-induced thymine-cytosine pyrimidine-pyrimidone (6-4) photoproducts. *Nucleic Acids Res.* 13: 5317–5325.

Gibbs, P.E.M. and Lawrence, C.W. (1993) U-U and T-T cyclobutane dimers have different mutational properties. *Nucleic Acids Res.* 21: 4059–4065.

Gibbs, P.E.M., Kilbey, B.J., Banerjee, S.K. and Lawrence, C.W. (1993a) The frequency and accuracy of replication past a thymine cyclobutane dimer are very different in *Saccharomyces cerevisiae* and *Escherichia coli. J. Bacteriol.* 175: 2607–2612.

Gibbs, P., Horsfall, M., Borden, A., Kilbey, B.J. and Lawrence, C.W. (1993b) Understanding spectra of UV induced mutations: studies with individual photoproducts. *In*: A. Shima, M. Ichakasi, Y. Fujiwara and K. Tahebe (eds): *Frontiers of Photobiology*. Elsevier Science Publishers B.V., pp 357–361.

Groy, T.L., Kim, S.-T. and Rose, S.D. (1991) Structures of two spiro[cyclopropane-1,3'-indoline]-*cis-syn*-1,3-dimethyluracil cyclobutane photodimers. *Acta Cryst.* C47: 1898–1902.

Hartman, R.F. and Rose, S.D. (1992a) A possible chain reaction in photosensitized splitting of pyrimidine dimers by a protonated oxidased flavin. *J. Org. Chem.* 57: 2302–2306.

Hartman, R.F. and Rose, S.D. (1992b) Efficient photosensitized pyrimidine dimer splitting by a reduced flavin requires the deprotonated flavin. *J. Am. Chem. Soc.* 114: 3559–3560.

Hartman, R.F., Rose, D.S., Pouwels, P.J.W. and Kaptein, R. (1992) Flavin-sensitized photochemically induced dynamic nuclear polarization detection of pyrimidine dimer radicals. *Photochem. Photobiol.* 56: 305–310.

Hartzfeld, D.G. and Rose, S.D. (1992) Efficient pyrimidine dimer radical anion splitting in low polarity solvents. *J. Am. Chem. Soc.* 115: 850–854.

Haseltine, W.A., Gordon, L.K., Lindan, C.P., Grafstrom, R.H., Shaper, N.L. and Grossman, L. (1980) Cleavage of pyrimidine dimers in specific DNA sequences by a pyrimidine dimer DNA-glycosylase of *M. luteus. Nature* 285: 634–641.

Heelis, P.F., Hartman, R.F. and Rose, S.D. (1993a) Detection of the excited singlet state of deprotonated, reduced flavin. *Photochem. Photobiol.* 57: 1053–1055.

Heelis, P.F., Hartman, R.F. and Rose, S.D. (1993b) Detection of radical ion intermediates in flavin-photosensitized pyrimidine dimer splitting. *Photochem. Photobiol.* 57: 442–446.

Henriksen, T., Dahlback, A., Larsen, S.H.H. and Moan, J. (1990) Ultraviolet-radiation and skin cancer. Effect of an ozone layer depletion. *Photochem. Photobiol.* 51: 579–582.

Hettich, R.L., Buchanan, M.V. and Ho, C.-H. (1990) Characterization of photo-induced pyrimidine cyclobutane dimers by laser desorption Fourier transform mass spectrometry. *Biomed. Environ. Mass Spectrom.* 19: 55–62.

Horsfall, M.J. and Lawrence, C.J. (1994) Accuracy of replication past the T-C (6-4) adduct. *J. Mol. Biol.* 235: 465–471.

Hruska, F.E., Voituriez, L., Grand, A. and Cadet, J. (1986) Molecular structure of the *cis-syn* photodimer of d(TpT)(cyanoethyl ester). *Biopolymers* 25: 1399–1417.

Inaka, K., Yosui, A., Deruiter, P.E. and Eker, A.P.M. (1993) Crystallization and preliminary X-ray diffraction studies of photolyase (photoreactivating enzyme) from the cyanobacterium *Anacystis-nidulans. J. Mol. Biol.* 233: 167–169.

Jiang, N. and Taylor, J.-S. (1993) *In vivo* evidence that UV-induced C→T mutations at dipyrimidine sites could result from the replicative bypass of *cis-syn* cyclobutane dimers or their deamination products. *Biochemistry* 32: 472–481.

Johns, H.E., Pearson, M.L., LeBlanc, J.C. and Helleiner, C.W. (1964) The ultraviolet photochemistry of thymidylyl-(3'-5')-thymidine. *J. Mol. Biol.* 9: 503–524.

Jordan, S.P., Alderfer, J.L., Chanderkar, L.P. and Schuman Jorns, M. (1989) Reaction of *Escherichia coli* and yeast photolyases with homogeneous short-chain oligonucleotide substrates. *Biochemistry* 28: 8149–8153.

Kamiya, H., Murata, N., Murata, T., Iwai, S., Matsikage, A., Masutani, C., Hanaoka, F. and Ohtsuka, E. (1993) Cyclobutane thymine dimers in *ras* proto-oncogene hot spot acivate the gene by point mutation. *Nucleic Acids Res.* 21: 2355–2361.

Kan, L.-S., Voituriez, L. and Cadet, J. (1988) Nuclear magnetic resonance studies of *cis-syn*, *trans-syn*, and (6-4) photodimers of TpT and *cis-syn* photodimers of TpT cyanoethyl ester. *Biochemistry* 27: 5796–5803.

Kan, L.-S., Voituriez, L. and Cadet, J. (1992) The Dewar valence isomer of the (6-4) photoadduct of thymidylyl-(3'-5')-thymidine monophosphate: formation, alkaline lability and conformational properties. *J. Photochem. Photobiol. B: Biol.* 12: 339–357.

Kao, J.F.L., Nadji, S. and Taylor, J.-S. (1993) Structure determination of a third cyclobutane photodimer of thymidylyl-(3'-5')-thymidine: the *trans-syn* II product. *Chem. Res. Toxicol.* 6: 561–567.

Kemmink, J., Eker, A.P.M., Koning, T.M.G., van der Marel, G.A., van Boom, J.H. and Kaptein, R. (1988) Photoreactivation of the thymine dimer containing DNA octamer d(GCGT^TGCG).d(CGCAACGC) by the photoreactivating enzyme from *Anacystis nidulans*. *J. Photochem. Photobiol. B: Biol.* 1: 323–328.

Kim, J.-K., Wallace, J.C. and Alderfer, J.L. (1993) Substituent effects on the puckering mode of the cyclobutane ring and the glycosyl bond of *cis-syn* photodimers. *Biopolymers* 33: 713–721.

Kim, S.-T. and Rose, S.D. (1990) Activation barriers in photosensitized pyrimidine dimer splitting. *J. Phys. Org. Chem.* 3: 581–586.

Kim, S.-T., Hartman, R.F. and Rose, S.D. (1990) Solvent dependence of pyrimidine dimer splitting in a covalently linked dimer-indole system. *Photochem. Photobiol.* 52: 789–794.

Kim, S.-T. and Sancar, A. (1991) Effect of base, pentose and phosphodiester backbone structures on binding and repair of pyrimidine dimers by *Escherichia coli* DNA photolyase. *Biochemistry* 30: 8623–8630.

Kim, S.-T. and Rose, S.D. (1992) Pyrimidine dimer splitting in covalently linked dimer-arylamine systems. *J. Photochem. Photobiol. B: Biol.* 12: 179–191.

Kim, S.-T., Heelis, P.F. and Sancar, A. (1992a) Energy transfer (deazaflavin $\to$ FADH$_2$) and electron transfer (FADH$_2 \to$ T $<$ $>$ T) kinetics in *Anacystis nidulans* photolyase. *Biochemistry* 31: 11244–11248.

Kim, S.-T., Sancar, A., Essenmacher, C. and Babcock, G.T. (1992b) Evidence from photoinduced EPR for a radical intermediate during photolysis of cyclobutane thymine dimer by DNA photolyase. *J. Am. Chem. Soc.* 114: 4442–4443.

Kim, S.-T. and Sancar, A. (1993) Photochemistry, photophysics, and mechanism of pyrimidine dimer repair by DNA photolyase. *Photochem. Photobiol.* 57: 895–904.

Kim, S.-T., Sancar, A., Essenmacher, C. and Babcock, G.T. (1993a) Time-resolved EPR studies with DNA photolyase: excited-state FADH^0 abstracts an electron from trp-306 to generate FADH$^-$, the catalytically active form of the cofactor. *Proc. Natl. Acad. Sci. USA* 90: 8023–8027.

Kim, S.-T., Malhotra, K., Smith, C.A., Taylor, J.-S. and Sancar, A. (1993b) DNA photolyase repairs the *trans-syn* cyclobutane dimer. *Biochemistry* 28: 7065–7068.

Koechler, D.R., Awadallah, S.S. and Glickman, B.W. (1991) Sites of preferential induction of cyclobutane pyrimidine dimers in the nontranscribed strand of lac I correspond with sites of UV-induced mutation in *Escherichia coli*. *J. Biol. Chem.* 356: 11766–11773.

Koning, M.G., van Soest, J.J.G. and Kaptein, R. (1991) NMR studies of dipyrimidine cyclobutane photodimers. *J. Biochem.* 195: 29–40.

Kovács, A., Gál, P. and Závodszky, P. (1993) A simple method to asses *in vivo* repair of ultraviolet radiation-induced lesions of specific DNA sequences of restriction sites. *Radiat. Res.* 136: 397–403.

Lawrence, C.W., Gibbs, P.E.M., Borden, A., Horsfall, M.J. and Kilbey, B.J. (1993) Mutagenesis induced by single UV photoproducts in *E. coli* and yeast. *Mutat. Res.* 299: 157–163.

LeClerc, J.E., Borden, A. and Lawrence, C.W. (1991) The thymine-thymine pyrimidine-pyrimidone (6-4) ultraviolet light photoproduct is highly mutagenic and specifically induces 3' thymine-to-cytosine transitions in *Escherichia coli*. *Proc. Natl. Acad. Sci. USA* 88: 9685–9689.

Lee, B.J., Sakashita, H., Ohkubo, T., Ikehara, M., Doi, T., Morikawa, K., Kyogoku, Y., Osafune, T., Iwai, S. and Ohtsuka, E. (1994) Nuclear magnetic resonance study of the interaction of T4 endonuclease V with DNA. *Biochemistry* 33: 57–64.

Lemaire, D.G.E. and Ruzsicska, B.P. (1993a) Quantum yields and secondary photoreactions of the photoproducts of dTpdT, dTpdC and dTpdU. *Photochem. Photobiol.* 57: 755–769.

Lemaire, D.G.E. and Ruzsicska, B.P. (1993b) Kinetic analysis of the deamination reaction of cyclobutane dimers of thymidylyl-3',5'-2'-deoxycytidine and 2'-deoxycytidylyl-3',5'-thymidine. *Biochemistry* 32: 2525–2533.

Li, Y.F., Kim, S.-T. and Sancar, A. (1993) Evidence for lack of DNA photoreactivating enzyme in humans. *Proc. Natl. Acad. Sci, USA* 90: 4389–4393.

Lipman, R.S.A. and Schuman Jorns, M. (1992) Direct evidence for singlet-singlet energy transfer in *Escherichia coli* DNA photolyase. *Biochemistry* 31: 786–791.

Lippke, J.A., Gordon, L.K., Brash, D.E. and Haseltine, W.A. (1981) Distribution of UV light-induced damage in a defined sequence of human DNA: detection of alkaline-sensitive lesions at pyrimidine nucleoside-cytosine sequences. *Proc. Natl. Acad. Sci. USA* 78: 3388–3392.

Liu, F.-T. and Yang, N.C. (1978) Photochemistry of cytosine derivatives. 1. Photochemistry of thymidylyl-(3′-5′)-deoxycytidine. *Biochemistry* 17: 4865–4876.

Matsunaga, T., Mori, T. and Nikaido, O. (1990) Base sequence specificity of a monoclonal antibody binding to (6-4) photoproducts. *Mutat. Res.* 235: 187–194.

Matsunaga, T., Hieda, K. and Nikaido, O. (1991) Wavelength dependent formation of thymine dimers and (6-4) photoproducts in DNA by monochromatic ultraviolet light ranging from 150 to 365 nm. *Photochem. Photobiol.* 54: 403–410.

Matsunaga, T., Hatakeyama, Y. and Ohta, M. (1993) Establishment and characterization of a monoclonal antibody recognizing the Dewar valence isomer of (6-4) photoproducts. *Photochem. Photobiol.* 57: 934–940.

McMordie, R.A.S. and Begley, T.P. (1992) Mechanistic studies on DNA photolyase. 5. Secondary deuterium isotope effects on the cleavage of the uracil photodimer radical cation and anion. *J. Am. Chem. Soc.* 114: 1886–1887.

McMordie, R.A.S., Altmann, E. and Begley, T.P. (1993) Mechanistic studies on DNA photolyase. 9. Is the cleavage of the cyclobutane pyrimidine photodimer radical anion a pericyclic reaction? *J. Am. Chem. Soc.* 115: 10370–10371.

Mitchell, D.L. (1988) The induction and repair of lesions produced by the photolysis of (6-4) photoproducts in normal UV-hypersensitive human cells. *Mutat. Res.* 194: 227–237.

Mitchell, D.L., Allison, J.P. and Nairn, R.S. (1990a) Immunoprecipitation of pyrimidine (6-4) pyrimidone photoproducts and cyclobutane pyrimidine dimers in UV-irradiated DNA. *Radiat. Res.* 123: 299–303.

Mitchell, D.L., Brash, D.E. and Nairn, R.S. (1990b) Rapid repair kinetics of pyrimidine (6-4) pyrimidone photoproducts in human cells are due to excision rather than conformational change. *Nucleic Acids Res.* 18: 963–971.

Mitchell, D.L., Nguyen, T.D. and Cleaver, J.E. (1990c) Nonrandom induction of pyrimidine-pyrimidine (6-4) photoproducts in ultraviolet-irradiated human chromatin. *J. Biol. Chem.* 265: 5353–5356.

Mitchell, D.L., Jen, J. and Cleaver, J.E. (1992) Sequence specificity of cyclobutane pyrimidine dimers in DNA treated with solar (ultraviolet B) radiation. *Nucleic Acids Res.* 20: 225–229.

Mizuno, T., Matsunaga, T., Ihara, M. and Nikaido, O. (1991) Establishment of a monoclonal antibody recognizing cyclobutane dimers in DNA: a comparative study with 64M-1 antibody specific for (6-4) photoproducts. *Mutat. Res.* 254: 175–184.

Moan, J. and Peak, M.J. (1989) Effects of UV radiation on cells. *J. Photochem. Photobiol. B: Biol.* 4: 21–34.

Mori, T., Matsunaga, T., Hirose, T. and Nikaido, O. (1988) Establishment of a monoclonal antibody recognizing ultraviolet light-induced (6-4) photoproducts. *Mutat. Res.* 194: 263–270.

Mori, T., Wani, A.A., D'Ambrosio, S.M., Chang, C.-C. and Trosko, J.E. (1989) In situ pyrimidine dimer determination by laser cytometry. *Photochem. Photobiol.* 49: 523–526.

Mori, T., Matsunaga, T., Chang, C.-C., Trosko, J.E. and Nikaido, O. (1990) In situ (6-4) photoproducts determination by laser cytometry and autoradiography. *Mutat. Res.* 236: 99–105.

Mori, T., Nakane, M., Hattori, T., Mutsunaga, T., Ihara, M. and Nikaido, O. (1991) Simultaneous establishment of monoclonal antibodies specific for either cyclobutane pyrimidine dimer or (6-4) photoproduct from the same mouse immunized with ultraviolet-irradiated DNA. *Photochem. Photobiol.* 54: 225–232.

Morikawa, K., Matsumoto, O., Tsujimoto, M., Katayanagi, K., Ariyoshi, M., Doi, T., Ikehara, M., Inaoka, T. and Ohtsuka, E. (1992) X-ray structure of T4 endonuclease V: an excision repair enzyme specific for pyrimidine dimer. *Science* 256: 523–526.

Nguyen, H.T. and Minton, K.W. (1988) Ultraviolet-induced dimerization of non-adjacent pyrimidines. *J. Biol. Mol.* 200: 681–693.

Niggli, H.J. (1990) Comparative studies on the correlation between pyrimdine dimer formation and tyrosine activity in cloudman S91 melanoma cells after ultraviolet-irradiation. *Photochem. Photobiol.* 52: 519–524.

Niggli, H.J. (1992) Determination of cytosine-cytosine photodimers in the DNA of cloudman S91 melanoma cells using high pressure liquid chromatography. *Photochem. Photobiol.* 55: 793–796.

O'Day, C.L., Burgers, P.M.J. and Taylor, J.-S. (1992) PCNA-induced DNA synthesis past *cis-syn* and *trans-syn* I thymine dimers by calf thymus DNA polymerase δ *in vitro*. *Nucleic Acids Res.* 20: 5403–5406.

Okamura, T., Sancar, A., Heelis, P.F., Begley, T.P., Hirata, Y. and Mataga, N. (1991) Picosecond laser photolysis studies of the photorepair of pyrimdine dimers by DNA photolyase. 1. Laser photolysis of photolyase-2-deoxyuridine dinucleotide photodimer complex. *J. Am. Chem. Soc.* 113: 3143–3145.

Olsen, W.M., Huitfeldt, H.S. and Eggset, G. (1989) UVB-induced (6-4) photoproducts in hairless mouse epidermis studied by quantitative immunohistochemistry. *Carcinogenesis* 10: 1669–1673.

Park, H.W., Sancar, A. and Deisenhofer, J. (1993) Crystallization and preliminary crystallographic analysis of *Escherichia-coli* DNA photolyase. *J. Mol. Biol.* 231: 1122–1125.

Pfeifer, G.P., Drouin, R., Riggs, A.D. and Holmquist, G.P. (1991) *In vivo* mapping of a DNA adduct at nucleotide resolution: Detection of pyrimidine (6-4) pyrimidone photoproducts by ligation-mediated polymerase chain reaction. *Proc. Natl. Acad. Sci. USA* 88: 1374–1378.

Pfeifer, G.P., Drouin, R. and Holmquist, G.P. (1993) Detection of DNA adducts at the DNA sequence level by ligation mediated PCR. *Mutat. Res.* 288: 39–46.

Ramsey, R.S. and Ho, C.-H. (1989) Determination of pyrimidine dimers in DNA by high performance liquid chromatography/gas chromatography and electron capture detection. *Anal. Biochem.* 182: 424–431.

Rao, S.N. and Kollman, P.A. (1993) Theoretical simulations on d(CGCGAATTCGCG)$_2$ with *cis-syn* thymine-thymine cyclobutane dimer. *Bull. Chem. Soc. Jpn.* 66: 3132–3134.

Roza, L., van der Wulf, K.J.M. and MacFarlane, S.J. (1988) Detection of cyclobutane thymine dimers in DNA of human cells with monoclonal antibodies raised against a thymine dimer-containing tetranucleotide. *Photochem. Photobiol.* 18: 627–633.

Ruiz-Rubio, M. and Bockrath, R. (1989) On the possible role of cytosine deamination in delayed photoreversal mutagenesis targeted at thymine-cytosine dimers in *E. coli*. *Mutat. Res.* 210: 93–102.

Rycyna, R.E. and Alderfer, J.L. (1985) UV irradiation of nucleic acids: formation, purification and solution conformational analysis of the '6-4 lesion' of dTpdT. *Nucleic Acids Res.* 13: 5949–5963.

Sage, E. (1993) Distribution of photolesions in DNA. Genetic consequences and the role of sequence context. *Photochem. Photobiol.* 57: 163–173.

Sancar, A. and Tang, M.-S. (1993) Nucleotide excision repair. *Photochem. Photobiol.* 57: 905–921.

Sancar, A. (1994) Structure and function of DNA photolyase. *Biochemistry* 33: 2–9.

Smith, C.A. and Taylor, J.-S. (1993) Preparation and characterization of a set of deoxyoligonucleotides 49-mers containing site specific *cis-syn*, *trans-syn* I, (6-4), and Dewar photoproducts of thymidylyl(3' → 5')-thymidine. *J. Biol. Chem.* 268: 11143–11151.

Strickland, P.T., Nikaido, O., Matsunaga, T. and Boyle, J. (1992) Further characterization of monoclonal antibody indicates specificity for (6-4) dipyrimidine photoproducts. *Photochem. Photobiol.* 55: 723–727.

Svoboda, D.L., Smith, C.A., Taylor, J.-S.A. and Sancar, A. (1993) Effect of sequence, adducts type, and opposing lesions on the binding and repair of ultraviolet photodamage by DNA photolyase and (A)BC excinuclease. *J. Biol. Chem.* 268: 10694–10700.

Tabaczynski, W.A., Lemaire, D.G.E., Ruzsicska, B.P. and Alderfer, J.L. (1993) An NMR and conformational investigation of the *trans-syn* cyclobutane photodimers of TpdU. *Biopolymers* 33: 1365–1375.

Taylor, J.-S. and Cohrs, M.P. (1987) DNA, light and Dewar pyrimidones: the structure and biological significance of TpT^3. *J. Am. Chem. Soc.* 109: 2834–2835.

Taylor, J.-S., Brockie, I.R. and O'Day, C.L. (1987) A building block for the sequence-specific introduction of *cis-syn* thymine dimers into oligonucleotides. Solid phase synthesis of TpT[c,s]pTpT. *J. Am. Chem. Soc.* 109: 6735–6742.

Taylor, J.-S. and Brockie, I.R. (1988) Synthesis of a *trans-syn* thymine dimer building block. Solid phase synthesis of CGTAT[*t,s*]TATGC. *Nucleic Acids Res.* 16: 5123–5136.

Taylor, J.-S. (1990) DNA, sunlight, and skin cancer. *J. Chem. Educ.* 67: 835–841.

Taylor, J.-S. and O'Day, C.L. (1990) *Cis-syn* thymine dimers are not absolute blocks to replication by DNA polymerase I of *Escherichia coli in vitro*. *Biochemistry* 29: 1624–1632.

Taylor, J.-S., Lu, H.-F. and Kotyk, J.J. (1990) Quantitative conversion of the (6-4) photoproduct of TpdC to its Dewar valence isomer upon exposure to simulated sunlight. *Photochem. Photobiol.* 51: 161–167.

Taylor, J.-S. and Nadji, S. (1991) Unravelling the origin of the major mutation induced by ultraviolet light, the C-T transition at dTpdC sites. A DNA synthesis building block for the *cis-syn* cyclobutane dimer of dTpdU. *Tetrahedron* 47: 2579–2590.

Thomas, D.C., Okumoto, D.S., Sancar, A. and Bohr, V.A. (1989) Preferential DNA repair of (6-4) photoproducts in the dihydrofolate reductase gene of Chinese hamster ovary cells. *J. Biol. Chem.* 264: 18005–18010.

Todo, T., Takemori, H., Ryo, H., Ihara, M., Matsunaga, T., Nikaido, O., Sato, K. and Nomura, T. (1993) A new photoreactivating enzyme that specifically repairs ultraviolet light-induced (6-4) photoproducts. *Nature* 361: 371–374.

Tressman, I., Kennedy, M.A. and Liu, S.-K. (1994) Unusual kinetics of uracil formation in single- and double-stranded DNA by deamination of cytosine in cyclobutane pyrimidine dimers. *J. Mol. Biol.* 235: 807–812.

Varghese, A.J. (1972) Photochemistry of nucleic acids and their constituents. *In*: A.C. Giese (ed.): *Photophysiology*, Vol. III, Academic Press, Inc., New York, pp 207–274.

Wang, S.Y. (1976) Pyrimidine bimolecular photoproducts. *In*: S.Y. Wang (ed.): *Photochemistry and Photobiology of Nucleic Acids*, Vol. 1 Chemistry, Academic Press, Inc., New York, pp 295–356.

Wang, C.-I. and Taylor, J.-S. (1992) In vitro evidence that UV-induced frameshift and substitution mutation at T tracks are the result of misalignment-mediated replication past a specific thymine dimer. *Biochemistry* 31: 3671–3681.

Witmer, M.R., Altmann, E., Young, H., Begley, T.P. and Sancar, A. (1989) Mechanistic studies on DNA photolyase. 1. Secondary deuterium isotope effects on the cleavage of 2'-deoxyuridine dinucleotide photodimers. *J. Am. Chem. Soc.* 111: 9264–9265.

Yang, D.-Y. and Begley, T.P. (1993) Mechanistic studies on DNA photolyase VIII: Studies on the fragmentation of the radical anion and cation of a uracil-alkene photoadduct. *Tetrahedron Lett.* 34: 1709–1712.

Yeh, S.-R. and Falvey, D.E. (1992) Model studies of DNA photorepair: energetic requirements for the radical anion mechanism determined by fluorescence quenching. *J. Am. Chem. Soc.* 114: 7313–7314.

Young, T., Nieman, R. and Rose, S.D. (1990) Photo-CINDP detection of pyrimidine dimer radical cations in anthraquinonesulfonate-sensitized splitting. *Photochem. Photobiol.* 52: 661–668.

Ziegler, A., Leffell, D.J., Kunala, S., Sharma, H.W., Gailani, M., Simon, J.A., Halperin, A.J., Baden, H.P., Shapiro, P.E., Bale, A.E. and Brash, D.E. (1993) Mutation hotspots due to sunlight in the p53 gene of nonmelanoma skin cancers. *Proc. Natl. Acad. Sci. USA* 90: 4216–4220.

Synthesis of active compounds

Interface between Chemistry and Biochemistry
ed. by P. Jollès and H. Jörnvall
© 1995 Birkhäuser Verlag Basel/Switzerland

Chemical and enzymatic synthesis of glycopeptides

M. Schultz and H. Kunz

Institut für Organische Chemie, Johannes Gutenberg-Universität, D-55099 Mainz, Germany

Summary. Progress recently made in the synthesis of biologically relevant N- and O-glycopeptides is illustrated by examples. In this context, developments in the preparation of complex saccharide side chains and in the subsequent coupling to peptide portions is described. Special emphasis is given to the synthesis of Lewis antigen-type structures. Furthermore, modern methods in solid phase peptide syntheses utilizing glycosylated building blocks are presented. Recent advances in glycopeptide syntheses employing enzymatic methods in deprotection steps as well as in peptide/saccharide chain elongation are reported.

Introduction

In the vast field of natural product synthesis the preparation of glycoconjugates is motivated by both chemical and biological objectives. Characterized by an astonishing structural diversity glycoconjugates occur most ubiquitously in the vegetable and animal kingdoms (Hart et al., 1989). Thus, their chemical synthesis has been taken as a challenge since Emil Fischer started to work on peptide and carbohydrate chemistry as separate fields. When this took place at the beginning of this century, the variety of biological information concerning the function of glycoproteins was far from the knowledge currently available. With the rapid development of bio-analytical methods during the past two decades, glycoconjugates have been proved to be of more than just biophysical or nutritional importance. It is now generally accepted that glycoconjugates are strongly involved in phenomena such as cell-cell-, cell-virus(bacteria)-interaction, and transmembrane signalling. For example, their participation in processes such as tumor-development and inflammation became evident and led to an increased activity in this field (Lis and Sharon, 1993).

From investigations in the area of the blood group determining Lewis-type structures it can be demonstrated what fascinating insight into the versatile significance of carbohydrates in living organisms is possible (for a review see Lasky, 1992):

- the sialylated-Lewis[x]-derivative and its sulfated analogue are involved in the recruitment of neutrophils during inflammatory processes (Philips et al., 1990; Polley et al., 1991; Yuen et al., 1992)

- the Lewis^x-structure is likely to be tumor-associated (Itzkowitz et al., 1986)
- the Lacto-Lewis^x-ligand of a parasite has been reported to be involved in down-regulation of an animal-host T-cell response (Vellupiallai and Harn, 1994).

The listed phenomena indicate some applications for glycoconjugates. As antigens they facilitate the production of highly specific antibodies, thus serving for diagnostic and therapeutic purposes. They are furthermore valuable model compounds for structure elucidation and enzyme-substrate interactions. Serious problems are imposed on the production of glycoconjugates from natural sources, since only inefficient access is possible due to microheterogenity and the lack of sufficient amount of material. Furthermore, bio-technological approaches have just begun to deliver recombinant glycoconjugates (Rasmussen, 1991). Finally, the implementation of chemical syntheses is essential to achieve the preparation of unnatural derivatives and analogues.

The present chapter will report on some recent progress made in the synthesis of structurally and stereochemically defined glycopeptides. For the sake of brevity, the field of glycolipids will be mentioned only in passing (for a recent review see Hakomori, 1993).

Synthesis of saccharide side-chains

To obtain a glycopeptide consisting of saccharide-portion and peptide-backbone, either N- or O-glycosidically linked, two different strategies are conceivable.

The first method involves the formation of the central carbohydrate-amino-acid linkage prior to the elongation of peptide and saccharide. In this case, the critical glycosylation-step can be carried out in an early stage of the synthesis. However, the required number of orthogonally stable protective groups is increased, since the hydroxy functions of the saccharide-portion and the functional groups of the amino acid have to be masked in such a way that deblocking and elongation are feasible in the desired fashion.

To minimize these problems, the whole saccharide is advantageously synthesized separately. Hence, the protective group pattern of the saccharide can be simplified to a few or even one principle (in most cases acyl-type protection) and the carbohydrate is finally equipped with an anomeric protective group accessible for activation. Subsequent to glycosylation of the corresponding amino-acid, the N- or the C-terminus is selectively deblocked. The target peptide sequence is then synthesized via common peptide condensation methods either in solution or on solid-phase.

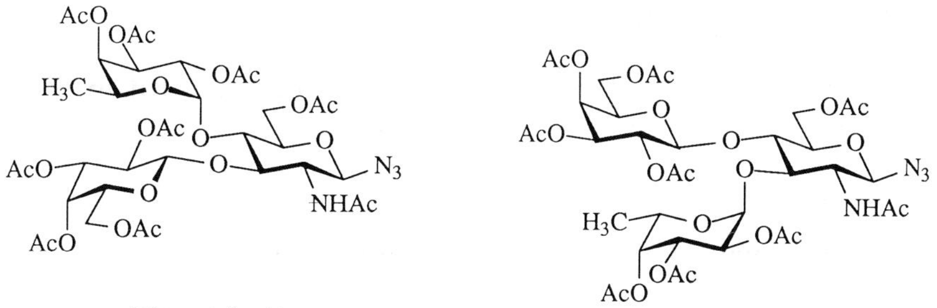

Figure 1. (1) Lewis[a] azide   (2) Lewis[x] azide

Lewis-type side-chains

Application of glycosyl azides

Since many natural glycoproteins contain saccharides N-linked to asparagine, it is desirable to make this linkage region efficiently accessible. Application of the anomeric azide as a stable protection which can readily be converted into the anomeric amine is a useful synthetic strategy. N-glycopeptides carrying oligosaccharides, such as the peracetylated Lewis-type antigenic determinants (1) and (2), respectively, have been prepared via this approach (Kunz and Unverzagt, 1988; Kunz and März, 1992; v. d. Bruch and Kunz, 1994).

The Lewis-type structures can be attained starting from glucosamine (3). The key glucosamine azide (4) is formed by peracetylation and subsequent reaction of the glycosylchloride with sodium-azide (Kunz and Waldmann, 1985; Kunz et al., 1989). Application of phase-transfer conditions in this step circumvents the use of silver-azide as the more dangerous reagent (Spinola and Jeanloz, 1970). From (4) the O-acetyl groups are removed with sodium methanolate in methanol and the 4- and 6-position are simultaneously masked as p-methoxybenzylidene acetal (Fig. 2). Employing the benzylidene acetal for this purpose would

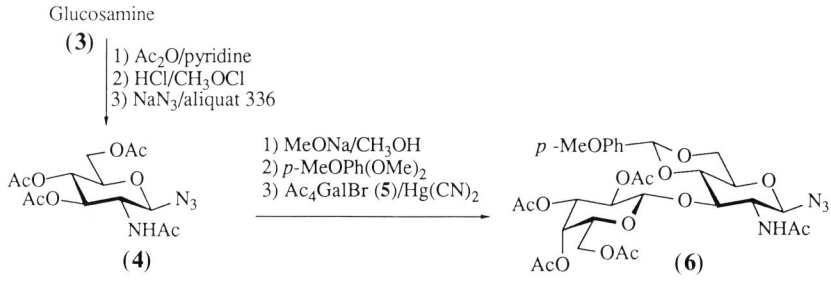

Figure 2.

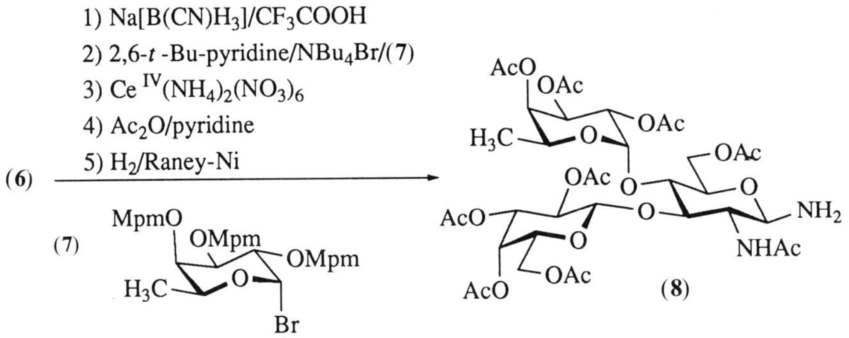

Figure 3.

finally lead to a benzyl ether protection in 6-position, removeable only by hydrogenolysis, thus conflicting with the azido-function.

To attain the series of Lewis[a]-conjugates the 3-OH function has to be galactosylated next. From various glycosyldonors and conditions employed, the use of the peracetylated galactosylbromide (5) activated by mercuric cyanide in a mixture of dichloromethane/nitromethane gives the highest yield of disaccharide (6). These conditions in particular are compatible with the acid-labile p-methoxybenzylidene group.

Cautious, selective acetal opening using sodium-cyanoborohydride and trifluoroacetic acid (Garegg et al., 1982) produces an acceptor suitable for subsequent fucosylation. The required fucose-donor (7) is synthesized in analogy to a method described in the literature (Lönn, 1985). In the subsequent fucosylation step, acidic conditions can be avoided by application of the *in situ* anomerization procedure using tetrabutylammonium bromide (Lemieux and Haymi, 1965).

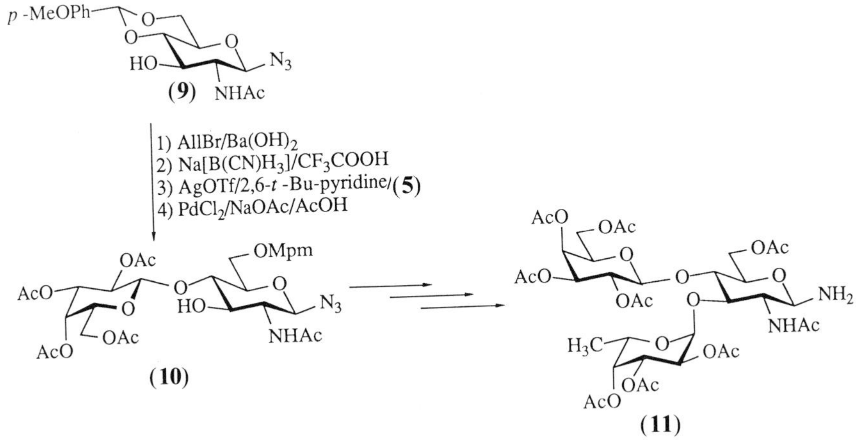

Figure 4.

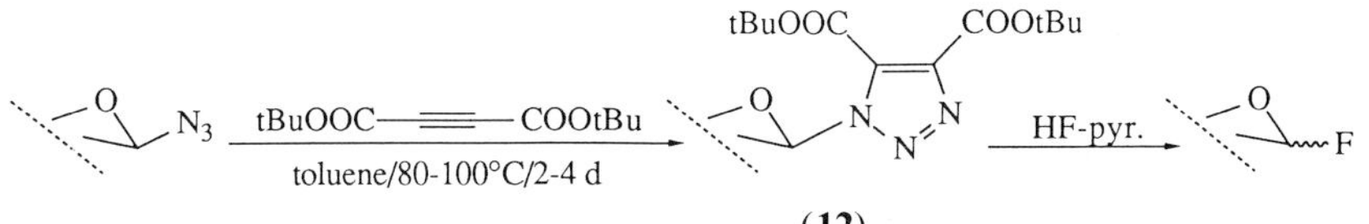

Figure 5.

Cleavage of the p-methoxybenzyl ether groups utilizing ceric ammonium nitrate (Johansson and Samuelsson, 1984) and subsequent acetylation gives the azide (**1**) which is hydrogenated in the presence of Raney-Nickel to yield quantitatively the glycosylamine (**8**). If the reaction is carried out in isopropanol no O → N-acetyl migration or anomerization is observed.

Starting from the glucosamine derivative (**9**) the Lewis^x determinant can be obtained by synthesizing an appropriately protected lactosamine intermediate (**10**) which is fucosylated at 3-OH (v. d. Bruch and Kunz, 1994). The inverse strategy, described in the literature (Lönn, 1985), fails in this particular case because of the low reactivity of 4-OH and the pronounced acid lability of the protective group pattern. The successful reaction sequence starts with a bariumhydroxide-mediated allylation at 3-OH of compound (**9**) (Jaquinet and Sinay, 1977). The regioselective acetal opening is followed by a silvertriflate-promoted galactosylation using peracetylated galactosylbromide (**5**). Once more, the reaction has to be conducted carefully. Di-tert-butyl pyridine as sterically hindered base was added in order to prevent acidic cleavage of the ether-type protection (Nilsson and Norberg, 1988).

In this case, glycosylation with the Helferich promotor (Hg(CN)$_2$) surprisingly yields the orthoester in 82% yield. To liberate 3-OH, the

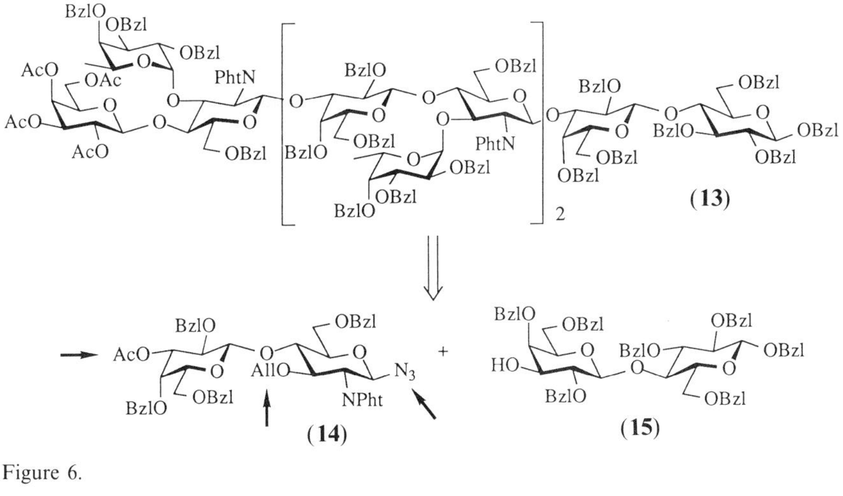

Figure 6.

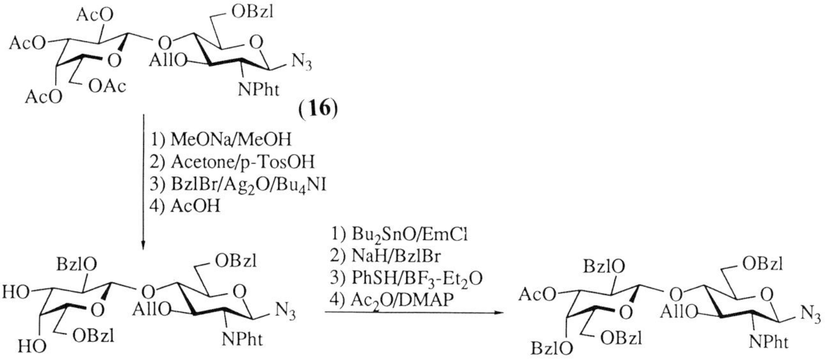

Figure 7.

allylether is isomerized via a Pd(II) mediated isomerization followed by acetic-acid promoted cleavage of the corresponding enol-ether (Ogawa et al., 1983). The outcome of this reaction is influenced by the lability of the p-methoxyphenyl (Mpm) ether. Exchange of the 6-OH protection, however, requires two additional steps. Similar to the strategy outlined in Figure 3, fucosylation of 3-OH utilizing p-methoxybenzylated fucosyl bromide (**7**), subsequent removal of the Mpm ethers and reduction of the azido function complete the pathway leading to Lewis^x amine (**11**).

Preparation of Lewis^a and Lewis^x via the azido methodology prove this approach to be an efficient strategy in which the more common benzyl ether protection can elegantly be substituted by oxidatively cleavable Mpm-groups. Therewith, the use of the anomeric azido function permits direct access to the corresponding glycosylamines ready for coupling the aspartic-acid building blocks (see below).

Moreover, the applicability of this procedure is expandable since glycosyl azides can be activated and used as donors in oligosaccharide synthesis (Bröder and Kunz, 1990). In this strategy, the crucial step consists in the transformation of the glycosyl azide to a potent glycosyl donor via a triazole intermediate formed in a 1,3-dipolar cycloaddition (Bröder and Kunz, 1990; Harmon et al., 1971). Among the glycosyl triazoles tested, only the di-tert-butyl esters (**12**) can successfully be converted into the glycosyl fluorides employed as glycosyl donors (Bröder and Kunz, 1993).

This reactivity of the di-tert-butyl esters can be explained by assuming a preceeding cleavage of the ester groups. The resulting di-carboxylic acid is de-carboxylated and thus, the leaving-group potential of the residual triazole-system is decisively increased. Although elevated temperature and elongated reaction times during triazole-formation are required, the general applicability of the method has been demonstrated

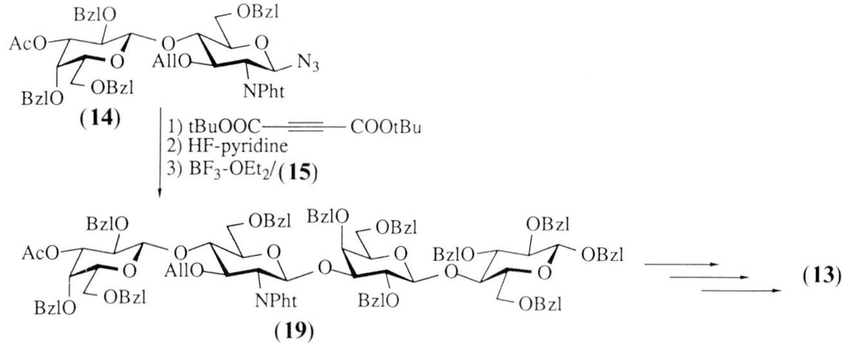

Figure 8.

in several examples and is highlighted by the synthesis of the trimeric Lewis^x-lactose structure (13) (Bröder and Kunz, unpublished). In the retrosynthetic analysis the molecule is de-fucosylated and split into a lactose and a lactosamine building block. To render a repetitive chain-extension the lactosamine unit (14) has to be equipped with orthogonally stable protective groups at the desired linkage positions, as illustrated in Figure 6.

Starting from disaccharide (16) the acetyl-groups are removed and 3,4-O-isopropylidene protection is introduced. After mild benzylation (Yamazaki et al., 1990), de-isopropylidenation and stannane-mediated ethoxymethyl (Em) etherification of 3'-OH (Kluge et al., 1972), 4'-OH is benzylated and the ethoxymethyl ether is exchanged to acetyl.

The glycosyl azide (14) can now be transformed into the corresponding triazole by reacting it with acetylene dicarboxylic acid di-tert-butylester in toluene. Treatment of the triazole with hydrogen fluoride/pyridine-complex according to Figure 5 gives the glycosyl fluoride in high overall yield. The latter is coupled to the lactose-derivative (15) utilizing boron trifluoride etherate. After deacetylation of tetrasaccharide (19) and repeated glycosylation with building block (14), the corresponding neo-lacto-octasaccharide (not shown) is obtained.

The synthesis is completed after de-allylation using an iridium-catalyst (Oltvoort et al., 1981) followed by simultaneous fucosylation of the 3-OH positions thus liberated, using tribenzyl fucosyl bromide. The application of common catalysts as Pd(II) or Wilkinson-catalyst failed (Bröder and Kunz, unpublished).

The previous example shows that the azido-triazole-fluoride sequence meets the requirements of a convergent saccharide synthesis. Moreover, since the glycosylation conditions employed for activation of the glycosyl fluoride are compatible with the anomeric azido function, synthesis of N-glycopeptide precursors is possible via the introduced strategy.

Trichloroacetimidate approach

The synthesis of a tetrameric Lewis^x-saccharide using the well-established trichloroacetimidate procedure (Schmidt and Michel, 1980) recently has been described (Schmidt and Toepfer, 1992). In contrast to the above-outlined strategy, the Lewis^x-monomer itself is employed as building block. Furthermore, the chosen strategy starts with the fucosylation of a suitable glucosamine acceptor. To attain an extendable trisaccharide, 1-O-tert-butyldimethylsilyl-2-azido-2-deoxyglucopyranoside (**20**) has been used as the starting compound (Bommer et al., 1991).

3-O-α-Fucosylation employing benzyl-protected fucopyranosyl trichloroacetimidate (**21**) in the presence of catalytic amounts of trimethylsilyl triflate yields disaccharide (**22**). Selective opening of the benzylidene acetal and introduction of peracetylated galactose results in the formation of Lewis^x-derivative (**23**). Then, the anomeric silyl group is cleaved using tetrabutylammonium fluoride, and base promoted addition to trichloroacetonitrile gives donor (**25**). On the other hand, de-acetylation and 4,6-O-benzylidenation leads to acceptor (**24**). From (**24**) and (**25**), hexasaccharide (**26**) is formed in high yield at $-40°C$ in acetonitrile upon treatment with trimethylsilyl triflate. A repeated application of this reaction-sequence starting from a dimeric-Lewis^x-lactose unit allows efficient access to the completely deprotected (Lewis^x)$_4$-lactose-sphingosine structure (**27**).

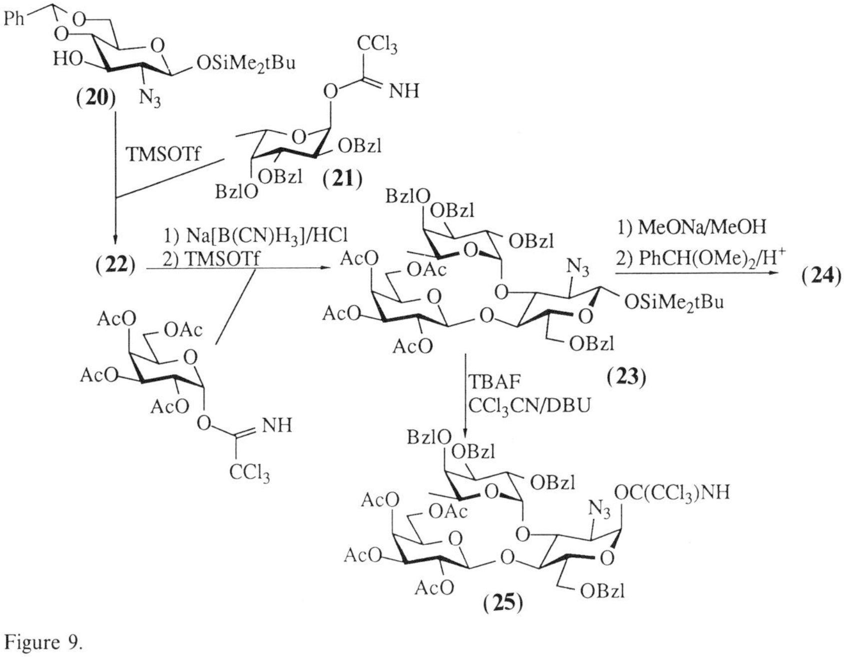

Figure 9.

(24)
+
(25)

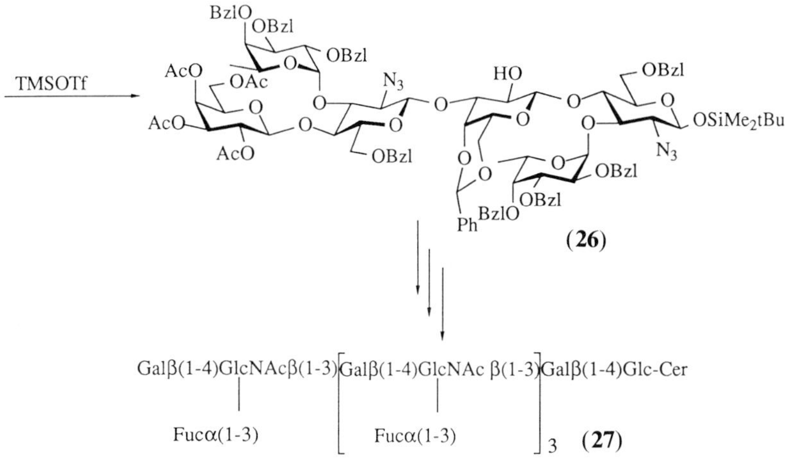

(26)

(27)

Figure 10.

Glycosyl fluoride approach

As mentioned before, sulfated Lewis-type molecules seem to reveal biological properties comparable to the sialylated derivatives (Yuen et al., 1992). Hence, the synthesis of both, sulfated Lewis^x and Lewis^a oligosaccharides has been described (Nicolaou et al., 1993). The employed strategy for sulfated Lewis^x again is based on the fucosylation of a lactosamine unit. The resulting Lewis^x unit (**28**) is equipped with a sulfuric acid function afterwards.

In the reaction sequence, all glycosidic linkages are formed via silverperchlorate/tindichloride-promoted activation of glycosyl fluorides (Mukaiyama et al., 1981).

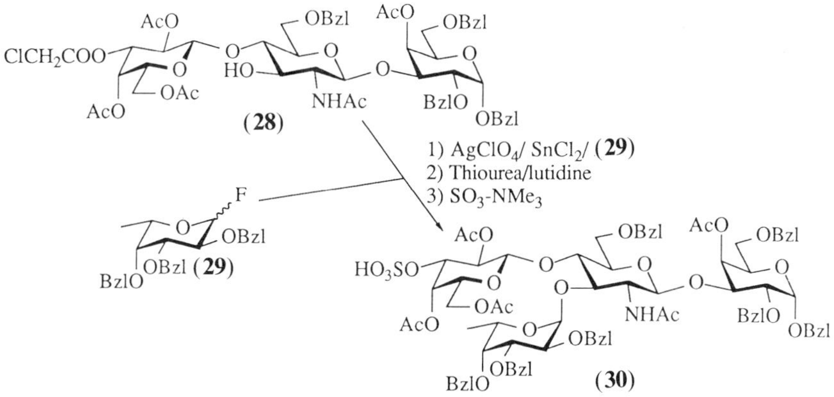

(28)

(29)

(30)

Figure 11.

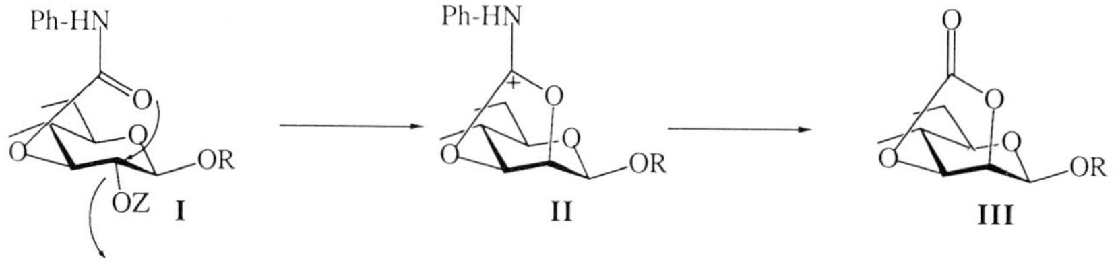

Figure 12.

β-Mannoside-containing side-chains

The construction of the β-mannosidic linkage is difficult for two reasons. First, neighboring group assistance of a 2-O-acyl group leads to a preferred trans-attack while the anomeric effect simultaneously favors the formation of the α-configurated product. In nature, mannose is widely found β-glycosidically linked to chitobiose, thus constituting a crucial part of N-glycoproteins. On that account the synthesis of this problematic linkage has seen increasingly more attempts (Toshima and Tatsuta, 1993).

Earlier investigations took advantage of the possibility to epimerize 2-OH via an oxidation-reduction sequence (Augé et al., 1980). A different concept employs silver-silicate mediated heterogeneous catalysis (Paulsen and Lockhoff, 1981). Both approaches, however, resulted in the formation of significant amounts of undesired anomers or epimers, respectively. To avoid such side-reactions, a new methodology was developed based on an intramolecular S_N2-type epimerization at C-2 of glucose carried out subsequent to the synthesis of β-glucosylglycosides

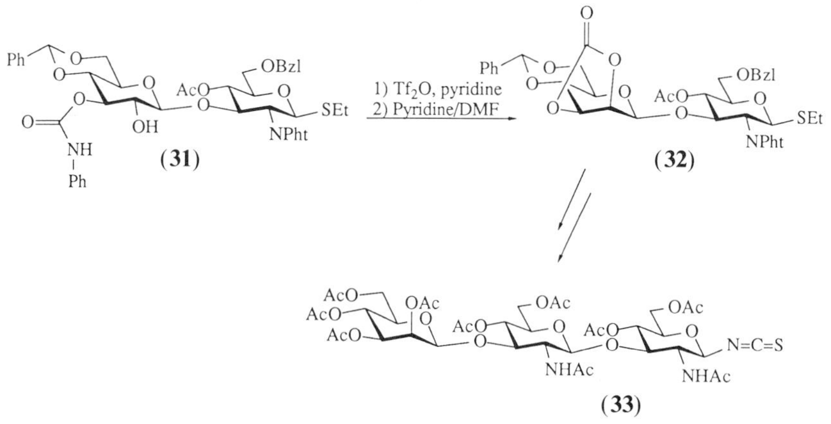

Figure 13.

(Kunz and Günther, 1988). This strategy, outlined in Figure 12, circumvents the repulsive interaction of the ring oxygen lone pair with an external nucleophile attacking axially at C-2. As a suitable neighboring group at C-3, selectively creating the 2,3-cis-diol, the N-phenylcarbamoyl group was chosen.

The efficient applicability of this method could be illustrated by the synthesis of the core-region trisaccharide of N-glycoproteins (Günther and Kunz, 1990). After introduction of the triflate as a leaving group in (**31**), the crude intermediate is converted to the 2,3-cis-carbonate (**32**) at 70°C in DMF/pyridine. This disaccharide is then coupled to a further glucosamine unit. Exchange of the protective groups for acetyl, Lewis acid-promoted oxazoline formation and treatment with potassium isothiocyanide (Khorlin et al., 1980) yield potential N-glycopeptide precursor (**33**).

Attempts to convert the intermediate oxazoline into the corresponding glycosyl azide by treating it with trimethylsilyl azide failed, in contrast to results described in the literature (Nakabayashi et al., 1988).

N-Glycopeptides: Coupling and peptide-chain extension

In the field of N-glycopeptides, the most widespread methodology employs the condensation of glycosylamines with aspartic-acid derivatives to form the N-glycosidic linkage. Two popular methods for introducing the anomeric amino group exist:

Glycosylazides as precursors
As already outlined in the course of the preceding sections, anomeric azides can easily be reduced to the corresponding glycosylamines and subsequently coupled to aspartic acid derivatives. Moreover, the azido-function serves as a useful protective group during saccharide synthesis. The azido group has also been introduced prior to coupling via the oxazoline precursor. Reduction can then be accomplished using neutral washed Raney-Nickel (Kunz et al., 1991) or Lindlar-catalyst (Nakabayashi et al., 1988). Usually, no side-reactions, such as anomerization or migration of acyl-groups, are observed during these transformations.

Reducing sugars as precursors
Glycosylamines are obtained from reducing (oligo-)saccharides after exhaustive treatment with saturated ammonium bicarbonate solution (Likhosherstov et al., 1986). In this way, even the complex heptasaccharide (**34**) can be converted into the corresponding β-configurated glycosylamine (**35**) in high yield (Cohen-Anisfeld and Lansbury, Jr., 1993).

To ensure a high coupling yield, it is important to carefully check the reaction mixture for surplus ammonia and to determine the extent of

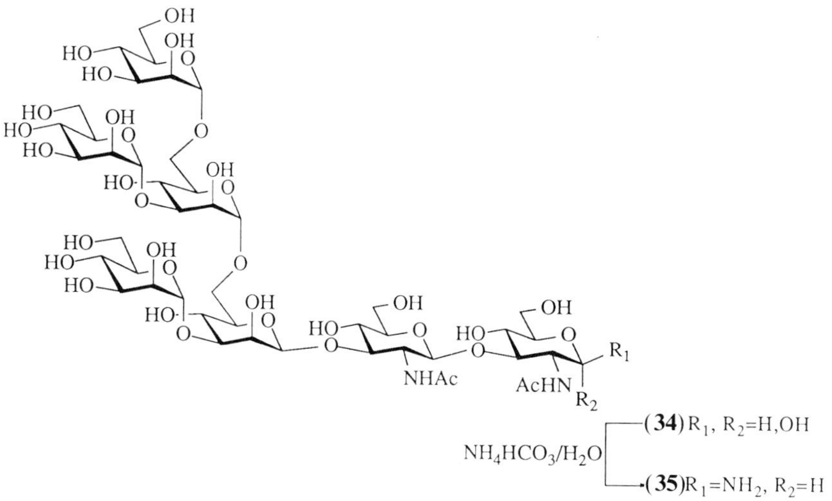

Figure 14.

(34) $R_1, R_2 = H, OH$
(35) $R_1 = NH_2, R_2 = H$
NH_4HCO_3/H_2O

conversion to the desired glycosylamine. Therefore, the reaction has to be followed thoroughly by HPLC analysis especially in the case of precious carbohydrates. If peracetylated sugars are used as starting material, the conversion is considerably less efficient, possibly due to the formation of 1-acetamides.

Peptide coupling and extension in solution

In the case of glycosylamines, reagents applicable for peptide condensation generally are useful for N-glycosylation, too. In this sense utilizing carbodiimides or dihydro-quinoline-derivatives (EEDQ, IIDQ) as condensing reagents in the synthesis of a series of carbohydrate-asparagine-conjugates has been reported. For example, glycosylamine (**36**), representing a protected Lewis[a] derivative, and N-trichloroethyloxycarbonyl/O-allyl-protected aspartic acid (**37**) are coupled in a carbodiimide-promoted reaction in high yield (Kunz and März, 1992). To attain an extended peptide-chain in these molecules, particular attention has to be paid to the selection of an adapted protective group strategy that allows orthogonal, mild deprotection and further elongation in the presence of the sensitive glycosidic linkages (Kunz, 1987).

In this sense, based on the Lewis[a]-Asn block (**38**), chain elongation is achieved in both directions starting with deprotection of the N- or C-terminus (Kunz and März, 1992). Whereas the trichloroethoxycarbonyl group is eliminated using zinc in acetic-acid (Windholz and Johnston, 1967), the allylester can be removed in a palladium-catalyzed reaction under neutral conditions (Kunz and Waldmann, 1984).

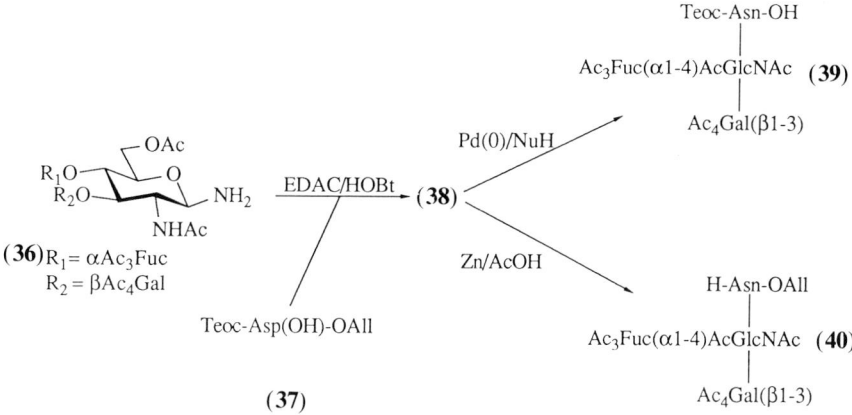

Figure 15.

Fragment condensation employing standard conditions is used for further chain extension. In order to avoid side-reactions and to enhance the solubility in organic solvents hydroxy amino acids are used in their tert-butyl protected forms. To simplify work-up and purification water soluble 1-ethyl-2-(3-dimethylamino-propyl)-carbodiimide (EDAC) serves as condensation reagent in the presence of 1-hydroxy benzotriazole. Full deprotection of glycoconjugate (41) can be accomplished in two steps, since only two types of protective groups, tert-butyl and acetyl, are present after the final coupling step.

Thus, Lewis^a-peptide T conjugate (42) is obtained. Peptide T (Kowalski et al., 1987) is a partial sequence of the heavily glycosylated envelope gp 120 of the HIV I virus. It is presumably involved in the T-cell/virus recognition.

Using the same methodology a glycopeptide carrying two Lewis^x-antigen side-chains has been prepared (v. d. Bruch and Kunz, 1994). As in the previous example the preformed acetyl-protected glycosyl azide (2)

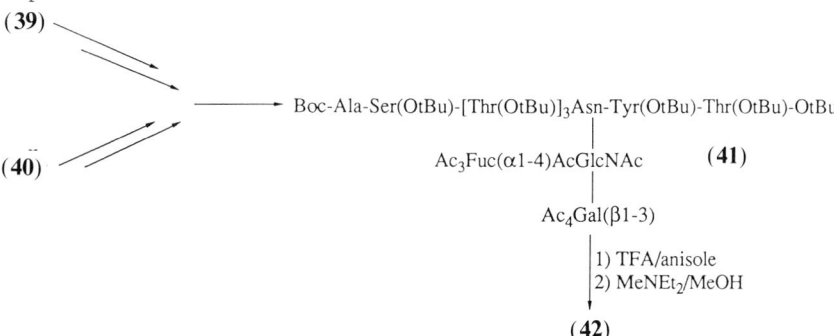

Figure 16.

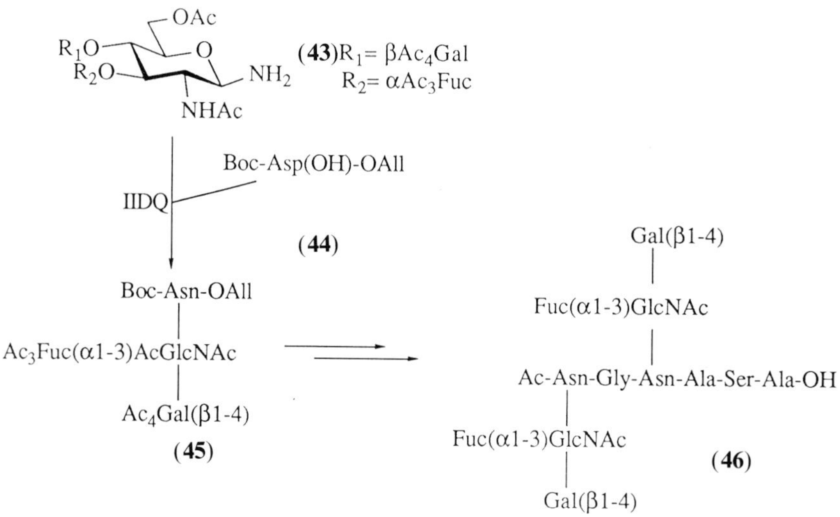

```
            ┌─OAc
    R₁O─────┐
            │─O
    R₂O─────┘         (43) R₁ = βAc₄Gal
            │  NH₂         R₂ = αAc₃Fuc
          NHAc
```

```
                Boc-Asp(OH)-OAll
         IIDQ ├──────────────
                    (44)                    Gal(β1-4)
                                                │
         Boc-Asn-OAll                      Fuc(α1-3)GlcNAc
              │                                 │
Ac₃Fuc(α1-3)AcGlcNAc    ─────→         Ac-Asn-Gly-Asn-Ala-Ser-Ala-OH
              │                                 │
         Ac₄Gal(β1-4)                    Fuc(α1-3)GlcNAc
             (45)                               │                  (46)
                                          Gal(β1-4)
```

Figure 17.

is converted into the corresponding amine (43) and subsequently condensed with a tert-butyloxycarbonyl/allyl protected aspartic-acid derivative utilizing the mixed-anhydride method (IIDQ).

The glycosylasparagine conjugate (45) illustrates the acid-stability of acyl-protected carbohydrates (Kunz and Unverzagt, 1988), since the N-terminus can be deprotected without affecting the other parts of the molecule by using HCl in diethylether. C-terminal deprotection can be achieved by a Rh(I) catalyzed cleavage of the allyl ester (Waldmann and Kunz, 1983). After introduction of a glycine-residue as spacer molecule, both Lewis^x-carrying parts are condensed employing water soluble carbodiimide. Considering the bulky structure, the diglycosylated tripeptide is obtained in fairly high yield. Then, another tripeptide-spacer is attached to the C-terminus and the resulting (Lewis^x)₂-hexapeptide is deblocked in two steps. Finally, the obtained hapten (46) is coupled to ε-amino groups of lysine-residues of bovine-serum albumin (BSA) or keyhole limpet hemocyanin (KLH) to furnish immunoactive Le^x antigens.

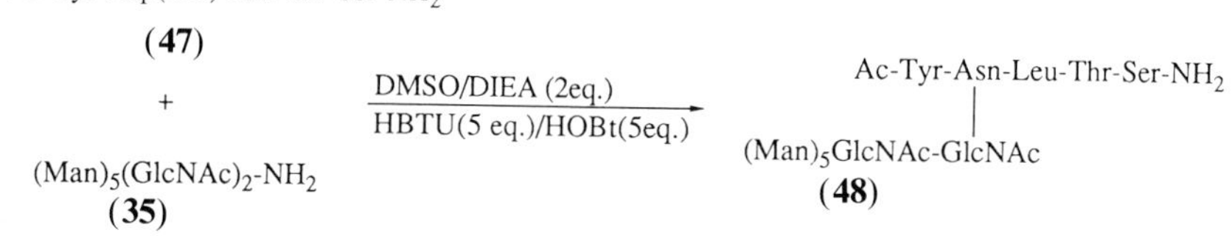

```
Ac-Tyr-Asp(OH)-Leu-Thr-Ser-NH₂
         (47)
                                                        Ac-Tyr-Asn-Leu-Thr-Ser-NH₂
           +              DMSO/DIEA (2eq.)                       │
                       ──────────────────→
                       HBTU(5 eq.)/HOBt(5eq.)           (Man)₅GlcNAc-GlcNAc
(Man)₅(GlcNAc)₂-NH₂                                            (48)
         (35)
```

Figure 18.

The presented concept of a simulated biantennary neo-glycoprotein considers the binding-site-structure of antibodies and related receptors and has previously been reported to result in significantly enhanced antigenicity (Lee and Lee, 1987).

Just recently, it has been demonstrated that complex N-glycopeptides with an extended peptide-chain are directly accessible by condensing peptides containing aspartic-acid residues unblocked in the side-chain and deprotected glycosylamines (Cohen-Anisfeld and Lansbury, Jr., 1993). Suitably protected peptides can advantageously be pre-synthesized separately on solid-phase.

The best results are obtained when a mixture of 5 eq. of uronium-salt (HBTU)/HOBt (Dourtoglou et al., 1978) in combination with 2 eq. of diisopropylethylamine is applied using neat DMSO as solvent and 1 eq. of both glycosylamine and peptide. Due to the lability of partially deprotected conjugates like (**48**) towards acids, special attention has to be paid to this point in the final deprotection steps.

Chain-extension on solid-phase

There is no principal obstacle to the employment of preformed N- or O-glycosylated amino acid conjugates as building blocks in solid phase synthesis. However, in the presence of sensitive glycosidic bonds any methodology that utilizes strong acids in the final cleavage of the peptide-resin linkage is not applicable.

To meet these preconditions, a new solid-phase anchoring principle especially adapted to glycopeptide chemistry has been developed (Kunz et al., 1988; Kunz and Dombo, 1988). Instead of the common acid-labile alkoxy-benzyl ester type peptide polymer linkage, an allylic ester anchor is introduced. This methodology takes advantage of the practically neutral conditions prevailing during the Pd(0)-catalyzed allyl ester cleavage (Kunz and Waldmann, 1984). Thus, neither the glycosidic linkages nor acid-labile (tert-butyl-type) protective groups are affected while the product is detached from the resin. Moreover, protected glycopeptides show improved properties during purification procedures.

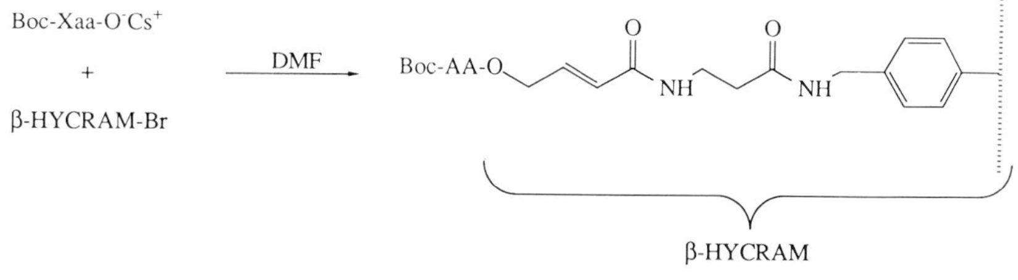

Figure 19.

To allow a convenient amino acid analysis during solid-phase synthesis as well, β-alanine has been inserted between the amino methyl resin and the hydroxy crotonic acid residue (Kosch et al., 1992). The starting amino-acid is coupled to the allyl halide side chain of the polymer via the corresponding cesium salt.

The merits of the allylic anchor are illustrated by the synthesis of the fully deprotected lactosamine peptide T conjugate (**50**) on a gram scale using Fmoc amino protection (Kunz et al., 1992; Kosch et al., 1994).

Again, the final coupling is carried out using Boc protected alanine to simplify deprotection. The glycopeptide in (**49**) is removed from the resin by a palladium(0)-catalyzed allyl transfer to N-methylaniline without affecting any of the numerous sensitive linkages.

From these results it can be concluded that the application of glycosylated amino acids as building blocks in solid-phase-peptide synthesis should probably be an efficient strategy for the construction of extended glycopeptides with a variety of different sequences. In fact, such multiple column syntheses, already established in peptide chemistry, have recently been carried out in the field of glycopeptides (Peters et al., 1992). The underlying methodology takes advantage of the well-known reactivity of amino acid pentafluorophenylesters (Meldal and Jensen, 1990). Thus, preformed Fmoc-Ser-α-(Ac$_3$GalNAc)OPfp and Fmoc-Thr-α-(Ac$_3$GalNAc)OPfp were utilized as building-blocks in solid phase synthesis employing highly acid-labile linkers. In this way, 40 different O-glycopeptides from human intestinal mucin and porcine submaxillary mucin have been synthesized. Moreover, the compatibility of these methods with sensitive saccharide structures has been demonstrated in the synthesis of the trisaccharide-cyclized-hexapeptide (**52**) from blood-clotting factor IX (Reimer et al., 1993).

However, concerning the obtained yields in the latter example (Fig. 21) it turns out that the use of larger unprotected saccharide blocks

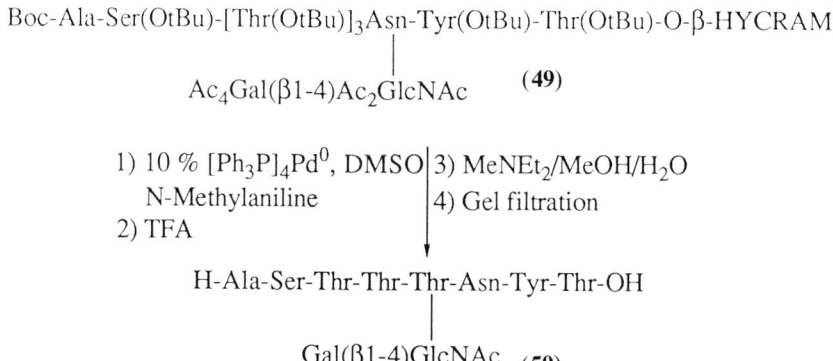

Figure 20.

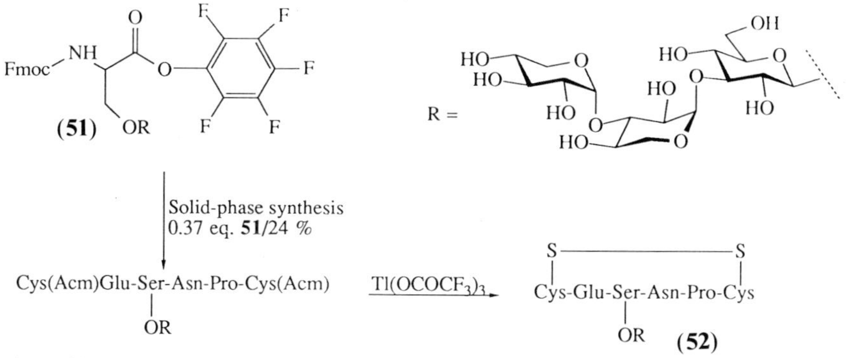

Figure 21.

leads to difficulties in the preparation of Pfp-esters like (**51**) as well as in the coupling reactions. According to these results, the use of acylated saccharides for such purposes seems to be more favorable.

Recent progress in the formation of 3-O-glycopeptides of serine or threonine

Glycosylation procedures which have been proved to be efficient in the formation of intersaccharidic bonds in principle should also be applicable for the glycosylation of appropriately protected serine/threonine building blocks. However, only a few methodologies comply with existing limitations and requirements in O-glycopeptide synthesis.

Trichloroacetimidates
The potential of this method (Schmidt and Michel, 1980) has recently been demonstrated by the synthesis of several complex glycopeptides in the series of proteoglycans (Goto and Ogawa, 1992, 1993). Subsequent to a sophisticated saccharide construction, the anomeric methoxybenzyl ether in (**53**) is cleaved by oxidation and the liberated hydroxy function is converted to the trichloroacetimidate (**54**). Coupling to N-benzyloxy-carbonyl serine benzyl ester is then successfully accomplished in a boron-trifluoride promoted reaction.

Cleavage of the levuoyl esters (OLev) is selectively achieved with hydrazine-acetate and sulfatation is accomplished with the triethylamine-sulfur trioxide complex. Finally, deprotection yields hexasaccharide serine conjugate (**56**).

Allylcarbamates
Despite those exceptional results, it should be noted that the utilized amino acid protective groups would not allow a chain extension in the

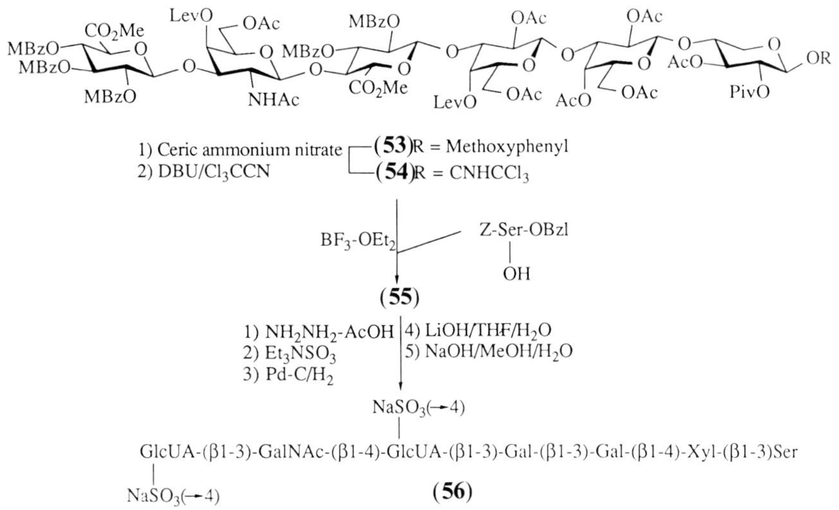

Figure 22.

peptide part. Rather, the use of such combinations as Fmoc/OtBu, Aloc/OtBu, is required for this purpose. However, tert-butyl type protections are usually not stable towards strong Lewis-acids employed for the activation of trichloroacetimidates or glycosyl fluorides. Thus, alternative glycosylation methods are desirable.

In this sense, the electrophile-induced activation of the glycosyl allyl carbamates facilitates a mild and selective glycosylation procedure (Kunz and Zimmer, 1993). The donor compounds, either acyl or ether protected, are readily prepared from anomerically deprotected saccharides and allyl isocyanate. Activation of the double-bond efficiently

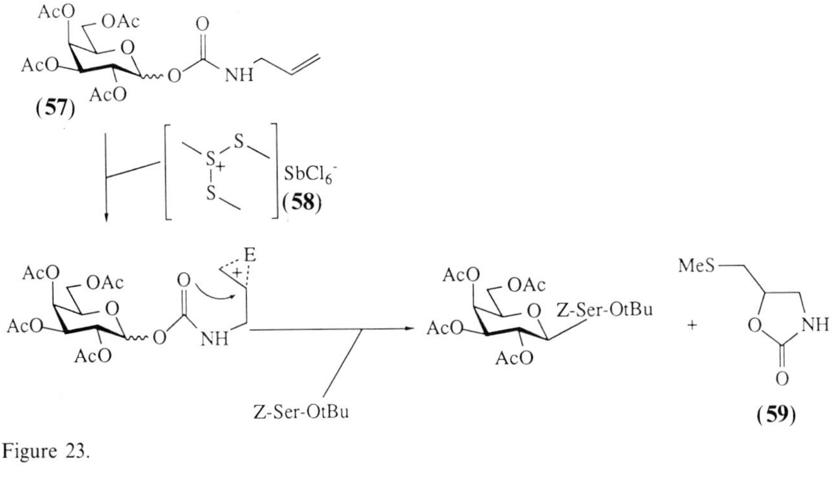

Figure 23.

proceeds if the allyl carbamates like (**57**) are treated with the soft electrophile (E) (**58**). Cyclisation of the five-membered carbamate is followed by nucleophilic attack of the serine/threonine derivative as the acceptor. The isolation of the cyclization product (**59**) strongly supports the suggested mechanism.

Several other glycosyl acceptors have successfully been reacted with acyl- and benzyl-protected allyl carbamates demonstrating the accessibility of α- or β-configurated glycosides by this procedure.

Thioglycosides
1-Thioalkyl glycosides are well-established donors for glycopeptide synthesis, as was demonstrated in the preparation of trisaccharide serine derivative (**60**) (Paulsen et al., 1988).

In this case the reaction is carried out in the presence of dimethylmethylthiosulfonium triflate (DMTST) as promotor. In another example, thioalkyl glycoside (**61**) serves as precursor for the corresponding bromo sugar which can be coupled to a serine residue in a silver triflate-promoted reaction (Nakahara et al., 1990).

Since the conversion of a complex saccharide structure into the corresponding thioglycoside prior to glycosylation can turn out to be problematic, the thioalkyl group usually has to serve as the anomeric protection throughout saccharide synthesis. However, its stability with

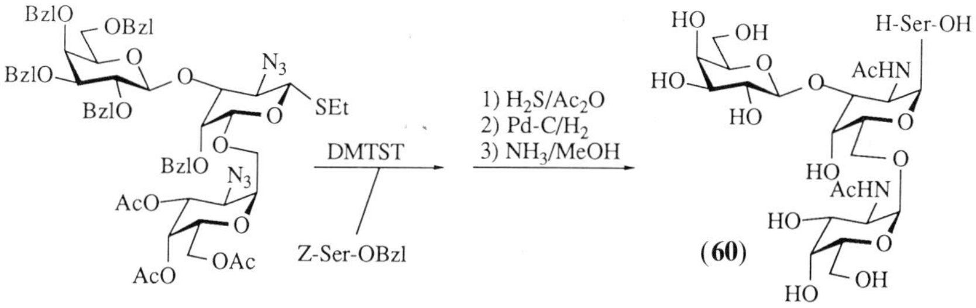

Figure 24.

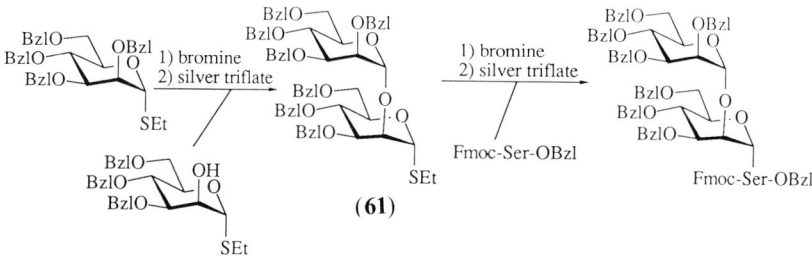

Figure 25.

Enzymes as tools in glycopeptide synthesis

Since enzyme catalyzed reactions typically proceed in a mild and selective fashion, their application in the synthesis of glycoconjugates is considered promising and became a rapidly growing area in recent years (Ichikawa et al., 1992). Meanwhile, almost all types of linkages occurring in glycopeptides have been synthesized by the *in vitro* utilization of enzymes. Moreover, an enzymatic deprotection methodology based on hydrolases has been developed and successfully applied in the synthesis of complex glycopeptides.

Enzymatic deprotection of glycopeptides

For a selective C- or N-terminal deprotection of glycopeptides conditions outside a certain pH-range have to be avoided. Otherwise, in many cases anomerization and/or cleavage of the glycosidic bond in acidic media as well as β-elimination and racemization under basic conditions are observed. Therefore, hydrolases have been recognized to be ideal tools for the *in vitro* cleavage of protective groups applied in glycopeptide chemistry.

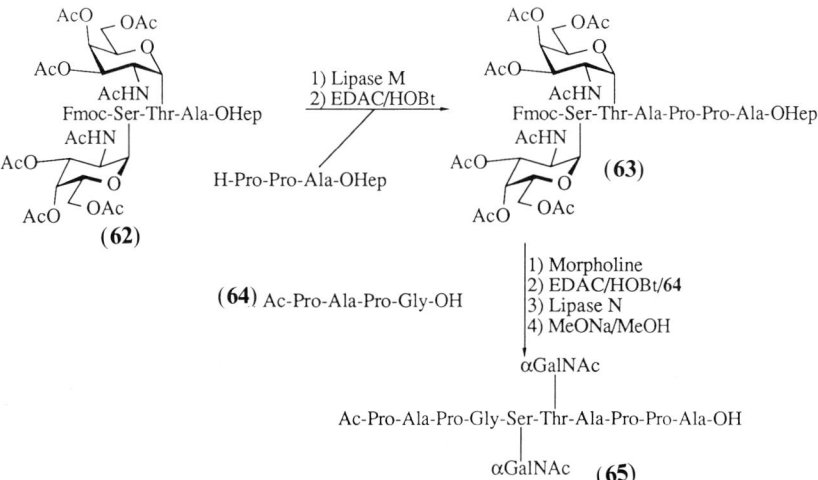

Figure 26.

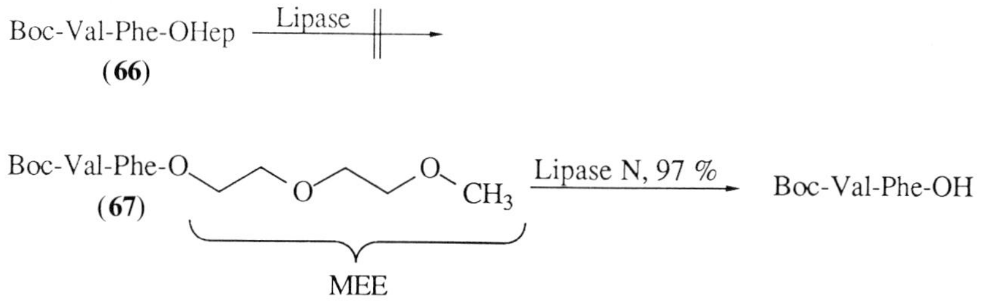

Boc-Val-Phe-OHep
(66)
Lipase
Boc-Val-Phe-O
(67)
O
O
CH3
MEE
Lipase N, 97 %
Boc-Val-Phe-OH

Figure 27.

Lipases for C-terminal deprotection
Since *in vivo* lipases hydrolyze triglycerides containing long fatty acid chains, it has been suggested that amino acid esters with higher alkyl chains should be preferred substrates for these easily accessible and inexpensive biocatalysts. In fact, the corresponding heptylesters of a series of dipeptides are efficiently cleaved after treatment with lipases at pH 7 and 37°C (Braun et al., 1990). Furthermore, after a thorough screening of available lipases the method has successfully been extended to the C-terminal deprotection of O-glycopeptides, e.g., the structural element of the tumor associated-T_n-antigen (**62**) (Braun et al., 1993).

A consequent application of the lipase-catalyzed removal of heptylesters finally resulted in the synthesis of the doubly glycosylated decapeptide (**65**), representing a partial sequence of the MUC I repeating unit (Gendler et al., 1988; Kunz et al., unpublished). None of the peptide bonds nor the acyl-protections of the carbohydrate residue have been affected during the enzyme-catalyzed steps.

However, the substrate activity can be further enhanced by introducing glycol-type esters as substrates (Kunz et al., 1994). Surprisingly, and in contradiction to current opinion on lipase action, the hydrophilic 2-[(2-methoxy)ethoxy]ethyl (MEE)-esters proved to be better substrates compared to n-alkyl esters. This can be illustrated by the hydrolysis of dipeptide esters (**66**) and (**67**) (Fig. 27).

The suitability of the MEE esters in the field of glycopeptide chemistry has been demonstrated by the cleavage of MEE ester (**68**) after treatment with lipase M from *mucor javanicus*.

Aloc-Ser-Val-Phe-OMEE
AcHN
AcO
O
AcO
OAc
(68)
Lipase M
Aloc-Ser-Val-Phe-OH
AcHN
AcO
O
AcO
OAc
(69)

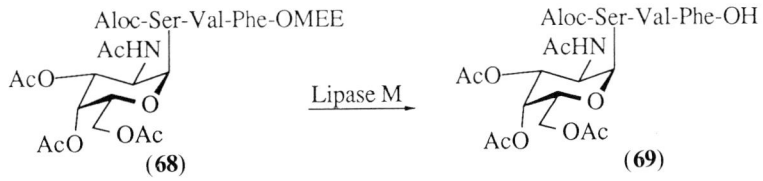

Figure 28.

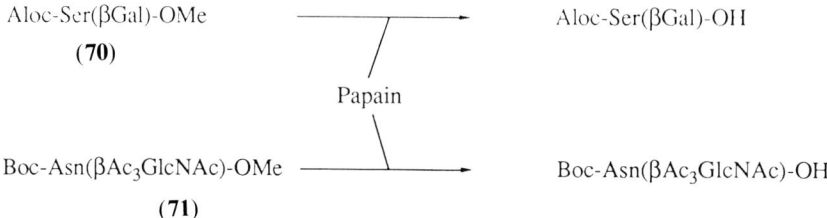

Figure 29.

Once again, only the intended reaction took place, exclusively leaving the residual linkages unaffected. The reasons for the significant differences in the reactivity of alkyl versus glycol esters is not yet completely understood. It should be noted though that the latter reveal a distinctly increased solubility in the employed aqueous buffer system.

Proteases for C-terminal deprotection
Although the protease-catalyzed cleavage of amino-acid esters obviously remains restricted to compounds without peptide-bonds, biocatalysts such as papain have been shown to be valuable tools for the synthesis of C-terminal deprotected amino acid glycosides (Ishii et al., 1990; Cantacuzène et al., 1991). Both N- and O-glycopeptide methyl esters (**70**) and (**71**) have been transformed into the corresponding acids without the emergence of any undesired by-product.

Glycopeptide chain elongation using proteases

Apart from basic research and some industrial applications, the use of proteases as C-N ligases to form peptides has not yet been applied to greater extent. Though inexpensive and easy to handle, proteases suffer from the major drawback that they can only serve for peptide synthesis if the unfavorable equilibrium is shifted by kinetic and thermodynamic manipulations. However, recently it was shown that a gene-technologically transformed thiosubtilisin was able to efficiently catalyze the formation of glycopeptide amides like (**72**) in aqueous solution (Wong et al., 1993).

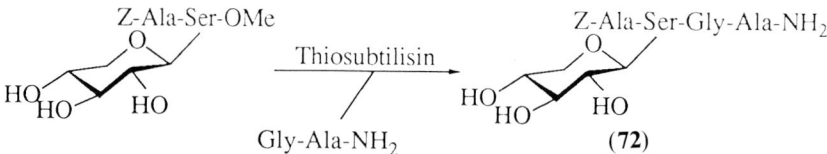

Figure 30.

In this case, site-directed mutagenesis led to subtilisin variants with clearly enhanced stability in aqueous or organic solvents compared to the wild-type protein.

Saccharide chain elongation in glycopeptides using glycosyltransferases

Some of the examples presented in the previous sections illustrate how laborious saccharide synthesis can be. To circumvent the underlying multi-step procedures, enzymes for the regio- and stereoselective construction of glycosidic linkages have been applied with increasing intensity during the past decade.

Glycosyltransferases are the natural and perfectly designed tools for the synthesis of the variety of glycosidic-linkages occurring. Hence, glycosyltransferases should hold a great potential for the *in vitro* synthesis of glycoconjugates. In fact, the regio- and stereoselective formation of saccharide structures in oligosaccharides, glycolipids and glycopeptides by the use of glycosyltransferases has been reported (Ichikawa et al., 1992).

The methodology becomes particularly competitive in the field of complex saccharide structures containing linkages, which are demanding in chemical construction, e.g., sialosides (Ito et al., 1993). In this sense, synthesis of sialyl lactosamine asparagine derivative (**74**) has been conducted by successive application of β-1,4-galactosyl- and α-2,6-sialyltransferase (Augé et al., 1989). During the galactosylation process regeneration of UDP-Gal was accomplished by a system of immobilized enzymes. Subsequent sialylation of the intermediate lactosamine derivative was attained using sialyltransferase and neuraminic acid cytidyl monophosphate.

The enzymatic approach to N-glycopeptides of type (**74**) is distinctly improved if both enzymatic glycosylations are performed in a "one-pot" procedure without regeneration of the nucleotide sugars (Unverzagt et al., 1990). In the latter strategy, the addition of alkaline phosphatase simultaneously serves to destroy potentially inhibiting UDP or CMP. If a chitobiose-asparagine derivative is reacted with galactosyltransferase the corresponding trisaccharide can be obtained (Thiem and Wiemann, 1990).

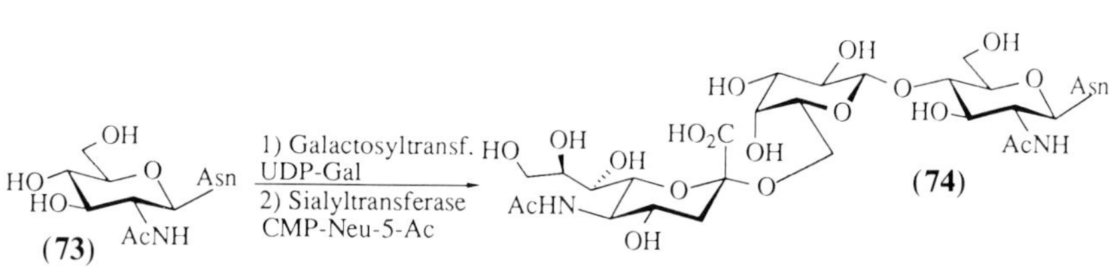

Figure 31.

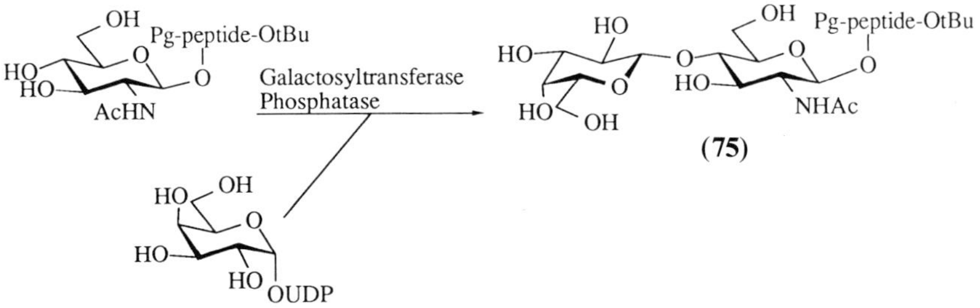

Figure 32.

While the selectivity of glycosyltransferases is a synthetically useful phenomenon, their substrate specificity usually constitutes a limitation. However, in case of galactosyltransferase several acceptor substrates have been shown to be useful for preparative application (Wong et al., 1991). Among those are O-glycopeptides derived from N-acetylglucosamine (Schultz and Kunz, 1993). The synthetic availability of such structures has been improved by utilizing a 1-thio-2-carbamoyl glucosamine donor. Subsequent to protective group exchange and removal of O-acetyl groups, O-glycopeptides of type (75) can be obtained by the galactosyltransferase-catalyzed elongation of the saccharide chain.

In a comparable conversion, a xylose-serine conjugate has been shown to be accepted as substrate (Wong et al., 1993).

Glycosyltransferases have just recently been shown to act on an adapted solid-phase substrate (Schuster et al., 1994). In the presented example the successive application of galactosyl- sialyl-, and fucosyl

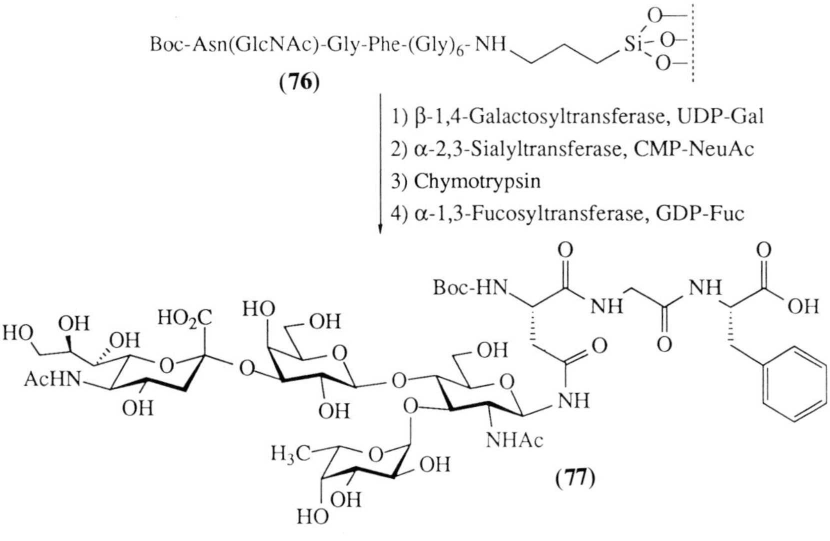

Figure 33.

transferase leads to the formation of sialyl Lewis^x glycopeptide (**77**). While galactosylation and sialylation have been performed on a resin-linked Boc-Asn(GlcNAc)-Phe-(Gly)$_6$ conjugate (**76**), the final fucosylation has been accomplished in solution subsequent to the chymotrypsin-catalyzed cleavage of the phenylalanine glycine bond.

According to the presented results the application of glycosyltransferases is likely to remain a rapidly growing area in the future.

Conclusion

The progress accomplished in the field of glycopeptide chemistry provides selective synthetic methods which allow access to very complex glycopeptide structures on preparative scale. Methods like solid phase synthesis and the use of enzymes facilitate access to a great variety of biologically interesting glycopeptides. However, the overall effort required for the synthesis of oligosaccharide peptide conjugates is still immense. Especially oligosaccharide synthesis constitutes a laborious task and a limitation of the availablity of glycoconjugates. Therefore, further methodological developments are considered necessary in order to improve the preparation of glycopeptides.

References

Augé, C., Warren, C.D., Jeanloz, R.W., Kiso, M. and Anderson, L. (1980) The synthesis of O-β-D-Mannopyranosyl-(1- > 4)-O-(2-acetamido-2-deoxy-β-D-glucopyranosyl)-(1- > 4)-2-acetamido-2-deoxy-D-glucopyranose. *Carbohydr. Res.* 82: 71–95.

Augé, C., Gautheron, C. and Pora, H. (1989) Enzymatic synthesis of a sialylglycopeptide, α-NeupAc(2- > 6)-β-D-Galp(1- > 4)-β-D-GlcpNAc(1- > 4N)-L-Asn. *Carbohydr. Res.* 193: 288–293.

Bommer, R., Kinzy, W. and Schmidt, R.R. (1991) Synthesis of the octasaccharide moiety of the dimeric Le^x antigen. *Liebigs Ann. Chem.* 425–433.

Braun, P., Waldmann, H., Vogt, W. and Kunz, H. (1990) Selective enzymatic removal of protecting functions: Heptyl esters as carboxy protecting groups in peptide synthesis. *Synlett* 37–38.

Braun, P., Waldmann, H. and Kunz, H. (1993) Chemoenzymatic synthesis of O-glycopeptides carrying the tumor associated T$_N$-antigen structure. *Bioorg. Med. Chem.* 1: 197–207.

Bröder, W. and Kunz, H. (1990) Glycoside and saccharide synthesis using N-glycosyl triazoles as hydrolytically stable glycosyl donors. *Synlett* 251–252.

Bröder, W. (1992) *Ein neues anomeres Schutz- und Aktivierungsverfahren unter Verwendung von Glycosylaziden in der Synthese von neo-lacto-Oligosacchariden und trimerem Lewis-X-Antigen.* Dissertation, Universität Mainz.

Bröder, W. and Kunz, H. (1993) A new method of anomeric protection and activation based on the conversion of glycosyl azides to glycosyl fluorides. *Carbohydr. Res.* 249: 221–241.

Cohen-Anisfeld, S.T. and Lansbury, P.T. (1993) A practical, convergent method for glycopeptide synthesis. *J. Am. Chem. Soc.* 115: 10531–10537.

Dourtoglou, V., Ziegler, J.C. and Gross, B. (1978) O-Benzotriazolyl-N,N-tetramethyluronium hexafluorophosphate: A new and effective reagent for peptide coupling. *Tetrahedron Lett.* 1269.

Garegg, P.J., Hultberg, H. and Wallin, S. (1982) A novel ring-opening of carbohydrate benzylidene acetals. *Carbohydr. Res.* 108: 97–101.

Gendler, S., Taylor-Papadimitriou, J., Duhig, T., Rothbard, J. and Burchell, J. (1988) A highly immunogenic region of human polymorphic epithelial mucin expressed by carcinomas is made up of tandem repeats. *J. Biol. Chem.* 263: 12820–12823.

Goto, F. and Ogawa, T. (1992) Synthesis of a sulfated glycopeptide corresponding to the carbohydrate-protein linkage region of proteoglycans. *Tetrahedron Lett.* 33: 5099–5102.

Goto, F. and Ogawa, T. (1993) Recent aspects of glycoconjugate synthesis: A synthetic approach to the linkage region of proteoglycans. *Pure and Appl. Chem.* 65: 793–801.

Günther, W. and Kunz, H. (1990) Synthesis of a β-mannosyl chitobiosyl asparagine conjugate – a central element of the core region of N-glycoproteins. *Angew. Chem. Int. Ed. Engl.* 29: 1068–1069.

Hakomori, S.-I. (1993) Carbohydrates as adhesion molecules. *New Chem. Trans.* 21: 583–595.

Harmon, R.E., Earl, R.A. and Gupta, S.K. (1971) 1,3-Dipolar additions of glycosyl azides to substituted acetylenes. *J. Org. Chem.* 36: 2552–2556.

Hart, G.W., Haltiwanger, R.S., Holt, G.D. and Kelly, W.G. (1989) Nucleoplasmic and cytoplasmic glycoproteins. *Annu. Rev. Biochem.* 58: 785–838.

Ichikawa, Y., Look, G.C. and Wong, C.-H. (1992) *Analyt. Biochem.* 202: 215–238.

Ishii, H., Funabashi, K., Mimura, Y. and Inoue, Y. (1990) Papain-catalyzed hydrolysis of N-protected glycosylated amino acid ester. *Bull. Chem. Soc. Jpn.* 63: 3042–3043.

Ito, Y., Gaudino, J.J. and Paulson, J.C. (1993) Synthesis of bioactive sialosides. *Pure and Appl. Chem.* 65: 75–762.

Itzkowitz, S.H., Yuan, M., Fukushi, Y., Palekar, A., Phelps, P.C., Shamsuddin, A.M., Trump, B.F., Hakomori, S.-I. and Kim, Y.S. (1986) Lewis[x] and sialylated Lewis[x] related antigen expression in human malignant and nonmalignant colonic tissues. *Cancer Res.* 46: 2627–2632.

Jacquinet, J.-C. and Sinay, P. (1977) Synthesis of blood group substances. 6. Synthesis of O-α-L-fucopyranosyl-(1 → 2)-O-β-galactopyranosyl-1(1 → 4)-O-[α-L-fucopyranosyl-(1 → 3)]-2-acetamido-2-deoxyglucopyranose – the postulated Le^d antigenic determinant. *J. Org. Chem.* 42: 720–724.

Johansson, R. and Samuelsson, B. (1984) Regioselective reductive ring-opening of 4-methoxybenzylidene acetals of hexopyranosides. Access to a novel protecting-group strategy. Part 1. *J. Chem. Soc. Perkin. Trans. I* 2371–2374.

Khorlin, A.Y., Zurabyan, S.E. and Macharadze, R.G. (1980) Synthesis of glycosylamides and 4-N-glycosyl-L-asparagine derivatives. *Carbohydr. Res.* 85: 201–208.

Kluge, A.F., Untch, K.G. and Fried, J.H. (1972) Synthesis of prostaglandin models and prostaglandins by conjugate addition of a functionalized organo-copper reagent. *J. Am. Chem. Soc.* 94: 7827–7832.

Kosch, W., März, J. and Kunz, H. (1994) Synthesis of glycopeptide derivatives of peptide T on a solid phase using an allylic linkage. *Reactive Polymers* 22: 181–194.

Kowalski, M., Potz, J., Basiripour, L., Dorfman, T., Goh, W.C., Terwilliger, E., Dayton, A., Rosen, C., Haseltine, W. and Sodroski, J. (1987) Functional regions of the envelope glycoprotein of human immunodeficiency virus type I. *Science* 237: 1351–1355.

Kunz, H. and Waldmann, H. (1984) The allyl group as mild and selective cleavable carboxy protective group for the synthesis of sensitive O-glycopeptides. *Angew. Chem. Int. Ed. Engl.* 23: 71–72.

Kunz, H. and Waldmann, H. (1985) Construction of N-glycopeptide disaccharides. Synthesis of the linkage region of the transmembrane neuraminidase of an influenza virus. *Angew. Chem. Int. Ed. Engl.* 24: 883–885.

Kunz, H. (1987) Synthesis of glycopeptides – partial structures of biological recognition components. *Angew. Chem. Int. Ed. Engl.* 25: 294–308.

Kunz, H. and Dombo, B. (1988) Solid-phase synthesis of peptides and glycopeptides on polymeric supports carrying allylic anchoring groups. *Angew. Chem. Int. Ed. Engl.* 27: 711–713.

Kunz, H. and Günther, W. (1988) β-Mannoside synthesis via inversion of configuration on β-glucosides by intramolecular nucleophilic substitution. *Angew. Chem. Int. Ed. Engl.* 29: 1068–1069.

Kunz, H. and Unverzagt, C. (1988) Protective group dependent stability of intersaccharidic bonds. Synthesis of fucosylchitobiose-glycopeptides. *Angew. Chem. Int. Ed. Engl.* 27: 1697–1699.

Kunz, H., Dombo, B. and Kosch, W. (1988) Solid phase synthesis of peptides and glycopeptides on resins with allylic anchoring groups. *Proc. Eur. Peptide Symp.* 20: 154–156.

Kunz, H., Waldmann, H. and März, J. (1989) Synthese von N-Glycopeptid-Partialstrukturen der Verknüpfungsregion sowohl der Transmembran-Neuramidase eines Influenza-Virus als auch des Faktors B des menschlichen Komplementsystems. *Liebigs Ann. Chem.* 45–59.

Kunz, H., Pfrengle, W., Rück, K. and Sager, W. (1991) Stereoselective synthesis of L-amino acids via strecker and Ugi reactions on carbohydrate templates. *Synthesis* 1039–1042.

Kunz, H. and März, J. (1992) Synthesis of selectively deprotectable asparagine glycoconjugates with a Lewis[a] antigen side chain. *Synlett* 589–593.

Kunz, H., Kosch, W. and März, J. (1992) Solid phase synthesis on polymeric support with allylic anchoring groups. *In*: J.A. Smith and J.E. Rivier (eds): *Peptides*, Escom. Leiden, pp 502–504.

Kunz, H. and Zimmer, J. (1993) Glycoside synthesis via electrophile-induced activation of N-allyl carbamates. *Tetrahedron Lett*. 34: 2907–2910.

Kunz, H., Kowalczyk, D., Braun, P. and Braum, G. (1994) Enzymatic hydrolysis of hydrophilic diethylenglycol and polyethylenglycol esters of peptides and glycopeptides by lipases. *Angew. Chem. Int. Ed. Engl.* 33: 336–339.

Lasky, L.A. (1992) An endothelial ligand for L-selectin is a novel mucin-like molecule. *Science* 258: 964.

Lee, R.T. and Lee, Y.C. (1987) Preparation of cluster glycosides of N-acetylgalactosamine that have subnanomolar binding constants towards the mammalian hepatic Gal/GalNAc specific receptor. *Glyconjugate J*. 4: 317.

Lemieux, R.U. and Haymi, J.I. (1965) The mechanism of the anomerization of the tetra-O-acetyl-D-glucopyranosyl chloride. *Can. J. Chem.* 43: 2162–2173.

Likhosherstov, L.M., Novikova, O.S., Derevitskaja, V.A. and Kochetkov, N.K. (1986) A new simple synthesis of amino-sugar β-D-glycosylamines. *Carbohydr. Res*. 146: C1–C5.

Lis, H. and Sharon, N. (1993) Protein glycosylation: Structural and functional aspects. *Eur. J. Biochem*. 218: 1–27.

Lönn, H. (1985) Synthesis of a tetra- and a nonosaccharide which contain α-fucopyranosyl groups and are part of the complex type of the carbohydrate moiety of glycoproteins. *Carbohydr. Res*. 139: 105–113.

Meldal, J. and Jensen, K.J. (1990) Pentafluorophenyl esters for the temporary protection of the α-carboxy group in solid-phase glycopeptide synthesis. *J. Chem. Soc., Chem. Commun*. 483.

Mukaiyama, T., Muria, Y. and Shoda, S. (1981) An efficient method for the glycosylation of hydroxy compounds using glycopyranosyl fluoride. *Chem. Lett*. 431.

Nakabayashi, S., Warren, C.D. and Jeanloz, R.W. (1988) Amino sugars 135. The preparation of a partially protected heptasaccharide-asparagine intermediate for glycopeptide synthesis. *Carbohydr. Res*. 174: 219–289.

Nakahara, Y., Ijima, H., Sibayama, S. and Ogawa, T. (1990) A highly stereoselective synthesis of di- and trimeric sialosyl-T_n epitope: A partial structure of glycophorin A[1]. *Tetrahedron Lett*. 31: 6897–6900.

Nicolaou, K.C., Bockovich, N.J. and Carcanague, D.R. (1993) Total synthesis of sulfated Le[x] and Le[a]-type oligosaccharide selectin ligands. *J. Am. Chem. Soc.* 115: 8843–8844.

Nilsson, M. and Norberg, T. (1988) Synthesis of a dimeric Lewis[x]-hexasaccharide derivative corresponding to a tumor-associated glycolipid. *Carbohydr. Res*. 183: 71–82.

Nilsson, S., Lönn, H. and Norberg, T. (1991) Synthesis of two-tumor-associated oligosaccharides: Di- and trifucosylated para-lacto-N-hexaose. *Glycoconj. J.* 8: 9–15.

Ogawa, T., Nakabayashi, S. and Kitajima, T. (1983) Synthetic studies on cell surface glycans. Part XXII. Synthesis of a hexasaccharide unit of a complex type of glycan chain of a glycoprotein. *Carbohydr. Res*. 144: 225–236.

Oltyoort, J.J., van Boeckel C.A.A., de Koning, J.H. and van Boom, J.J. (1981) Use of the cationic iridium complex 1,5-cyclooctadienebis[methyldiphenylphosphine] iridium hexafluorophosphate in carbohydrate chemistry: Smooth isomerization of allyl ethers to 1-propenyl ethers. *Synthesis* 305–308.

Paulsen, H. and Lockhoff. O. (1981) Neue effektive β-Glycosidsynthese für Mannose-Glycoside. Synthesen von Mannose-haltigen Oligosacchariden. *Chem. Ber.* 114: 3102–3114.

Paulsen, H., Rauwald, W. and Weichert, U. (1988) Building units of oligosaccharides. LXXXVI. Glycosydation of oligosaccharide thioglycosides to O-glycoprotein segments. *Liebigs Ann. Chem.* 75–88.

Peters, S., Bielfeldt, T., Meldal, M., Bock, K. and Paulsen, H. (1992) Multiple-column solid-phase glycopeptide synthesis. *J. Chem. Soc. Perkin Trans I* 1163–1171.

Phillips, M.L., Nudelman, E., Gaeta, F.C.A., Perez, M., Singhal, A.K., Hakomori, S.-I. and Paulson, J.C. (1990) ELAM 1 mediates cell adhesion by recognition of a carbohydrate ligand, Sialyl-Lewis[x]. *Science* 250: 1130–1132.

Polley, M.J., Phillips, M.L., Wagner, E., Nudelman, E., Shinghal, A.K., Hakomori, S.-I. and Paulson, J.C. (1991) CD 62 and endothelial cell-leukocyte adhesion molecule 1 (ELAM 1) recognize the same carbohydrate ligand, sialyl-Lewis^x. *Proc. Natl. Acad. Sci. USA* 88: 6224–6228.

Rasmussen, J.R. (1991) *In*: V. Ginsburg and P.W. Robbins (eds): *Biology of Carbohydrates*, JAI Press, London, pp 179–285.

Reimer, K.B., Meldal, M., Kusumoto, S., Fukase, K. and Bock, K. (1993) Small-scale solid-phase O-glycopeptide synthesis of linear and cyclized hexapeptides from blood-clotting factor IX containing O-(α-D-Xyl-1- > 3-α-D-Xyl-1- > 3-β-D-Glc)-L-Ser. *J. Chem. Soc. Perkin Trans I* 925.

Schmidt, R.R. and Michel, J. (1980) Simple synthesis of α- and β-O-glycosyl imidates; preparation of glycosides and disaccharides. *Angew. Chem. Int. Ed. Engl.* 19: 731–732.

Schmidt, R.R. and Toepfer, A. (1992) An efficient synthesis of the Le^x antigen family. *Tetrahedron Lett.* 33: 5161–5164.

Schultz, M. and Kunz, H. (1993) Synthetic O-glycopeptides as model substrates for glycosyltransferases. *Tetrahedron Asymmetry* 4: 1205–1220.

Schuster, M., Wang, P., Paulson, J.C. and Wong, C.-H. (1994) Solid-phase chemical-enzymatic synthesis of glycopeptides and oligosaccharides. *J. Am. Chem. Soc.* 116: 1135–1136.

Spinola, M. and Jeanloz, R.W. (1970) Synthesis of a di-N-acetylchitobiose asparagine derivative. *J. Biol. Chem.* 245: 4158–4168.

Suzuki, K., Maeta, H., Suzuki, T. and Matsumoto, T. (1989) Cp$_2$ZrCi$_2$-AgBF$_4$ in benzene: A new reagent system for rapid and highly selective α-mannoside synthesis from tetra-O-benzyl-D-mannosyl fluoride. *Tetrahedron Lett.* 30: 6879.

Thiem, J. and Wiemann, T. (1990) Synthesis of galactose-terminated oligosaccharides by use of galactosyltransferase. *Angew. Chem. Int. Ed. Engl.* 29: 80–82.

Toshima, K. and Tatsuta, K. (1993) Recent progress in O-glycosylation methods and its application to natural products synthesis. *Chem. Rev.* 93: 1503–1531.

Unverzagt, C., Kunz, H. and Paulson, J.C. (1990) High-efficiency synthesis of sialyloligosaccharides and sialoglycopeptides. *J. Am. Chem. Soc.* 112: 9308.

v.d. Bruch, K. and Kunz, H. (1994) Synthesis of N-glycopeptide clusters with Lewis^x antigen side chains and their coupling to carrier proteins. *Angew. Chem. Int. Ed. Engl.* 33: 101–103.

Velupillai, P. and Harn, D.A. (1994) Oligosaccharide-specific induction of interleukin 10 production by B220$^+$ cells from schistosome infected mice: A mechanism for regulation of CD4$^+$ T-cell subsets. *Proc. Natl. Acad. Sci. USA* 91: 18–22.

Waldmann, H. and Kunz, H. (1983) Allylester als selektiv abspaltbare Carboxylschutzgruppen in der Peptid- und N-Glycopeptisynthese. *Liebigs Ann. Chem.* 1712–1725.

Windholz, T.B. and Johnston, D.B.R. (1967) Trichlorethoxycarbonyl. A generally applicable protecting group. *Tetrahedron Lett.* 27: 2555–2557.

Wong, C.-H., Ichikawa, Y., Krach, T., Gautheron-Le Narvor, C., Dumas, D.P. and Look, G. (1991) Probing the acceptor specificity of β-1, 4-galactosyltransferase for the development of enyzmatic synthesis of novel oligosaccharides. *J Am. Chem. Soc.* 113: 8137–8145.

Wong, C.-H., Schuster, M., Wang, P. and Sears, P. (1993) Enzymatic synthesis of N- and O-linked glycopeptides. *J. Am. Chem. Soc.* 115: 5893–5901.

Yamazaki, F., Kitajiama, T., Nukada, T., Ito, Y. and Ogawa, T. (1990) Synthetic studies on cell surface glycans 66. Synthesis of an appropriately protected core glycotetraoside a key intermediate for the synthesis of "bisected" complex-type glycans of a glycoprotein. *Carbohydr. Res.* 201: 15–30.

Yuen, C.-T., Lawson, A.M., Chai, W., Larkin, M., Stoll, M.S., Stuart, A.C., Sullivan, F.X., Ahern, T.J. and Feizi, T. (1992) Sulfated blood group Lewis^a. *Biochemistry* 31: 9126.

Interface between Chemistry and Biochemistry
ed. by P. Jollès and H. Jörnvall
© 1995 Birkhäuser Verlag Basel/Switzerland

Peptides as active probes

A. Undén and T. Bartfai

*Department of Neurochemistry and Neurotoxicology Stockholm University,
S-106 91 Stockholm, Sweden*

Summary. The use of peptides as probes of peptide binding sites of neuropeptide receptors, and of peptidases and proteases is discussed. The rapidly expanding use of peptide antigens as probes of protein structure and valuable diagnostics and vaccines is described. We also discuss the use of synthetic peptide motifs in studies on the molecular details of protein-protein and protein-nucleic acid interactions. Covalently modified peptides such as phosphopeptides exemplifies the use of synthetic peptides in the study of posttranslational modifications of proteins.

Introduction

Naturally occurring peptides are small polymeric molecules which show great variety at the level of primary structure, possess little secondary structure because of their limited size, and seldom have well determined tertiary structure. Despite, or perhaps thanks to, their flexible structures in solution, peptides are abundant in multicellular organisms, acting as signal substances at specific peptide hormone or neuropeptide receptors. Peptides serve as substrates of a great variety of peptidases and proteases as protein degradation proceeds through degradation of peptides. Hence naturally occurring peptides provide interesting probes of recognition/binding sites of receptors and of the active sites of peptidases and proteases.

The development of chemical synthesis of peptides in solution and later, more importantly, on solid phase has enabled the broad use of peptides in chemistry and biology – including those with non-naturally occurring amino acids and posttranslational modifications. Peptides can often be used as surrogate antigens instead of proteins. These peptides, of which peptide antigens are the most important, are now widely used as diagnostic agents and vaccines instead of the corresponding viral and bacterial proteins. The peptide antigens are probes of antigen recognition sites of antibodies, and antipeptide antibodies can also in some cases be used as probes of the topology of the protein.

By introducing different reporter groups such as biotin, affinity labelling groups, radioactive isotopes or fluorescent groups, peptide probes can be employed for studies of active sites, in ligand binding studies and in studies of enzyme kinetics.

In many protein-protein interactions, only a limited number of amino acid residues are critical for the binding between the two interacting proteins. Therefore, in many cases it is possible to use fragments of proteins as surrogates for the whole protein in studies of such interactions. This has been utilised for instance in studies of the SH_2 domains connected with the tyrosine kinase signalling cascade (Pawson and Schlessinger, 1993), when using peptides derived from extracellular matrix proteins such a fibronectin and laminine (Robley, 1993) and in studies of the recognition of RNA by viral proteins such as Tat (Frankel, 1992).

The widespread application of solid phase peptide synthesis has led to an explosion in the number of peptides used in science and medicine. The number of commercially available peptides grew about 15-fold during 1983–1993, making it a difficult task to write a review with the above title. We will therefore only concentrate on the principles of the use of peptides as probes and provide a few illustrative examples.

Peptides as probes of receptor sites

Structure activity relation of peptide hormones as probes for receptor structure

Peptide receptors are membrane-bound proteins which interact with peptide ligands at the extracellular side of the plasma membrane. The peptides are allosteric modulators of the receptor that act to catalyze the activation of G-proteins or alternatively, peptide hormones increase tyrosine kinase activity in the intracellular domain of many growth hormone peptide receptors or promote dimerisation of the receptor-peptide complexes and thereby transmit the signal. The reason for peptide-receptor interactions being far less understood than peptide-enzyme or peptide-antibody interactions can to a large extent be attributed to the difficulties associated with spectroscopic and crystallographic studies of membrane proteins and the allosteric nature of the peptide-receptor interaction.

The majority of the studies of peptide hormone receptors have used the peptide hormone and fragments of analogues thereof to study their properties as ligands at the receptors in different binding- and bioassays. Until recently, very little was known about the protein chemistry of peptide receptors, and structure activity relationship (SAR) studies were the only means by which these receptors could be studied.

In SAR studies, the primary structure of the peptide hormone is dissected by sequential substitutions by "neutral" amino acid residues like L-alanine, by making N- or C-terminal deletions, by substitutions with D-amino acids or by modifications of the peptide bonds. These

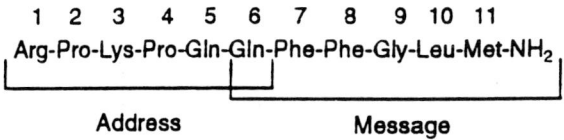

Figure 1. Amino acid sequence of substance P. The amino acid residues at positions 1–6 are important for receptor subtype specificity and high affinity binding and to the NK1 receptor (address), but not critical for agonist action which is carried by the C-terminal hexapeptide (message) cf. Table 1.

aspects have been covered in excellent reviews (Dutta, 1991; Fauchére, 1986).

A broad generalisation of the outcome of a large number of such SAR studies of peptide-peptide receptor interaction is that the peptide ligand carries a very limited number of functional groups (amino acid side chains and/or peptide bonds) that are absolutely essential for receptor binding, and even conservative substitutions within these groups result in analogues with low or no affinity at all for the receptor. In many cases, such as galanin, the opioid peptides and NPY, the amino acid residue that are most critical for peptide recognition are primarily located in the N- or C-terminal regions of the peptide, respectively. In a large number of cases of peptide hormones that interact with G-protein coupled receptors, the presence of at least one aromatic amino acid residues is in many cases absolutely essential for agonist action.

The relative importance of different amino acid residues and peptide bonds in a peptide ligand for its interactions with the appropriate receptors varies greatly. Seeking to rationalise these different interactions, Schwyzer (Sargent and Schwyzer, 1986) suggested that the peptide hormone sequence in many cases could be divided into an "address" and a "message" part. The "message" carries the amino acids that are essential for agonist action, and contributes to the affinity of the peptide for the receptor. The "address" residues are not essential for agonist action but can contribute to the free energy of the ligand-receptor interactions, increasing it by several orders of magnitude, thus accounting for the nanomolar or subnanomolar dissociation constant of peptide receptor complexes (K_D). This is illustrated in Figure 1 and Table 1 for substance P, a peptide hormone of the tachykinin peptide family. The three major mammalian tachykinin peptides are substance P, neurokinin A and neurokinin B, Table 1 (Escher and Regoli, 1989; Schwyzer, 1987). The C-terminal sequence Gln-Phe-Xaa-Gly-Leu-Met-NH$_2$ is the "message" and this hexapeptide of substance P is a full agonist that can activate SP receptors (NK1 receptors) in guinea pig ileum. As shown in Table 1, the N-terminal hexapeptide is required for high affinity receptor binding and provides the structural basis for discrimination between different tachykinin receptors.

Table 1. Relative affinities of substance P (SP) analogs and tachykinins for rat brain NK1 receptors

Peptide	NK1 receptor Relative affinities[a]
Substance P 100[b]	
1 2 3 4 5 6 7 8 9 10 11	
(R P K P Q Q F F G L M–NH_2)	
Neurokinin A	0.5[b]
(H K T D S F V G L M–NH_2)	
Neurokinin B	0.03[b]
(D M H D F F V G L M–NH_2)	
[Lys^1]SP	160[b]
[Asp^3]SP	60[b]
[Met^5]SP	110[b]
[Met^6]SP	100[b]
[Gly^{1,2}]SP	25[c]
[Gly^{2,3}]SP	19[c]
[Gly^{3,4}]SP	18[c]
[Gly^{4,5}]SP	33[c]
[Gly^{5,6}]SP	4[c]
[Met^7]SP	0.02[b]
[Met^8]SP	4[b]
[Nle^{11}]SP	4[b]
SP-Methyl ester	0.7[b]
SP(4–11)	3[d]
SP(5–11)	0.5[d]
SP(6–11)	0.3[d]

Analogues of substance P and related peptides as ligands to the NK1 receptor. (a) Relative affinities are expressed in percent of IC_{50} values from displacement of ^{125}I-Bolton Hunter substance P from rat synaptosomes excluding cerebral cortex and cerebellum, determined in radioligand competition studies against ^{125}I-Bolton-Hunter labeled substance P; (b) Lavielle et al. (1986); (c) Rivera-Baeza and Undén (1991); (d) Wang et al. (1993)

As will be discussed below, this concept can be expanded to describe a wide range of other interactions apart from the classical peptide hormone receptor interaction.

The flexibility of peptides and peptide hormones

The interpretation of SAR studies as a means of mapping peptide-receptor interactions is complicated, as a functional group of a peptide may not be important for ligand recognition *per se*, but rather acts to stabilise secondary or teriary structures of the peptide hormone when presented to and recognised by the receptor.

Peptides are flexible molecules with very little ordered structure in water solution. Probably the only peptide hormones with less than 50 amino acid residues that have more or less defined structures in water solution and that binds to G-protein coupled receptors, is the 36 amino acid long neuropeptide Y (NPY) and homologous peptides of the pancreatic polypeptide (PP) family (Tatemoto, 1982; Glover et al.,

1984). NPY, which actually is a small globular protein, is stabilised by hydrophobic interactions between an α-helix (C-terminal half) and a polyproline helix-like structure (N-terminal half) forming a small hydrophobic core.

It appears though, that the ordered secondary and tertiary structure of NPY is an exception and that other peptide hormones are extremely flexible molecules. In order to dissect the entropy factors (i.e., those interactions within the peptide hormone which lower the entropy barrier for recognition) from the importance of individual functional groups for the interaction with the receptor, attempts have been made to study by spectroscopic methods, the ability of peptides to form ordered structures in artificial media such as organic solvents. The relevance of such studies for the receptor recognition event is difficult to evaluate as no means to directly investigate the conformation of the peptide during its interaction with membrane bound receptor are available.

Conformational analysis of peptides using chemically modified analogues

In the absence of spectroscopic data of peptide-receptor interactions, attempts have been made to investigate the importance of peptide conformation indirectly by restricting the conformational flexibility of the peptide and analyse the outcome of such modifications in terms of affinity, receptor subtype specificity and agonist/antagonist action.

Local conformational constraints can be introduced by a wide range of amino acid derivatives and peptide bond surrogates (Toniolo, 1990). C^α-alkyl amino acids such as α-amino isobutyric acid, Aib (1, Fig. 1) have frequently been employed to restrict the allowed ψ and ϕ angles to essentially the area in the Ramachandran plot that corresponds to right- and left-handed α-helices and 3_{10} helices (Paterson et al., 1981; Degrado, 1988). Several microbial peptide antibiotics such as alamethicin, which forms transmembrane ion channels in lipid bilayer membranes, contain Aib (Karle and Balaram, 1990). N-methylation of a peptide bond removes a hydrogen bond donator and reduces the conformational space by increasing the size of the peptide bond. Cis peptide bonds are rare in proteins and occur almost exclusively in peptide bonds where proline donates the amide nitrogen. As a model for N-methyl peptide bonds, the peptide N-methylglycinyl-N-methylglycine (sarcosyl-sarcosine) has been synthesised and the cis-isomer of the peptide bond was shown to be almost isoenthalpic with the trans isomer (Howard et al., 1973). Depending on the chirality of the amino acids in N-methylated peptide bonds, different cis-trans isomers are preferred. Homochiral amino acids tend to adopt cis peptide bonds while heterochiral pairs favour the trans isomer (Vitoux et al., 1986). An example where these effects possibly played a role is in an analogue of CCK 28–32 where

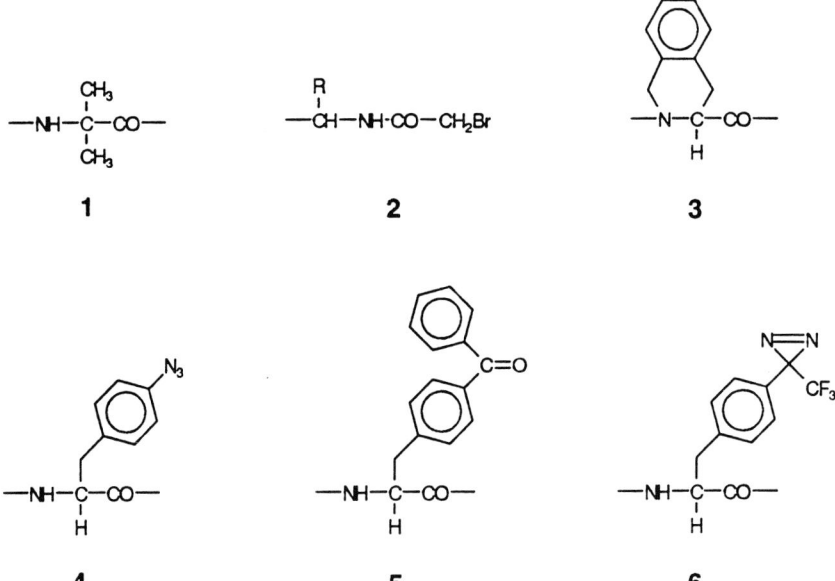

Figure 2. Functional groups which are frequently introduced into peptides as structural probes or affinity labels. (**1**) α-Aminoisobutyric acid (Aib) residue, (**2**) Peptide with N-terminus acylated by the alkylating probe α-bromoacetic acid, (**3**) Tetrahydro-isoquinoline carboxylic acid (Tic) residue; a rigid phenylalanine analog, (**4**) p-Azidophenylalanine residue, (**5**) p-Benzoylphenylalanine residue, (**6**) 4'(Trifluoromethyl-diazirinyl) phenylalanine residue.

norleucine in position 26 was substituted by N-methylnorleucine. A large increase in receptor subtype specificity for the CCK-B receptor was in this case obtained by N-methylation. ^{1}H-NMR studies suggested that the N-methylation favoured the formation of a cis-peptide bond (Hruby et al., 1990).

Increasing the rigidity of the amino acid side chains has been used as a strategy to study the conformational requirements of receptor recognition. The most studied of these constrained analogues is tetrahydro-isoquinolinecarboxylic acid (Tic) which can be considered a constrained analogue of phenylalanine where the N^α-nitrogen is linked to the phenyl ring by a methylene bridge (**3**, Fig. 2). By introducing Tic into opioid peptide analogues, a high specificity for the μ-opiate receptor was obtained (Kazmierski and Hruby, 1988).

Peptide bonds formed by two amino acid residues of different chiralities have a tendency to form β-turns (Rose et al., 1985). For mammalian luteinizing hormone releasing hormone pGlu^1-His^2-Trp^3-Ser^4-Tyr^5-Gly^6-Leu^7-Arg^8-Pro^9-Gly^{10}-NH$_2$, it was suggested that the type II' β-turn was favoured around Gly^6-Leu^7 (Momany, 1975). Type II' β-turns are for sterical reasons favoured by heterochiral pairs where the i+1 residues is a D-amino acid and the i+2 residue is an L-amino acid. This suggestion was supported by studies of analogues where Gly^6 was substituted by D-amino acids (Monahan et al., 1973). An L-Ala^6 analogue had 4% and a D-Ala^6-analogue had 350% of the activity of that of the native hormone.

A classical example where the turn-promoting properties of D-amino acids were exploited is in the case of melanocyte stimulating hormone (α-MSH). Heat-alkali treatment of α-MSH, which results in racemisation of Phe^7, resulted in a hormone with prolonged activity. This led to the suggestion that a turn structure could be important for biological activity (Sawyer et al., 1980). Later, synthetic cyclic analogues where this turn was stabilised were shown to be super-agonists in some assay systems (Sawyer et al., 1982).

Several peptide hormone receptor antagonists, such as the substance P receptor antagonists spantide, are analogues of the naturally occurring peptide carrying D-amino acid residues in different positions. In the light of recently published data on peptide receptors (see below) it can be speculated that the side chains of D-amino acid residues are directed towards those surfaces of the receptor that interact with non-peptide receptor antagonists, i.e., surfaces which normally do not bind the native agonist.

Peptide hormones and many peptide toxins act at the interface between water and the phospholipid membrane. Based on the analysis of the primary structure of many such peptides, it has been suggested that peptides have anisotropic distribution of polar and nonpolar residues that interact with the different compartments. These ideas were exploited by Kaiser and co-workers (Kaiser and Kézdy, 1983) who showed that many peptides, when analysed by helical wheel plots (Schiffer and Edmundsson, 1976), have preferentially polar residues and nonpolar residues on different sides of the helix. This suggests that gross hydrophilic/hydrophobic properties play an important role in peptide-receptor interaction and suggests that an α-helix is induced when the hydrophobic residues interact with the phospholipid membrane. This type of two-step interaction is likely to facilitate the hormone receptor interactions as they result in the docking of two relatively ordered structures, thus reducing the entropy barriers for ligand recognition. This has been tested in studies with synthetic analogues of peptide toxins and peptide hormones displaying amphiphilic structures in helical-wheel plots. The results largely corroborate the assumption that the membrane promotes the formation of amphiphilic α-helices for several flexible peptide ligands (Kaiser and Kézdy, 1983; Degrado, 1988).

A more radical approach to decrease the flexibility of biologically active peptides is to introduce global conformational restrictions by cyclisation (Hruby, 1982; Hruby et al., 1990; Rizo and Gierasch, 1992). Numerous peptide toxins derived from non-mammalian sources that bind mostly to ion channels with very high affinity are cyclic peptides with one or several disulphide bonds, but also several neuropeptides contain one or several disulphide bridges (e.g., endothelin). These toxins have proved invaluable as probes for the pharmacological characterisation of subtypes of potassium and calcium channels (Watson and Girdlestone, 1994).

Cyclisation of peptide hormones has produced analogues with increased potency as ligands to their respective receptors. Classical examples of conformationally restricted cyclic peptide hormones are the α-MSH superagonist (Sawyer et al., 1982) and cyclic somatostatin analogues (Verber et al., 1979).

As can be expected, none of the approaches described above is of general value and for most peptide hormones the importance of intramolecular interactions within the peptide is extremely difficult to evaluate with any method. For many peptide hormones such as the tachykinins, galanin, NPY and members of the secretin family, restriction of conformational flexibility has not resulted in any super agonists or high affinity antagonists, although several good examples of various analogues with globally and locally constrained conformation resulting in receptor subtype-specific ligands or in ligands with retained affinity can be found in the literature.

However, caution must be taken when conformationally restricted peptide analogues are assayed in systems of high protease activity. Decreased flexibility and modified peptide bonds will in many cases result in anlogues that have lower k_{cat}/K_m values as substrates to endogenous proteases, i.e., these analogues are proteolytically more stable and will be apparently better agonists than the natural peptide when assayed in complicated biological systems or *in vivo*.

Alkylating peptide probes

Affinity labelling of proteins by alkylating ligands has ever since early studies on peptidases been regarded as an important tool in delineating the active site of proteins (Means and Feeney, 1971; Brunner, 1993; Chowdhry and Westheimer, 1979). This approach becomes all the more attractive when the receptor-ligand binding is characterised by high affinity (subnanomolar), such as in the case of peptide hormones, since these conditions permit the use of low concentrations of the affinity labelling ligand and therefore, at least in theory, low levels of non-specific incorporation of the peptide probe.

Several alkylating functional groups such as α-halocarbonyl, epoxides, nitrogen mustards and Michael substrates are possible to introduce into peptides (Roberts and Vellaccio, 1983). One of the simplest approaches to create an alkylating peptide analogue, is to couple α-bromoacetic acid to a peptide carrying a free amino group. This approach was originally described by Walter et al. (1972) who introduced the bromoacetyl group at the N-terminus of oxytocin, thereby converting the hormone into an irreversible antagonist. Haloacetyl groups react preferentially with the strongly nucleophilic thiolate anion, but can in principle react with any nucleophilic group as shown by the labelling of the histidine by iodoacetate in the active site of ribonuclease.

Photoaffinity labelling with peptide ligands

Photoaffinity labelling is a popular technique with protein chemists studying the active site of proteins, since the reversible, high specificity and high affinity binding of the ligand is utilised in this technique before the alkylation of the protein by the ligand is initiated by irradiation, thus reducing non specific labelling.

One of the most widely used functional groups that can be photoactivated is the arylazides. Peptides carrying p-azidophenylalanine as photoactivable group have been employed in several studies (**4**, Fig. 2). One of the earliest examples of a successful use of p-azidophenylalanine was a study in which the model peptide benzyloxycarbonyl-Ala-Ala-Phe(4-N_3)-OH (but also similar peptides containing p-nitrophenylalanine) was used to photoaffinity label chymotrypsin (Escher and Schwyzer, 1974). A general problem in the synthesis of p-azidophenylalanine containing peptides has been to find a convenient strategy to incorporate this residue during the synthesis of the peptide. Seyer et al. (1989) coupled the unprotected amino acid p-azidophenylalanine directly to an angiotensin II receptor ligand which had a blocked N-terminus. As an alternate route the same research group coupled the methyl ester of p-azidophenylalanine to the C-terminal part of the same peptide. Since the azido group is stable under basic conditions the methyl ester could be cleaved by saponification. A more general approach to the incorporation of the p-azidophenylalanine group was taken by Landis et al. (1989) who incorporated N^α-Boc-p-aminophenylalanine where the p-amino group was protected by benzyloxycarbonyl group into μ and δ opioid receptor subtype selective peptides. This approach allows a precursor of p-azidophenylalanine to be incorporated during solid phase peptide synthesis in the same manner as any conventional protected amino acid. After synthesis of the peptide, the p-amino phenyl group is diazotised whereafter sodium azid is added, thereby forming the desired p-azidophenylalanine analogue. Peptide probes with the p-azidophenylalanine residue can be radio-labelled by standard procedures for radioiodination (Seyer et al., 1989).

Other photoreactive groups used to label peptide hormone receptors are different benzophenone derivatives. Upon irradiation the benzophenone group is activated and has the advantage of having high reactivity with hydrocarbon derivatives but low reactivity with water (Boyd et al., 1991). These benzophenone derivatives have been successfully used when labelling the substance P receptor (Boyd et al., 1991) and the CCK_B receptor (Thiele and Fahrenholz, 1993). Incorporation of the group is simple, and commercially available p-benzoylbenzoic acid can be activated to form an active ester and acylate primary amino groups. An even more elegant approach was taken by Boyd et al. (1991) who synthesised p-benzoylphenylalanine (Kauer et al., 1986) and incorporated this as a phenylalanine surrogate at position 8 of substance P.

Studies with affinity labeled peptides have also proved valuable when characterising the molecular weight and the presence of glycosylation of neuropeptide-receptors. However, to our knowledge, there are no reports of affinity labelling of peptide hormone receptors providing information about which amino acid residues of the receptor interact with the peptide. Perhaps this is not surprising since the peptide hormone receptors are present in low concentrations (fmoles/mg protein) in the plasma membrane which in combination with heterogeneous labelling makes the purification of labelled alkylated receptor fragments technically very difficult.

Peptide probes for purification of peptide receptors

Biotinyl peptides can be used in affinity chromatographic purification of solubilised receptors using avidin columns. The high-affinity biotin-avidin interaction is favourable when small quantities of peptide receptors are to be purified.

Early attempts to purify peptide hormone receptors of the G-protein coupled receptor family were made with limited success. The purity of the receptor preparations obtained was however low, and until recently no segments of any peptide receptor had been isolated and sequenced. Recently, Eppler and co-workers published the purification of a somatostatin receptor of the μ-opioid receptor after using biotinylated analogues for affinity purification of solubilised receptors (Hulmes et al., 1992; Eppler et al., 1992; Eppler et al., 1993). To our knowledge the purification of the somatostatin receptor from rat GH4C1 pituitary cells was the first example where purification of a G-protein coupled peptide receptor also lead to the elucidation of the amino acid sequence of the whole receptor. An additional outcome of this study was that the interacting G-protein, which was determined to be $G_{i\alpha}$, was co-purified with the receptor (Eppler et al., 1993).

Biotin can be incorporated into peptides with the same activation procedures as an N-protected amino acid, and acylate the N-terminus and/or the ω-amino groups on diamino acids such as lysine or ornitine. Biotin is compatible with the reaction conditions of Boc and Fmoc strategies, although care should be taken to avoid S-alkylation during acidolytic cleavage of protective groups.

Peptides as antigens

Continuous and discontinuous epitopes of proteins

If peptides derived from a protein are used as antigens, antibodies which recognise the holoprotein can be obtained. This was first demonstrated in 1963 (Anderer, 1963) when it was shown that peptide fragments from the Tobacco Mosaic Virus could elicit antibodies that neutralise the infectiousness of the virus. Proteins are macromolecules with a defined tertiary structure and it is therefore not obvious how small flexible peptides with little or no secondary or tertiary structure can mimic the antigenic properties of the intact protein antigen. Empirically it has been found that peptide antigens have a minimal length of 5–6 amino acid residues (the "minimal epitope size"), even though longer peptides often give better antibody responses (Sutcliff et al., 1983; Berzofsky, 1985). In order to recognise a protein the peptide epitopes used for immunisation have to be accessible to the antibody; consequently surface accessibility and atomic mobility (Tainer et al., 1984) are both very good predictors of B-cell antigenic sites of a protein.

Hypothetically, three kinds of epitopes on a protein have been postulated (Lewy and Weiner, 1993):

(1) Sequential epitopes where only the primary structure of the segments of a protein is important for the immune response.
(2) Continuous epitopes where a linear sequence is recognised in a conformation-dependent manner.
(3) Discontinuous epitopes where the amino acid residues recognised by the antibody are scattered over different positions but adjacent in the folded state.

Only sequential and continuous epitopes can be expected to be mimicked by small peptides and the epitopes have to be at least partly exposed to water in order to be recognised by the antibody. As to the role of the secondary structure in eliciting an antibody response and recognising protein antigens, both the relative unimportance and importance of secondary structure have been put forth in the literature.

Peptides as probes of antigenic sites on proteins

Antibodies to small synthetic peptides derived from a protein can be used as probes to map which parts of the protein are accessible and antigenic. In many cases the cDNA sequences of proteins from microorganisms have been obtained through molecular cloning. Having this sequence, the next step is often to determine which parts will give rise to antibodies if the host organism is exposed to the infectious agent *in vivo*.

The immunogenic domains of protein epitopes can be mapped by synthesising a large number of overlapping peptides and testing if these peptides react with antibodies present in individuals that have been exposed to the whole micro-organism. Furthermore, no conclusions about the physiological significance of the peptide sequences that react with the antisera can be drawn from these studies. As the antiserum is raised against the native protein it can be assumed that in many cases the peptide sequences that are most avidly recognised possess similarities in conformation with the peptide when present in the protein. This *per se* does not ascertain that this peptide represents an immuno dominant site on the holoprotein (cf. above).

Epitope mapping using synthetic peptides as surrogates for immunologically active sequences in proteins is one of the most widely used applications of peptide synthesis. The methods of traditional solid phase peptide synthesis, although several orders of magnitude faster than synthesis in homogenous solution, were initially not fast enough to produce the hundreds or thousands of different peptides needed to map the epitopes of a protein. During the last 5 years a great amount of work has been invested into developing methods by which a large number of short (5–10aa) peptides can be synthesised simultaneously on a small, usually submilligram, scale. These efforts have been focused on developing new solid phase carriers that can help separate a large number of different peptides during the synthesis. One of the simplest, cheapest and most attractive procedures is to synthesise peptides by Fmoc chemistry as spots on a sheet of cellulose paper (Frank et al., 1990). Several other approaches have been suggested for the specific purposes of "epitope-mapping" and these techniques have recently been reviewed (Jung and Beck-Sickinger, 1992).

Structural studies of peptide-antibody interaction

Antigen-antibody interactions are often of very high affinity ($K_D \approx 10^{-11}–10^{-13}$), suggesting multiple interactions between the antigen and the antibody (Webster et al., 1994).

Although a number of X-ray crystallographic studies of peptide – antibody complexes have been published, no valid generalisations in terms of surface area, minimal number of contacts and affinity can be made (Webster et al., 1994). It is of importance to identify and compare the structure of the peptide antigen in the protein, the structure (if any) of the peptide when bound in solution or conjugated to a carrier, and the structure of the peptide in a peptide-antibody complex. One of the few detailed studies which illustrates these problems is a study by Stanfield et al. (1990) in which a 19-residue peptide, derived from the C-helix of myohemerythrin, was used to raise a monoclonal antibody

(Feiser et al., 1987) that recognised both the peptide, the native myohemerythrin and apo-myohemerythrin. No stable structure in solution could be determined for the peptide by ^{1}H-NMR and CD analysis (Dyson et al., 1988). In the native protein this peptide was known from x-ray crystallographic studies to form an α-helix. In view of this it is noteworthy that crystallographic studies of the peptide – antibody complex show that this peptide is bound as a type II β-turn to the antibody.

Although a limited number of studies are available at this stage, it can be concluded that antipeptide antibodies may be regarded as probes of surface accessibility or orientation of the protein chain in membrane proteins, but not as probes of secondary structure of the peptide antigen in the native protein. This possible weakness of peptide antigens as probes of local protein secondary structure does not deduct from their tremendous practical use as vaccines, diagnostic agents, etc.

Phosphopeptides – an example of peptide antibodies as probes of posttranslational modifications of proteins

Posttranslational modifications of proteins and peptides play an important regulatory and structural role in biological processes. In order to monitor these processes, posttranslational modifications of a given protein can be monitored by antibodies raised against peptides carrying the site to be modified and by antibodies raised against the covalently modified peptide.

Phosphorylation of proteins is the most common posttranslational modification, and phosphorylation-dephosphorylation cycles of proteins play a key role in regulating a number of intracellular processes. It has been estimated that about a third of all intracellular proteins are phosphorylated; so far, 200 protein kinases and 100 protein phosphatases have been identified, and it has been esimated that 2–3% of all genes in eukaryotes encode protein kinases (Hubbard and Cohen, 1993). The possibility to generate antibodies that specifically recognise both the phosphorylated and dephosphorylated forms of a phosphoprotein is therefore of great importance.

As most phosphoproteins are not available in sufficient quantities to generate antibodies, the most convenient route of generating antibodies that recognise a phosphoprotein is to immunise with the corresponding phosphorylated peptides.

In studies of the substrate specificity and kinetics of protein kinases and protein phosphatases, phosphorylated peptides are usually used as surrogates for protein substrates. A major problem is that the methods used so far to synthesise phosphorylated peptides are considerably more complicated than those of synthesising unmodified peptides. Much work has therefore been invested to develop simple, general and preferably solid phase methods to synthesise phosphopeptides.

Three strategies have been suggested:

(1) Enzymatic phosphorylation using various protein kinases. This route is attractive as it is not dependent on the rather complicated chemical synthesis of phosphopeptides, but is limited by the substrate specificity of the protein kinases.

(2) The so-called global strategy. Peptides and even proteins can be phosphorylated by polyphosphoric acid. This method was used to phosphorylate a neurofilament-derived peptide containing several serine residues (Otvos and Hollosi, 1993). When treated for 3 days all serine residues of the peptide were phosphorylated, whereas interruption of the reaction after 12 h resulted in different phosphorylated forms of the peptide.

In order to specifically direct the phosphorylation to a particular residue this strategy can be carried out on synthetic peptides where all reactive functional groups except the selected hydroxyl groups are protected.

Serine and threonine residues can be phosphorylated directly on the resin using Fmoc chemistry by incorporating the serine/threonine with unprotected β-hydroxyl groups (Knapp et al., 1993), which then can be modified on the resin by phosphorylation followed by oxidation (Andrew et al., 1991), Figure 3. A similar strategy was described by Shapiro et al. (1994) who employed an Fmoc serine derivative with the side chain protected with the t-butylmethylsilyl group which could be selectively removed prior to phosphorylation (Fig. 3). To incorporate side chain protected serine and threonine residues with orthogonally removable protective groups has the advantage that several potential problems associated with the reactivity of unprotected hydroxyl groups can be avoided. The problems connected with global strategies are incomplete phosphorylation and side reactions on sensitive amino acid residues such as tryptophane, methionine and cysteine in the oxidation step when the dialkylphosphate triester is oxidized to dialkylphosphorotriester (Fig. 3). The bulky trityl protective group is reported to protect cysteine residues from oxidation during these reaction conditions (Andrew, 1991).

(3) Incorporation of protected phosphoamino acids into the peptide. For the synthesis of serine/threonine phosphopeptides this approach is complicated by the danger of β-elimination when the

---

Figure 3. Synthesis of a phosphopeptide antigen. The general strategy for the solid phase synthesis of the phosphopeptide is adopted from Andrew et al. (1991) and Shapiro et al. (1994). The conjugation of the peptide to a protein carrier is carried out by the glutaric aldehyde method as described by Czernik et al. (1991).

Peptides as active probes

I   Boc-Ala-Glu(OtBu)-Ser(tBu)-Ile-Ser(OSiMe₂tBu)-Ala-OCH₂-C₆H₄-O-Res.

↓ NBu₄F x 3H₂O

II Boc-Ala-Glu(OtBu)-Ser(tBu)-Ile-Ser(OH)-Ala-OCH₂-C₆H₄-O-Res.

↓ ((CH₃)₂CH)₂NP(OCH₂C₆H₅)₂ / tetrazole

III Boc-Ala-Glu(OtBu)-Ser(tBu)-Ile-Ser(OP(OBzl)₂)-Ala-OCH₂-C₆H₄-O-Res.

↓ tBuOOH

IV Boc-Ala-Glu(OtBu)-Ser(tBu)-Ile-Ser(OP(O)(OBzl)₂)-Ala-OCH₂-C₆H₄-O-Res.

↓ CF₃COOH

V   H₂N-Ala-Glu-Ser-Ile-Ser(OP(O)(OH)₂)-Ala-OH

↓ Protein-NH₂ / Glutaraldehyde / NaBH₄

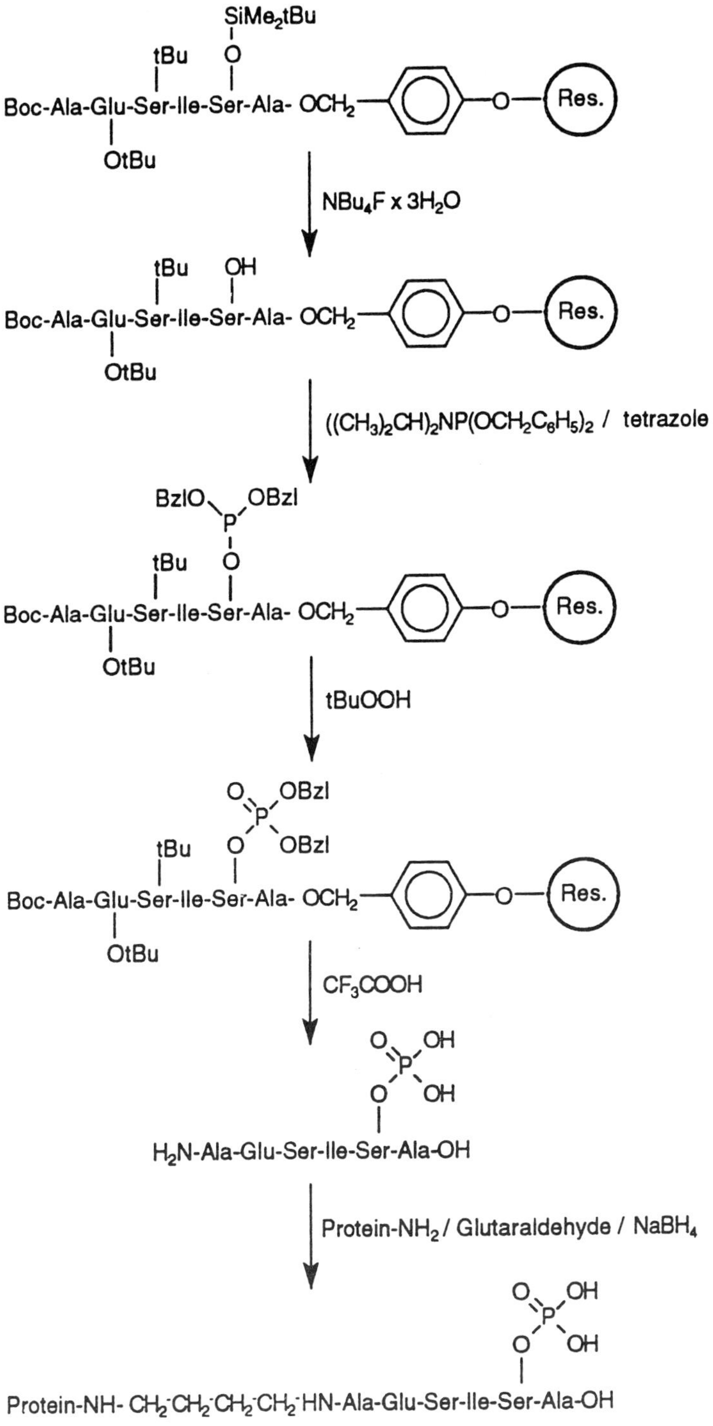

VI Protein-NH-CH₂CH₂CH₂CH₂-HN-Ala-Glu-Ser-Ile-Ser(OP(O)(OH)₂)-Ala-OH

peptide is treated with organic base during the removal of the Fmoc group. This reaction does not present a problem for phosphotyrosine-containing peptides and convenient and efficient methods for the solid phase synthesis of such peptides with Boc and Fmoc chemistry have recently been reported (Tian et al., 1993; Perich et al., 1994).

To raise antibodies against phosphopeptides follows the same immunisation procedures as for dephosphopeptides. A general procedure of raising antibodies against phosphoserine and phosphothreonine containing peptides synthesised by enzymatic methods and conjugated to a carrier protein by glutaraldehyde followed by reduction of the Schiff's base by sodium borohydride is given by Czernik et al. (1991). Such antibodies have been used in several projects to measure the phospho-dephospho state of proteins.

Peptides as probes of active sites of peptidases and proteases

Design of probes to study the active site of proteases: general aspects

Protease-peptide interactions are far better understood than peptide-receptor interactions. The first high-resolution X-ray crystallographic studies of protease-ligand complexes were carried out already in the late 1960s (Blow and Steitz, 1970). Much of the current interest in proteases is derived from the fact that several proteases such as renin, angiotensin converting enzyme and the HIV-1 and HIV-2 proteases are established or promising targets for drug development. As a consequence, much work has been invested into determining the X-ray structure of these and other proteases. Especially in the case of the HIV-1 protease, a large number of X-ray structures of the protein-inhibitor complexes has been determined (Fitzgerald, 1993).

Compared to peptide-receptor interactions, our understanding of peptide-enzyme interactions is far deeper. At peptide receptors, receptor-peptide ligand binding is reversible and does not, like enzymatic catalysis, involve the formation of new chemical bonds. The essence of the mechanisms of receptor-ligand interactions must be understood in terms of conformational changes, i.e. the making and breaking of a large number of non-covalent bonds, induced upon ligand binding.

Designing ligands/inhibitors of enzymes such as proteases, has been far more successful and is based on a more rational ground than the design of peptide-receptor ligands. When designing enzyme inhibitors the transition state concept of catalytic reactions has proved to be of enormous value. The basic idea was outlined almost 50 years ago by Linus Pauling (Pauling, 1948): *This picture of the nature of enzymes may well make us optimistic about the future of chemotherapeutics, for it predicts that for every enzyme . . . it would be possible to find an inhibiting molecule that is more closely complementary in its structure to the enzyme than the substrate itself, and which would accordingly be an effective inhibitor. The picture even presents us with ideas as to the nature of substances that would be effective inhibitors – they should resemble the activated complex.*

Much has been written about affinity label and photoaffinity label of enzymes (Brunner, 1993; Chowdhry and Westheimer, 1979). An example where photolabelling of enzymes has been successfully carried out with an active peptide probe was reported by Falchetto et al., 1991. 4'(Trifluoromethyl-diazirinyl) phenylalanine (**6**, Fig. 2). was incorporated into a 29-residue peptide derived from the calmodulin-binding segment of the plasma membrane Ca^{2+} pump using Fmoc strategy. This peptide inhibited the activity of a truncated form of the enzyme. After cross-linking, fragmentation, separation of labelled fragments and Edman degradation a short 8-residue fragment corresponding to residues 537–544 of the Ca^{2+} pump could be identified.

Miller and Kaiser (1988) labelled the peptide binding site of the cAMP-dependent protein kinase with peptides carrying p-benzoylphenylalanine (**5**, Fig. 2) and could identify two residues Gly^{125} and Met^{127} which were modified following activation.

Peptide probes of protease activity

As discussed above, several proteases have received attention as targets for drug design. Drug development requires simple and rapid assay systems where a large number of potential ligands can be screened. For proteases with simple substrate specificity such as trypsin or carboxypeptidases, model substrates that are both good substrates and form coloured products subsequent to hydrolysis by the enzyme can be designed. Enzymes such as the HIV proteases or renin have a more complex binding requiring recognition of several amino acid residues flanking the scissable peptide bond. In these cases, screening for inhibitors can be carried out by measuring the inhibition of the rate of hydrolysis of complex peptides that are good substrates to the protease. This procedure is however slow and relatively insensitive. To overcome these difficulties, model substrates that carry two reporter groups – one fluorescent group and one group that quenches the fluorescence – flanking the hydrolysed peptide bond can be used. Upon hydrolysis the concentration of the quenching group in the vicinity of the fluorescent group decreases dramatically as the two groups no longer are parts of the same molecule. The fluorescent signal can be followed and thus reflect the formation fo peptidolytic products. Several pairs of donor-

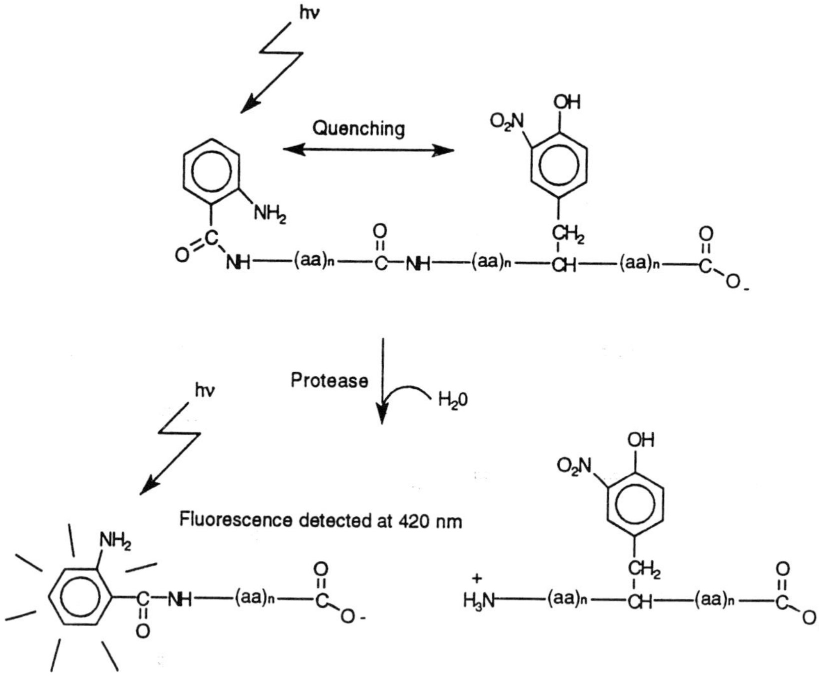

Figure 4. Peptides as probes of the activity of proteases with complex binding interactions. By introducing anthranilic acid residues as a fluorescent emitting group and p-nitrotyrosine as a quencher of the fluorescence, protease activity can be monitored as an increase in fluorescence intensity upon cleavage of the peptide substrate and removal of the quencher.

acceptor pairs have been suggested. The fluorophore 5-(2-aminoethyl) aminonaphthalene-1-sulfonic acid (Edans) and the quencher 4-(4-dimethylamino-phenylazo)benzoic acid (Dabsyl) form an efficient acceptor-donor pair (Matayoshi et al., 1990) but these have the disadvantage of presenting problems in solid phase peptide synthesis. The anthranilamide-nitrotyrosine donor acceptor pair suggested by Meldal and Breddam (1991) is, in contrast, easy to incorporate into peptides (Fig. 4). The levels of background fluorescence in a study of

Table 2. Background fluorescence for the anthranilic acid-p-nitrotyrosine donor-acceptor pair for model peptides where anthralinic acid-p-nitrotyrosine are separated by different numbers of amino acid residues. Data from Meldal and Breddam (1990)

Peptide substrate	F_0/F (%)
ABz-FGY(NO$_2$)D	0.3
ABz-AAFGY(NO$_2$)D	1.1
ABz-AAAAFGY(NO$_2$)D	3.4
ABz-DSGSKSAAAAFGY(NO$_2$)D	10.2
ABz-SDSDSGSKSAAAAFGY(NO$_2$)D	18.5

peptide substrates for subtilisin was only ≈3% if the donor/acceptor pair was separated by six amino acid residues. Even if as many as 18 amino acid residues separated the anthranilic acid residue from the p-nitrotyrosine only 18% of the residual fluorescence was observed (Tab. 2). This donor-quencher system has the advantage the p-nitrotyrosine can be incorporated into the peptide substrate as a conventional Fmoc amino acid C-terminally to the scissable bond and the peptide can be easily capped in the N-terminus by activated Boc-anthranilic acid.

Peptides as probes of protein-protein and protein nucleic acid interactions

Peptides as protein surrogates in studies of protein-protein interaction – definition of motifs

Protein-protein interactions can in certain cases be studied by replacing one of the interacting proteins with a peptide which adequately represents the particular motif of the interacting protein. When such a motif is well stabilised, such as a metal binding site is by the co-ordinated metal, or when the motif is a short sequence, such as the peptide specificity sequences of serine and threonine protein kinases, small peptides can be used to "map" the structural requirement of the protein substrate-enzyme interaction. In other cases however, a motif involves a larger portion of the protein and mimicking it by small peptides is more difficult.

The current data for several peptide/protein-protein/nucleic acid interactions support a model where one flexible ligand interacts with a structurally more ordered binding site. This generalisation does not exclude that the binding site is flexible and to a large extent disordered, but in most cases the overall structural framework must be intact for interaction with the protein or the peptide surrogate. The first step in the protein-protein or protein-nucleic acid interaction is in many cases a long-range acting charge-charge interaction that, after an initial interaction, forms a complex where the structure of the binding site largely determines the conformation of the interacting peptide surrogate or of the protein domain. The relative importance of various residues/functional groups follows a hierarchy where the binding site for some functional groups carry the "message" or the pharmacophore, while other groups form the most important interactions that encode the specificity (address) of the interaction. In addition to these interactions, there are also weaker interactions which are not particularly strong but nevertheless contribute to the overall affinity and specificity of the interaction. This simple, multiple-step model which is analogous to Schwyzer's model of how the peptide ligands bind to proteins, will be illustrated with a number of selected examples.

Peptide probes of protein-nucleic acid interaction: Tat and Rev

RNA can be folded into complex tertiary structures and many of these structures interact with proteins. The best studied examples of such interactions are the ribosomal RNA-protein complexes.

Tat is an 86-residue HIV virus protein that interacts with viral mRNA at a specific site; the so-called TAR (the trans-acting response element) site (Fig. 5). The TAR site has three unpaired nucleotides close to a six nucleotide long loop. Tat contains seven cysteines that form a metal-binding domain and an arginine-rich region that is thought to interact with nucleic acids. Small arginine-rich subdomains have also been found as a motif in other proteins that interact with RNA such as bacterial antiterminators, splicing factors and ribosomal proteins (Lazinski et al., 1988; Kenan et al., 1991). Synthesis and purification of recombinant Tat has proved difficult as the cysteine residues tend to oxidise and therefore several researchers have used peptides as probes to study the structural requirements of Tat-TAR interactions. Studies with synthetic peptides have shown that even very small peptide fragments such as Tat 49–57 (Tab. 3) bind to TAR with as high affinity as 6 nM, which is approximately the same as for the intact protein. CD studies showed that the Tat-derived peptide has little or no ordered structure in solution.

As shown in Table 3, the TAR-RNA binds to a wide range of synthetic peptides containing arginine in different positions. Although

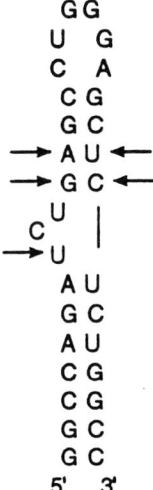

Figure 5. Structure of HIV TAR-RNA that binds the Tat protein and corresponding synthetic peptides. Nucleotides that are important for interaction with arginine residues on TAT binding peptides and proteins are indicated with arrows. Adapted from Frankel (1992). For peptide ligands to TAR see Table 3.

Table 3. Synthetic peptides as ligands to the TAR site

Peptide	Sequence	K_d (nM)
Tat (47–58)	YGRKKRRQRRRP	6
[Ala^{52}] Tat (47–58)	YGRKKARQRRRP	10
[Ala^{53}] Tat (47–58)	YGRKKRAQRRRP	12
[Ala^{54}] Tat (47–58)	YGRKKRRARRRP	5
[Ala^{55}] Tat (47–58)	YGRKKRRQARRP	12
[Ala^{56}] Tat (47–58)	YGRKKRRARARP	12
[Ala^{52}, Ala^{53}] Tat (47–58)	YGRKKAAQRRRP	>100
[Lys^{52}, Lys^{53}] Tat (47–58)	YGRKKKKQRRRP	13
[Ala^{55}, Ala^{56}] Tat (47–58)	YGRKKRRQAARP	>100
[Lys^{55}, Lys^{56}] Tat (47–58)	YGRKKRRQKKRP	4
SIV (76–91)	YEKSHRRRRTPKKAKA	6
Rev (34–50)	TRQARRNRRRRWRERQR	2

SIV; Tat protein from simian immunodeficiency virus. Data from Calnan et al. (1990)

the affinity of the synthetic peptides is high, the specificity is apparently low, and other studies suggested that larger peptides or intact Tat displays higher specificity and can better discriminate between mutant forms of TAR RNA, although these differences between small synthetic peptides and large peptides are not dramatic (Churcher et al., 1993; Gait and Karn, 1993).

The HIV-1 Rev protein is another protein that binds specifically to viral RNA, to Rev response element (RRE) (Gait and Karn, 1993). Like TAR, RRE also contains a distorted "bubble" structure with at least one non-Watson-Crick base pair. Rev is a 116-residue protein and, like the Tat protein, it contains an arginine-rich sequence that interacts with RNA. In this case RNA recognition of the interacting protein Rev is more complex in comparison to Tat. In several studies synthetic peptides of Rev have been used as surrogates of the Rev protein (Kjems et al., 1992; Tan et al., 1993). Tan et al., found that peptides derived from Rev that bound with high affinity to RRE contained an α-helical structure as measured by circular dichroism (CD). Although no high-resolution spectroscopic studies are available for Rev, CD studies show ≈50% α-helix content for the intact protein. SAR studies with Rev 34–50 (TRQARRNRRRRWRERQR) and different synthetic peptides with point substitutions showed that the Rev 34–59 bound with an affinity of 40 nM, which is one order of magnitude lower than for the native protein. Single amino acid residue substitution of T^{34}, R^{35}, R^{38}, R^{39}, N^{40} or R^{44} by L-alanine all resulted in a tenfold or larger reduction in the affinity of these peptides as ligands to the RNA binding site. To what extent the α-helical conformation of these peptides is important for binding is not clear as the peptide of lowest helical content, a 43 R ⇒ A analogue, had the highest affinity for RNA. The hypothesis that an α-helical conformation is of importance for RNA recognition could be further tested by stabilisation of α-helices using techniques developed from the field of peptide hormone research, such as introductions of C^α-alkyl residues or stabilisation by side chain interactions. What this study clearly shows though is that RNA binding by Rev is considerably more complex than that by Tat, that it involves a larger number of critical amino acid residues, and that the RRE-RNA discriminates better between different cationic peptide ligands than TAR-RNA does.

Peptide probes of protein-protein interaction: SH_2 domains-phosphotyrosine interaction

Several intracellular eukaryotic proteins carry domains that recognise structural motifs of other proteins. One of the best characterised domains is the so-called SH_2 domains on proteins that interact with proteins in the receptor-tyrosine kinase signal transduction cascade (Schlessinger and Ullrich, 1992). The SH_2 domains were first discovered and named after the *src* oncogene product, but soon analogous regions were found in several enzymes and proteins without enzyme activity (Pawson and Schlessinger, 1993). The SH_2 regions, which consist of around 100 amino acid residues, recognize phosphotyrosine residues formed by the action of protein tyrosine kinases.

The phosphotyrosine residue is the most important residue for interaction with SH_2 domains and it appears that the recognition site for this residue is, to a large extent, conserved in different proteins carrying SH_2 domains.

The X-ray structure of the phosphotyrosine binding site of *src* has been determined to 1.5 Å resolution when complexed with two low-affinity phosphotyrosine pentapeptides. The tyrosine phosphate group interacts with several hydrogen bonds and cationic residues in the *src*-protein, but one arginine residue that is conserved in all other SH_2-binding domains forms a salt bridge with two hydrogen bonds to two of the oxygens in the phosphate group (Waksman, 1992).

This study did not address the question to what extent other amino acid residues flanking the phosphotyrosine residue contribute to the interaction with the SH_2 domain. Songyang et al. (1993) used a phosphopeptide library to identify the sequence dependence of the phosphotyrosine binding site of different proteins carrying SH_2 domains. Different specificities were observed but the general motif for all these interactions were pTyr-X-Y-Hydrophobic, where X and Y are hydrophilic amino acids. In the case of the *src* subfamily, binding of the sequence pTyr-Glu-Glu-Ile was selected.

The crystal structure of the *src* domain complexed with an 11-residue phosphopeptide Glu-Pro-Gln-pTyr-Glu-Glu-Ile-Pro-Ile-Tyr-Leu has recently been determined (Waksman, 1993). The peptide interacted with the SH$_2$ domain primarily at four residures; pTyr-Glu-Glu-Ile which is identical to the tetrapeptide defined by binding studies using phosphopeptide libraries (cf. above).The SH$_2$ protein appears to bind the peptide in an extended form. The most important of the residues C-terminal of the phosphotyrosine is the isoleucine residue completely buried in a hydrophobic binding site and tightly bound by hydrocarbons of a number of mostly hydrophobic residues. It was concluded that the binding site for isoleucine was critical for the specificity of peptide binding and the authors made the analogy between the peptide-SH$_2$ interaction and fitting a "two-pronged plug (the peptide) into a two-holed socket (the SH$_2$ domain)". It is also interesting to note that in the absence of phosphopeptide the binding site on the domain was relatively disordered.

The SH$_2$ domain binding of phosphotyrosine peptides is a good illustration of how peptides can be used as probes to investigate the structural requirements of a protein to recognise another protein. Phosphopeptide libraries were used to investigate substrate specificity and the use of peptides complexed to the SH$_2$ domain allowed crystallographic studies where, in the terminology of peptide hormone interaction, the sturctural details of both the "message" (phosphotyrosine) and "address" (isoleucine) interaction could be characterised with high resolution.

Perspectives

The ease by which reasonably long peptides and covalently modified peptides can be obtained in large quantities makes the use of peptides a tool for future studies on antigenicity, protein-protein and protein-RNA/DNA interactions. The introduction of methods for making productive peptide libraries will enhance the use of peptides as probes of biological function. The X-ray structures for neuropeptide-receptor complexes will contribute to our understanding of the peptide recognition and agonist action at these membrane-bound receptors in the near future.

Use of peptidomimetics together with the use of peptide libraries is rapidly increasing in the pharmaceutical industry. In summary, a sharp increase in the number of synthetic peptides for scientific and therapeutic purposes is expected.

References

Anderer, F.A. (1963) Versuche zur Bestimmung der Serologisch Determinanten Gruppen des Tabakmosaikvirus. *Z. Naturforsch.* 18b: 1010.

Andrew, D.M. Kitchin, J. and Seale, P.W. (1991) Solid-phase synthesis of a range of O-phosphorylated peptides by post-assembly phosphitylation and oxidation. *Int. J. Peptide Protein Res.* 38: 469–475.
Berzofsky, J.A. (1985) Intrinsic and extrinsic factors in protein antigenic structure. *Science* 229: 932–940.
Blow, D.M. and Steitz, T.A. (1970) X-ray diffraction studies of enzymes. *Ann. Rev. Biochem.* 39: 63–100.
Boyd, N.D., White, C.F., Cerpa, R., Kaiser, E.T. and Leeman, S.E. (1991) Photoaffinity labeling the substance P receptor using a derivative of substance P containing p-benzoylphenlalanine. *Biochemistry* 30: 336–342.
Brunner, J. (1993) New photolabeling and crosslinking methods. *Annu. Rev. Biochem.* 62: 483–514.
Calnan, B.J., Biancalana, S., Hudson, D. and Frankel, A.D. (1991) Analysis of arginine-rich peptides from HIV Tat protein reveals unusual features of RNA-protein interaction. *Genes and Dev.* 5: 201–210.
Chodhry, V. and Westheimer, F.H. (1979) Photoaffinity labeling of biological systems. *Ann. Rev. Biochem.* 48: 293–325.
Churcher, M.J., Lamont, C., Hamy, F., Dingwall, C., Green, S.M., Lowe, A.D., Butler, P.J.G., Gait, M.J. and Karn, J. (1993) High affinity binding of TAR RNA by the human immunodeficiency virus type-1 *tat* protein requires base-pairs in the RNA stem and amino acid residues flanking the basic region. *J. Mol. Biol.* 230: 90–110.
Czernik, A.J., Girault, J.-A., Nairn, A.C., Chen, J., Synder, G., Kebabian, J. and Greengard, P. (1991) Production of phosphorylation state-specific antibodies. *Methods Enzymol.* 201: 264–283.
DeGrado, W.F. (1988) Design of peptides and proteins. *In*: C.B. Anfinsen, J.T. Edsall, F.M. Richards and D.S. Eisenberg (eds): *Advances in Protein Chemistry*. Academic Press, Inc., Harcourt Brace Jovanovich, London.
Dutta, A.S. (1991) Design and therapeutic potential of peptides. *Advances in Drug Research* 21: 147–286.
Dyson, H.J., Rance, M., Houghten, R.A., Wright, P.E. and Lerner, R.A. (1987) Folding of immunogenic peptide fragments of proteins in water solution. II. The nascent helix. *J. Mol. Biol.* 201: 201–217.
Eppler, C.M., Zysk, J.R., Corbett, M. and Shieh, H.-M. (1992) Purification of a pituitary receptor for somatostatin. *J. Biol. Chem.* 267: 15603–15612.
Eppler, C.M., Hulmes, J.D., Wang, J.-B., Johnson, B., Corbett, M., Luthin, D.R., Uhl, G.R. and Linden, J. (1993) Purification and partial amino acid sequence of a μ opioid receptor from rat brain. *J. Biol. Chem.* 268: 26447–26451.
Escher, E. and Schwyzer, R. (1974) p-Nitrophenylalanine, p-azidophenylalanine, m-azidophenylalanine, and o-nitro-p-azido-phenylalanine as photoaffinity labels. *FEBS Lett.* 46: 2947–2955.
Escher, E. and Regoli, D. (1989) Substance P and related peptides. Peptide hormones as prohormones. *In*: J. Martinez (ed.): *Processing, Biological Activity, Pharmacology*. Halsted Press, Ellis Horwood Ltd., pp 26–52.
Falchetto, R., Vorherr, T., Brunner, J. and Carafoli, E. (1991) The plasma membrane Ca^{2+} pump contains a site that interacts with its calmodulin-binding domain. *J. Biol. Chem.* 266: 2930–2936.
Fauchére, J.-L. (1986) Elements for the rational design of peptide drugs. *Advances in Drug Research* 15: 29–69.
Fieser, T.M., Tainer, J.A., Geysen, H.M., Houghten, R.A. and Lerner, R.A. (1987) Influence of protein flexibility and peptide conformation on reactivity of monoclonal anti-peptide antibodies with a protein α-helix. *Proc. Natl. Acad. Sci. USA* 84: 8568–8572.
Fitzgerald, P.M.D. (1993) HIV protease-ligand complexes. *Curr. Opin. Struct. Biol.* 3: 868–874.
Frank, R., Güler, S., Krause, S. and Lindenmaier, W. (1990) Facile and rapid "spot-synthesis" of large numbers of peptides on membrane sheets. *In*: E. Giralt and D. Andreu (eds): *Peptides*, ESCOM, Leiden, The Netherlands, pp 151–152.
Frankel, A.D. (1992) Peptide models of the Tat-TAR protein-RNA interaction. *Protein Science* 1: 1539–1542.
Gait, M.J. and Karn, J. (1993) RNA recognition by the human immunodeficiency virus Tat and Rev proteins. *TIBS* 18: 255–259.

Glover, I.D., Barlow, D.J., Pitts, J.E., Wood, S.P., Tickle, I.J., Blundell, T.L., Tatemoto, K., Kimmel, J.R., Wolmer, A., Strassburger, W. and Zhang, Y.-S. (1985) Conformation studies on the pancreatic polypeptide family. *Eur. J. Biochem.* 142: 379–385.

Howard, J.C., Momany, F.A., Andreatta, R.H. and Scheraga, H.A. (1973) Investigation of the cis and trans isomers of sarcosyl-sarcosine by nuclear magnetic resonance spectroscopy and conformational energy calculations. *Macromolecules* 6: 535–541.

Hruby, V.J. (1982) Conformational restrictions of biologically active peptides via amino acid side chain groups. *Life Sciences* 31: 189–199.

Hruby, V.J., Al-Obeidi, F. and Kazmierski, W. (1990) Emerging approaches in the molecular design of receptor-selective peptide ligands: conformational, topographical and dynamic considerations. *Biochem. J.* 268: 249–262.

Hubbard, M.J. and Cohen, P. (1993) On target with a new mechanism for the regulation of protein phosphorylation. *TIBS* 18: 172–177.

Hulmes, J.D., Corbett, M., Zysk, J.R., Böhlen, P. and Eppler, C.M. (1992) Partial amino acid sequence of a somatostatin receptor isolated from GH_4C_1 pituitary cells. *Biochem. Biophys. Res. Comm.* 184: 131–136.

Jung, G. and Beck-Sickinger, A.G. (1992) Multiple peptide synthesis methods and their applications. *Angew. Chem. Int. Ed. Engl.* 31: 367–383.

Kaiser, E.T. and Kézdy, F.J. (1983) Secondary structures of proteins and peptides in amphiphilic environments. *Proc. Natl. Acad. Sci. USA* 80: 1137–1143.

Kazmierski, W. and Hruby, V.J. (1988) A new approach to receptor ligand design: Synthesis and conformation of a new class of potent and highly selective μ opioid antagonist utilizing tetrahydroisoquinoline acid. *Tetrahedron* 44: 697–710.

Karle, I.L. and Balaram, P. (1990) Structural characteristics of α-helical peptide molecules containing Aib residues. *Biochemistry* 29: 6747–6756.

Kauer, J.C., Ericson-Viitanen, S., Wolfe, H.R. and DeGrado, W.F. (1986) *p*-Benzoyl-L-phenylalanine, a new photoactive amino acid. *J. Biol. Chem.* 261: 10695–10700.

Kenan, D.J. Query, C.C. and Keene, J.D. (1991) RNA recognition: towards identifying determinants of specificity. *TIBS* 16: 214–220.

Kjems, J., Brown, M., Chang, D.D. and Sharp, P.A. (1992) Structural analysis of the interaction between the human immunodeficiency virus Rev and the Rev response element. *Proc. Natl. Acad. Sci. USA* 88: 683–687.

Knapp, D.R., Oatis, Jr., J.E. and Papac, D.I. (1993) Small-scale manual multiple peptide synthesis system. Application to phosphopeptide synthesis. *Int. J. Peptide Protein Res.* 42: 259–263.

Landis, G., Lui, G., Shook, J.E., Yamamura, H.I., Burks, T.F. and Hruby, V.J. (1989) Synthesis of highly μ and δ opioid receptor selective peptides containing a photoaffinity group. *J. Med. Chem.* 32: 638–643.

Lavielle, S., Chassaing, G., Julien, S., Besseyre, J. and Marquet, A. (1986) Influence of the amino acids of substance P in the recognition of its receptor: Affinities of synthesized SP analogues for the specific ^{125}I-BHSP binding site on rat brain synaptosomes. *Neuropeptides* 7: 191–200.

Lazinski, D., Grzadzielska, E. and Das, A. (1989) Sequence-specific recognition of RNA hairpins by bacteriophage antiterminators requires a conserved arginine-rich motif. *Cell* 59: 207–218.

Levy, D.N. and Weiner, D.B. (1993) Synthetic peptide-based vaccines and antiviral agents including HIV/AIDS as a model system. *In*: M.V. Williams and D.B. Weiner (eds): *Biologically Active Peptides: Design, Synthesis, and Utilization.* Technomic Publishing Co., Inc., Lancaster, Basel, pp 219–267.

Matayoshi, E.D., Wang, G.T., Krafft, G.A. and Erickson, J. (1989) Novel fluorogenic substrates for assaying retroviral proteases by resonance energy transfer. *Science* 247: 954–958.

Means, G.E. and Feeney, R.E. (1971) *Chemical Modification of Proteins.* Holden-Day, Inc., San Francisco.

Meldal, M. and Breddam, K. (1991) Anthranilamide and nitrotyrosine as a donor-acceptor pair in internally quenched fluorescent substrates for endopeptidases: Multicolumn peptide synthesis of enzyme substrates for subtilisin Carlsberg and pepsin. *Analytical Biochemistry* 195: 141–147.

Miller, W.T. and Kaiser, E.T. (1988) Probing the peptide binding site of the cAMP-dependent protein kinase by using a peptide-based photoaffinity label. *Proc. Natl. Acad. Sci. USA* 85: 5429–5433.

Momany, F.A. (1975) Conformational energy analysis of the molecule, luteinizing hormone-releasing hormone. 1. Native decapeptide. *J. Am. Chem. Soc.* 98: 2990–2996.

Monahan, M.W., Amoss, M.S., Anderson, H.A. and Vale, W. (1973) Synthetic analogs of the hypothalamic luteinizing hormone releasing factor with increased agonist or antagonist properties. *Biochemistry* 12: 4616–4620.

Otvos, L., Jr., and Hollosi, M. (1993) Development of chemically modified peptides. *In*: M.V. Williams and D.B. Weiner (eds): *Biologically Active Peptides: Design, Synthesis, and Utilization.* Technomic Publishing Co., Inc., Lancaster, Basel, pp 155–186.

Paterson, Y., Rumsey, S.M., Benedetti, E., Némethy, G. and Schegara, H.A. (1981) Sensitivity of polypeptide to geometry. Theoretical conformation analysis of oligomers of α-aminoisobutyric acid. *J. Am. Chem. Soc.* 103: 2947–2955.

Pauling, L. (1948) Chemical achievement and hope for the future. *American Scientist* 36: 51–58.

Pawson, T. and Schlessinger, J. (1993) SH2 and SH3 domains. *Curr. Biol.* 3: 434–442.

Perich, J.W., Ruzzene, M., Pinna, L.A. and Reynolds, E.C. (1994) Efficient Fmoc/solid-phase peptide synthesis of O-phosphotyrosyl-containing peptides and their use as phosphatase substrates. *Int. J. Peptide Protein Res.* 43: 39–46.

Rivera Baeza, C. and Undén, A. (1991) Investigation of the importance of the N-terminal peptide bonds of substance P by synthetic pseudopeptide analogs. *Neuropeptides* 20: 83–86.

Rizo, J. and Gierasch, L.M. (1992) Constrained peptides: Models of bioactive peptides and protein substructures. *Annu. Rev. Biochem.* 61: 387–418.

Roberts, D.C. and Vellacio, F. (1983) Unusal amino acids in peptide synthesis. *In*: E. Gross and Meienhofer (eds): *The Peptides*, Vol. 5. Academic Press, New York, pp 341–447.

Robley, F.A. (1993) Biology and chemistry of extracellular matrix cell attachment peptides. *In*: M.V. Williams and D.B. Weiner (eds): *Biologically Active Peptides: Design, Synthesis, and Utilization.* Technomic Publishing Co., Inc., Lancaster, Basel, pp 307–324.

Rose, G.D., Gierasch, L.M. and Smith, J.A. (1985) Turns in peptides and proteins. *Adv. in Prot. Chem.* 37: 1–109.

Sargent, D.F. and Schwyzer, R. (1986) Membrane lipid phase as catalyst for peptide-receptor interaction. *Proc. Natl. Acad. Sci. USA* 83: 5774–5778.

Sawyer, T.K., Sanfilippo, P.J., Hruby, V.J., Engel, M.H., Heward, C.B., Burnett, J.B. and Hadley, M.E. (1980) 4-Norleucine, 7-D-phenylalanine-a-melanocyte-stimulating hormone: A higly potent a-melanotropin with ultralong biological activity. *Proc. Natl. Acad. Sci. USA* 77: 5754–5758.

Sawyer, T.K., Hruby, V.J., Darman, P.S. and Hadley, M.E. (1982) [half-Cys^4, half-Cys^{10}]-α-Melanocyte-stimulating hormone: A cyclic α-melanotropin exhibiting superagonist biological activity. *Proc. Natl. Acad. Sci. USA* 79: 1751–1755.

Schiffer, M. and Edmundson, A.-B. (1976) Use of helical wheels to represent the structure of proteins and to identify segments with helical potential. *Biophys. J.* 7: 121–135.

Schlessinger, J. and Ullrich, A. (1992) Growth factor signalling by receptor tyrosine kinases. *Neuron* 9: 383–391.

Schwyzer, R. (1987) Membrane-assisted molecule mechanism of neurokinin receptor subtype selection. *EMBO J.* 6: 2255–2259.

Seyer, R., Aumelas, A., Tence, M., Marie, J., Bonnafous, J.-C., Jard, S. and Castro, B. (1989) Synthesis of a biotinylated, iodinatable, and photoactivatable probe for angiotensin receptors. *Int. J. Peptide Protein Res.* 34: 235–245.

Shapiro, G., Swoboda, R. and Stauss, U. (1994) FMOC solid phase synthesis of serine phosphopeptides via selective protectin of serine and on resin phosphorylation. *Tetrahedron Letters* 35: 869–872.

Songyang, Z., Shoelson, S.E., Chaudhuri, M., Gish, G., Pawson, T., Haser, W.G., King, F., Roberts, T., Ratnofsky, S., Lechleider, R.J., Neel, B.G., Birge, R.B., Fajardo, J.E., Chou, M.M., Hanafusa, H., Schaffhausen, B. and Cantley, L.C. (1993) SH2 Domains recognize specific phosphopeptide sequences. *Cell* 72: 767–778.

Stanfield, R.L., Fieser, T.M., Lerner, R.A. and Wilson, I.A. (1990) Crystal structures of an antibody to a peptide and its complex with peptide antigen at 2.8 Å. *Science* 248: 712–719.

Sutcliffe, J.G., Shinnick, T.M., Green, N. and Lerner, R.A. (1983) Antibodies that react with predetermined sites on proteins. *Science* 219: 660–666.

Tainer, J.A., Getzoff, E.D., Alexander, H., Houghten, R.A., Olson, A.J., Lerner, R.A. and Hendrickson, W.A. (1984) The reactivity of anti-peptide antibodies is a function of the atomic mobility of sites in a protein. *Nature* 312: 127–134.

Tan, R., Chen, L., Buettner, J.A., Hudson, D. and Frankel, A.D. (1993) RNA recognition by an isolated α helix. *Cell* 73: 1031–1040.

Tatemoto, K. (1982) Neuropeptide Y: complete aminoacid sequence of the brain peptide. *Proc. Natl. Acad. Sci. USA* 79: 5485–5489.

Thiele, C. and Fahrenholz, F. (1993) Photoaffinity labeling of central cholecystokinin receptors with high efficiency. *Biochemistry* 32: 2741–2746.

Tian, Z., Gu, C., Roeske, R.W., Zhou, M. and van Etten, R.L. (1993) *Int. J. Peptide Protein Res.* 42: 155–158.

Toniolo, C. (1990) Conformationally restricted peptides through a short-range cyclization. *Int. J. Peptide Protein Res.* 35: 287–300.

Veber, D.F., Holly, F.W., Nutt, R.F., Bergstrand, S.J., Brady, S.F., Hirschmann, R., Glitzer, M.S. and Saperstein, R. (1979) Highly active cyclic and bicyclic somatostatin analogues of reduced ring size. *Nature* 280: 512–514.

Vitoux, B., Aubry, A., Cung, M.T. and Marraud, M. (1986) *N*-Methyl peptides. VII. Conformational perturbations induced by *N*-methylation of model dipeptides. *Int. J. Peptide Protein Res.* 27: 617–632.

Waksman, G., Kominos, D., Robertson, S.R., Pant, N., Baltimore, D., Birge, R.B., Cowburn, D., Hanafusa, H., Mayer, B.J., Overduin, M., Resh, M.D., Rios, C.B., Silverman, L. and Kuriyan, J. (1992) Crystal structure of the phosphotyrosine recognition domain SH2 of *v-src* complexed with tyrosine-phosphorylated peptides. *Nature* 358: 646–653.

Waksman, G., Shoelson, S.E., Pant, N., Cowburn, D.and Kuriyan, J. (1993) Binding of a high affinity phosphotyrosyl peptide to the *Src* SH2 domain: Crystal structures of the complexes and peptide-free forms. *Cell* 72: 779–790.

Walter, R., Schwartz, J.L., Meuhter, A., Dousa, T. and Hoffman, P.F. (1972) Bromoacetyl-oxytocin an irreversible inhibitor of neurohypophysal hormone-stimulated adenylate cyclase and a possible affinity label for hormone receptors. *Endocrinology* 91: 213–222.

Wang, J.-X., Bray, A.M., Dipasquale, A.J., Maeji, N.J. and Geysen, M.H. (1993) Systematic study of substance P analogs. *Int. J. Peptide Protein Res.* 42: 384–391.

Watson, S. and Girdlestone, D. (1994) TIPS receptor and ion channel nomenclature supplement. *TIPS*

Webster, D.M., Henry, A.H. and Rees, A.R. (1994) Antibody-antigen interactions. *Curr. Opin. Struct. Biol.* 4: 123–129.

Metalloproteins

Zinc metallochemistry in biochemistry

B.L. Vallee[1] and D.S. Auld[2]

[1]*Center for Biochemical and Biophysical Sciences and Medicine and* [1,2]*Department of Pathology, Harvard Medical School and Brigham and Women's Hospital, 250 Longwood Avenue, Boston, MA 02115, USA*

Summary. The chemically stable but stereochemically flexible, non-toxic nature of zinc combined with its amphoteric properties has permitted it to orchestrate a number of zinc-binding motifs critical to life processes. For zinc enzymes, *catalytic, cocatalytic,* and *structural* zinc sites exist. DNA-binding proteins have zinc *fingers, twists,* and *clusters* exist.

Introduction

The last three decades have established that zinc is an integral component of numerous functional proteins with wide ranging biological properties (Vallee and Auld, 1993a–c). The resultant new insight has had an impact on virtually all aspects of molecular and cell biology. Identification of these zinc proteins, the elucidation of their biochemistry together with information generated in nutrition, physiology, medicine, and pathology have converged to establish a large body of knowledge on zinc metabolism (Vallee and Falchuk, 1993).

Overview

Zinc is the only one of the pre-, post-, and transition elements known both to be non-toxic and indispensable to all forms of life. It is critical to the transmission of the genetic message, development, growth and differentiation. Zinc is prerequisite to the anabolism and catabolism of all essential foodstuffs and their intermediates, and it functions in hormone action by regulating the structures of their receptors. All these physiological roles have been clarified through the recognition and definition of biological zinc chemistry. As a telling example, the structure of metallothionein, in particular, is not only a unique prototype of natural products in biological chemistry but has no equivalent precedent in inorganic chemistry (Vallee and Auld, 1990b).

The feature(s) of zinc chemistry which accounts for the selective utilization of this element in biology has not been self-evident. Very little, if any, zinc is free in solution in biological systems and is usually bound to enzymes and other proteins. Zinc, having a filled D-shell, does not undergo reduction or oxidation. It is amphoteric, existing both as the aquo- and hydro-metal complex at pH values near neutrality. Zinc has Lewis acid properties and ligates N and O as readily as S. It is stereochemically flexible so that it can assume multiple coordination geometries, which contribute to its biochemical versatility. Four, five, and six coordinate complexes, however, are the ones encountered most frequently in enzymes and other, biological molecules, with geometries ranging from regular or distorted tetrahedral to trigonal bipyramidal, square pyramidal, and octahedral.

The visible absorption, magnetic circular dichroism, and electron paramagnetic resonance spectra of catalytically active, cobalt-substituted zinc metalloenzymes are unusual and unlike those of typical cobalt complex ions (Maret and Vallee, 1993). The three-dimensional structure of the molecule and the heterogeneity of the ligands jointly generate the typical coordination properties at active sites that have been defined as "entatic" (Vallee and Williams, 1968). This descriptive term signifies a state of tension or stress at the active site of an enzyme. Denaturation abolishes both enzymatic function and these atypical spectra considered to be indicative of a biologically active state in which the metal is "poised for catalysis."

Collectively, the physicochemical features of zinc are important means for the translation of chemical structure into multiple biological functions. Biology has evolved very varied permutations of zinc-protein coordination complexes with structures and conformations which serve both enzyme function and the expression of the genetic message. X-ray diffraction and NMR structure determination of 22 enzymes, five gene regulatory proteins and metallothionein have provided absolute standards of reference for zinc ligands which are characteristic of protein families. For zinc enzymes catalytic, cocatalytic, and structural zinc sites have been recognized, while for the gene regulatory proteins, zinc fingers, twists, and clusters have become proverbial (Vallee and Auld, 1990b, 1992a, 1993a–c).

Zinc enzymes

In its catalytic role the metal participates directly in enzymatic action (Fig. 1A). Removal by chelating or other agents inactivates the enzyme. Cocatalytic zinc atoms enhance catalytic function in conjunction with an active site zinc atom in the same enzyme (Fig. 1B). Structural zinc atoms essentially stabilize the protein (Fig. 1C).

Catalytic zinc sites
In 14 enzymes whose X-ray structure is known there is a single catalytic zinc atom. In such catalytic sites zinc forms complexes with any three

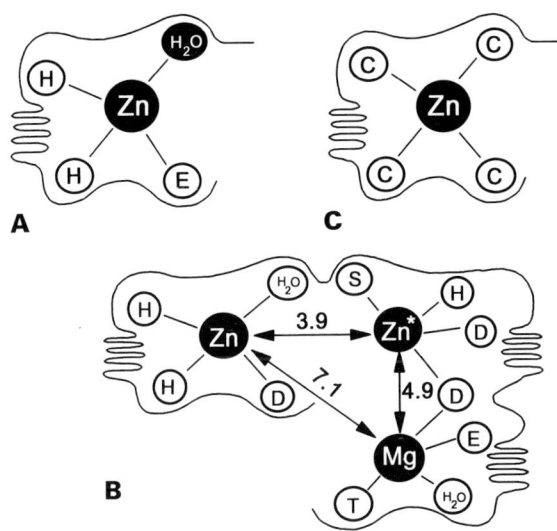

Figure 1. Zinc-binding sites in zinc enzymes: (A) catalytic, (B) cocatalytic, (C) structural.

nitrogen and oxygen donors of His and Glu, and the binding frequency is His ≫ Glu ≫ Cys. When this count includes the multizinc enzyme ADH and the catalytic-like zinc sites of cocatalytic enzymes, the binding frequency becomes His ≫ Glu > Asp > Cys (Fig. 2) (Vallee and Auld, 1992a, 1993a–c). A water molecule is the fourth ligand of the catalytic zinc atom (Vallee and Auld, 1989, 1990a).

Three histidine ligands are typical of the lyases human carbonic anhydrase I and II (Kannan et al., 1975; Liljas et al., 1972) and the hydrolases *Bacillus cereus* β-lactamase (Sutton et al., 1987), DD-carboxypeptidase of *Streptomyces albus* G (Dideberg et al., 1982), adenosine deaminase (Wilson and Quiocho, 1993), astacin (Gomis-Rüth et al., 1993b) alkaline protease of *Pseudomonas aeruginosa* (Baumann et al., 1993), adamalysin II of *Crotalus adamanteus* (Gomis-Rüth et al., 1993a) and the catalytic domain of fibroblast collagenase (Lovejoy et al., 1994). Two histidines are characteristic of the hydrolases bovine carboxypeptidase A and B (Quiocho and Lipscomb, 1971; Schmid and Herriott, 1976); thermolysin, the neutral protease of *B. thermoproteolyticus* (Matthews et al., 1972); *B. cereus* neutral protease (Pauptit et al., 1988); *Pseudomonas aeruginosa* elastase (Thayer et al., 1991); *B. cereus* phospholipase C (Hansen et al., 1993) and *Penicillium citrinum* nuclease P1 (Volbeda et al., 1991). One histidine occurs in the catalytic zinc site of ADH (Brändén et al., 1975) and cytidine deaminase (Betts et al., 1994). The latter sites are also unique in that they contain two cysteine ligands. Glutamate is the oxygen donor in six and aspartate in two of these enzymes. Overall, in catalytic sites zinc prefers the imidazolyl nitrogen by far.

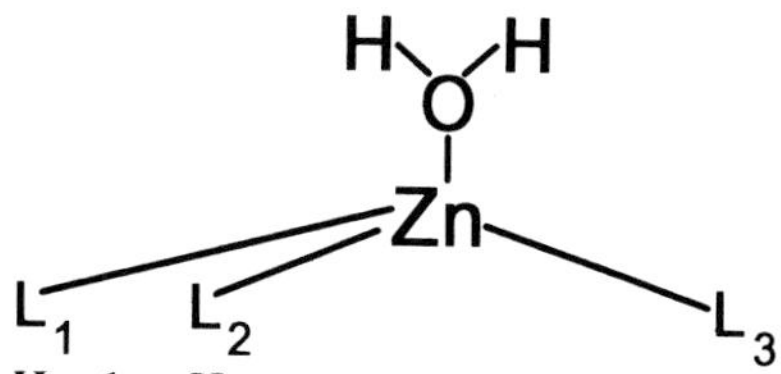

	L_1		L_2		L_3
Carbonic Anhydrase I, II	H	1	H	23	H
B-Lactamase	H	1	H	121	H
Adenosine Deaminase	H	1	H	196	H
Di D-Carboxypeptidase	H	2	H	40	H
Astacin	H	3	H	5	H
Adamalysin II	H	3	H	5	H
P. aeruginosa Alk. Protease	H	3	H	5	H
Cat.Domain Fibro.Collagenase	H	3	H	5	H
Thermolysin	H	3	H	19	E
B. cereus Neutral Protease	H	3	H	19	E
P. aeruginosa Elastase	H	3	H	19	E
Carboxypeptidase A, B	H	2	E	123	H
Phospholipase C	H	3	E	13	H
Nuclease P1	D	3	H	12	H
Alkaline Phosphatase	D	3	H	80	H
Cytidine Deaminase	C	2	C	26	H
Alcohol Dehydrogenase	C	20	H	106	C

Figure 2. Zinc ligand spacing in catalytic zinc sites.

H_2O, the fourth ligand of active-site catalytic zinc atoms, is likely the very objective of this motif's design: the zinc-bound water can be ionized, polarized or displaced in the catalytic process. The nature of the zinc ligands and the degree of ionization of the bound water significantly affects the net charge of the resulting active site complex; this can range from a di-cation for the lyase, carbonic anhydrase, to a mono-cation for many of the hydrolases to a neutral complex for ADH. Ionization of the activated water or its polarization brought about by a base form of an active-site amino acid provides hydroxide ions at neutral pH, and its displacement results in Lewis acid catalysis on the part of the catalytic zinc atom (Vallee and Auld, 1990a, 1993c). The structure of the active site implies that both the identity of the three protein ligands and their spacing underlie mechanistic pathways to activate water, determining details of the ensuing catalytic reactions, which is accomplished in conjunction with other active-site residues.

Ligand spacing and support structure
The regularity of amino acid spacing between the ligands of catalytic zinc atoms is striking (Vallee and Auld, 1989, 1990a,b). In 18 of the 19 catalytic zinc sites "short" spacers, consisting of from one to three

amino acids, separate the first two ligands, L_1 and L_2 (Fig. 2). When properly oriented, L_1 and L_2 can apparently form a bidentate zinc complex. It is equally characteristic that a "long" spacer, from 5 to 196 residues, separates L_3 from both L_1 or L_2, generally located in the C-terminal end of the protein. This long spacer arm could contribute to the completion of the active catalytic site, substrate-binding groups and hydrogen bonds to form the active center.

The zinc binding sites frequently have an α-helical or β-structural region of the protein that supplies the zinc ligands of the catalytic zinc sites (Vallee and Auld, 1992a). The length of the short spacer conditions the support structure. Thus, in carbonic anhydrase the short spacer consists of but one amino acid. As a consequence, the ligands are provided by the same side of the β-sheet. In contrast, when the spacer consists of three amino acids, as, for example, in thermolysin, an α-helix support structure becomes feasible and juxtaposes the ligands in an orientation suitable to establish a tetrahedral-like coordination sphere.

These structural features of zinc metalloenzymes further call attention to the importance of protein folding and conformation which are the basis of and maintain the structure and function of proteins. In zinc enzymes, the primary and secondary structure enveloping the short and long spacer arms might contain information on and give directions to the creation of zinc complexes with suitable coordination geometries and numbers. Zinc, and chromophoric replacement metals, may be useful for probing the folding process.

Such considerations will no doubt bear on the design and synthesis of enzyme model systems. Overall, the catalytic potentials of zinc enzymes seem closely related to the nature of the short and long amino acid spacers and the environment that they create for metal ligands. Their incorporation into synthetic designs should reflect the potentials of catalysis and specificity inherent in zinc enzymes.

Families of catalytic zinc sites
Results of the X-ray structure determinations of these 19 catalytic zinc sites serve as standards of reference for other enzymes with similar functional characteristics in families of both identical and closely related enzymes from many different species (Vallee and Auld, 1990b, 1993b,c). Both their structural identities (or the converse) and conformations of their active and structural enzymatic zinc ligands can be recognized by comparing their amino acid sequences with those of the enzymes whose three-dimensional structures have been determined.

Such findings have predicted the catalytic zinc binding sites of enzymes whose structures have not as yet been determined, e.g., monozinc aminopeptidases and leukotriene A_4 hydrolase (Vallee and Auld, 1990b). In the mono-zinc human intestinal, rat kidney, and *E. coli* aminopeptidases, a domain of ~300 amino acids contain two His and one Glu, similar to the zinc binding site of thermolysin. The three amino acids of the short spacer in thermolysin are identical to those of the mono-zinc aminopeptidases. The long spacer in thermolysin contains 19 amino acids, corresponding to 18 amino acids in aminopeptidases (Vallee and Auld, 1990b). This structural analysis of the zinc binding sites in those zinc enzymes predicted the presence of zinc an a hitherto unrecognized aminopeptidase activity in leukotriene A_4 hydrolase. It contains an amino acid segment that is homologous to the zinc binding domain of intestinal aminopeptidase. On this basis, the leukotriene A_4 hydrolase was then shown to contain 1 g-at zinc/mol protein, to exhibit aminopeptidase activity and to be inhibited by bestatin and captopril, specific peptidase inhibitors (Haeggström et al., 1990). Moreover, mutagenic replacements of leukotriene A_4 zinc ligands completely abolish its activity (Haeggström et al., 1990; Medina et al., 1991; Vallee and Auld, 1992a), the first proof of the identity of a hitherto unknown zinc enzyme binding site by mutagenesis.

A class of metalloendopeptidases having the same short spacer and His ligands seen in thermolysin and leukotriene A_4 hydrolase, but with the potential catalytic Glu adjacent to the second His (HxxEH), has also recently been proposed (Becker and Roth, 1992). This group comprises protease III from *E. coli* and the human and *Drosophila* insulin degrading enzymes. Mutagenic replacement of the two potential His ligands and the catalytic Glu group lead to mutants devoid of proteolytic activity. The Glu mutant still contains a stoichiometric amount of zinc while the His mutants contain none, consistent with their assigned roles. The third zinc ligand was not identified.

A considerable number of zinc endoproteases and homologous proteins have recently been sequenced which contain the predicted zinc binding site signature, HExxHxxGxxH (Stöcker et al., 1990; Auld, 1992; Vallee and Auld, 1992b) of astacin (Fig. 3). These enzymes include a number of early developmental proteins that are homologous to astacin (Gomis-Rüth et al., 1993b), the snake venom endoproteases represented by adamalysin II (Gomis-Rüth et al., 1993a), the large bacterial proteases represented by the alkaline protease of *P. aeruginosa* (Baumann et al., 1993) and all the collagenases and stromelysins represented by the structure of the catalytic domain of fibroblast collagenase bound to an inhibitor (Lovejoy et al., 1994). The two histidines, separated by a short spacer of three amino acids, are supplied by a central α-helix. The conserved Gly breaks the helix, causing a sharp turn in the peptide backbone, thus placing the third histidine in a zinc ligating position.

Outside of the catalytic zinc binding site region these proteins contain little primary structure identities. In fact, the catalytic domain of the fibroblast collagenase contains a second zinc site in agreement with zinc determinations of the matrilysin zymogen, its activated form and the catalytic

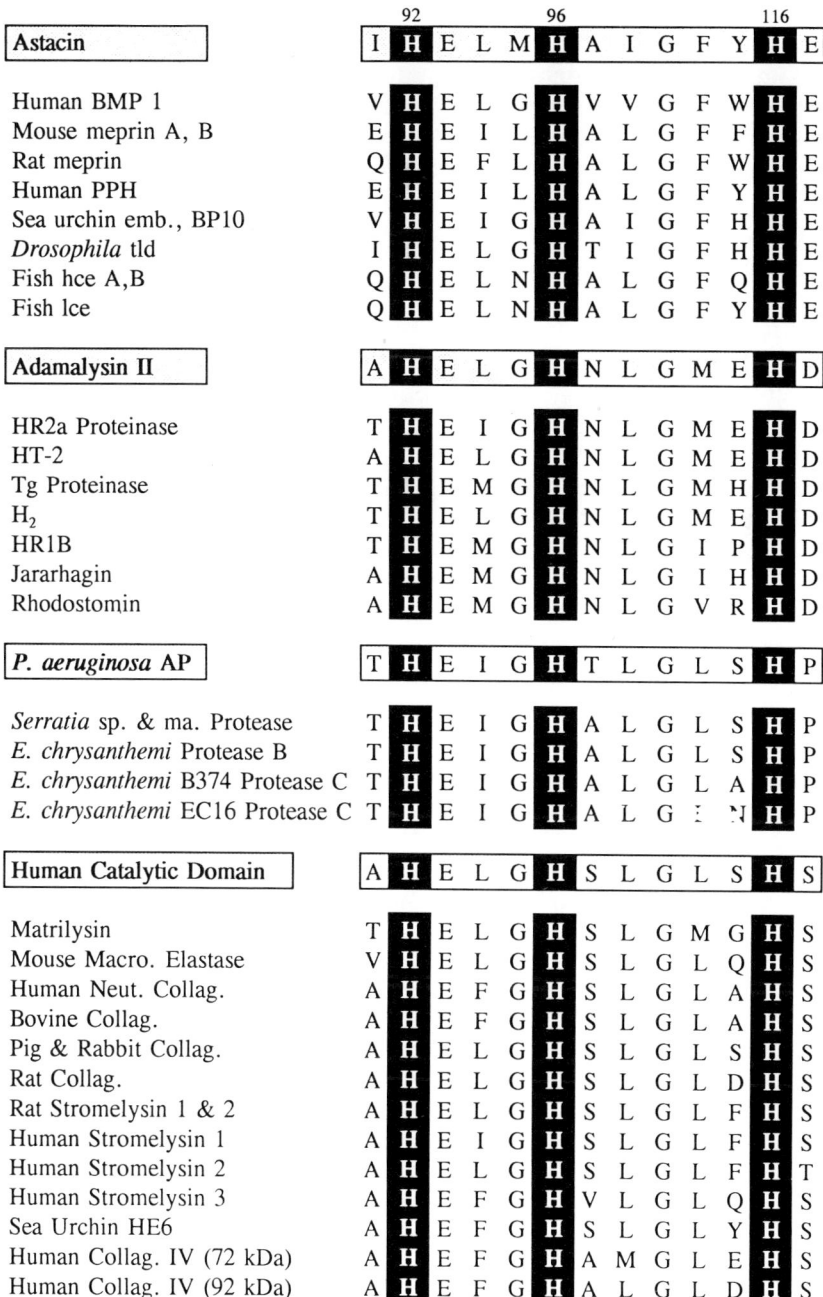

Figure 3. Zinc ligands of the extended astacin family of zinc endopeptidases. The lightly shaded boxes denote the X-ray standard of reference. Sequences were obtained from the National Biomedical Research Foundation Protein Identification Resource, The Swiss Prodata base and translations of coding regions of DNA-sequences from GenBank/EMBL data base.

domain of stromelysin-1 (Soler et al., 1994). Thus, it is quite remarkable that these enzymes are topologically similar molecules, consisting of an upper domain containing a central 5-stranded β-pleated sheet flanked by two long helices and a lower domain which is organized in a more irregular manner (Bode et al., 1993). The catalytic zinc binding site which contains most of the likely components critical to catalysis, i.e., the zinc and its associated ligands, the activated water and the catalytic Glu may have thus influenced the overall fold of these small proteases.

Cocatalytic zinc sites

These occur in multi-metal zinc enzymes where – prior to the X-ray structure determination – the function of these metal atoms was dubbed "modulating" or "regulatory" (Vallee, 1983; Vallee and Auld, 1992b). X-ray structures of these enzymes in numbers sufficient to allow the systematization of their zinc binding sites have become available only in the course of the last few years.

Thus far, in *E. coli* alkaline phosphatase, in addition to the active site zinc atoms, these have been identified as zinc and magnesium atoms (Kim and Wyckoff, 1991) as two zinc atoms in *B. cereus* phopholipase C (Hansen et al., 1993) and *Pencillium citrinum* P1 (Volbeda et al., 1991), and as zinc or magnesium in bovine lens leucine aminopeptidase (Burley et al., 1992). In all of these enzymes one zinc atom is catalytic (**Zn**) while the remaining pair of metal atoms, **Zn*** and **Z̄n** or Mg, are linked by a bridging amino acid, usually Asp, so that all three metal atoms are in close physical proximity to one another as exemplified for alkaline phosphatase in Figure 1B. In such zinc sites a single amino acid residue, either Asp or Glu, simultaneously binds to two zinc atoms or a zinc and a magnesium atom to form a bridge between two of the participating metal atoms. Alteration in the bond lengths or dissociation of either or both of these atoms from their respective metals during substrate interaction with the enzyme or formation of intermediates could profoundly affect catalysis due to the change in charge imposed upon the metals (Vallee and Auld, 1993a,b).

There are yet other features that set the coordination of their cocatalytic (**Zn*** and **Z̄n**/Mg) apart from that of **catalytic Zn** sites. While His and Cys predominate as ligands of catalytic and structural zinc atoms, Asp predominates in cocatalytic zinc sites where the frequency is Asp > His > Glu. Ser, Thr, Lys and Trp turn out to be ligands to the **Zn*** and **Z̄n**/Mg sites (Vallee and Auld, 1993a,b). The ionizable groups of such ligands have high pK_a values. Hence, at neutrality their affinity for binding zinc is relatively lower than that of the His-imidazole N, Glu and Asp, carboxylate O and Cys thiol S donor groups. Moreover, the amide carbonyl ligands of Trp should also coordinate zinc weakly. Thus, when the substrate enters the coordination sphere, ligand displacement and/or increased coordination would be plausible with a cocatalytic mechanism.

The cocatalytic metal-atoms may play alternate roles in different steps of catalysis. Thus, for the phosphate ester hydrolyzing enzymes interaction of the substrate with their cocatalytic zinc site may involve initial contact with all three metals as is observed in their final phosphate product complex (Kim and Wyckoff, 1991; Volbeda et al., 1991; Hansen et al., 1993). One of the zinc atoms may then polarize a nucleophile, e.g., either water or Ser-102 in the case of alkaline phosphatase, to attack an amide or a phosphoester bond. A second zinc atom could act transiently as the receptor for the amine or alcohol leaving group. In subsequent steps the role of these zinc atoms may be reversed. The first zinc atom may polarize a phospho-oxygen or carbonyl bond of an acyl intermediate for the attack of water promoted by the second zinc atom. The bridging Asp and H_2O ligands could have critical roles in this process. Thus, their dissociation from either or both metal atoms during catalysis could change the charge on the metal promoting the metal's action as a Lewis acid or allowing interaction with an electronegative atom of the substrate. Alternatively, the bridging ligand may participate transiently in the reaction as a nucleophile or general acid/base catalyst. In this manner the metal atoms and their associated ligands would play specific roles in each step of the reaction that works in concert to bring about catalysis.

Cocatalytic sites in zinc metalloenzymes would seem to represent an unusual biological utilization of zinc chemistry directed toward different ends of biological needs. Recent mutagenic and X-ray crystallographic studies of *E. coli* alkaline phosphatase have demonstrated that replacement of aspartic acid by histidine at position 153 results in conversion of the octahedral Mg binding site into a distorted tetrahedral $\bar{Z}n$ binding site (Murphy et al., 1993). In the native enzyme the aspartate carboxylate oxygens H-bond to Mg bound water molecules while in the histidine mutants the imidazole nitrogen is a ligand to the zinc. Mg activates the enzyme in a time-dependent fashion likely due to a slow displacement of the $\bar{Z}n$. The structure of the Mg bound mutant enzyme has yet to be determined. The discernment of the structure/function relationships in cocatalytic sites likely heralds yet other motifs to be discovered.

Structural zinc sites
Thus far, structural zinc atoms have been found only in alcohol dehydrogenase (Brändén et al., 1975) and aspartate transcarbamylase (Honzatko et al., 1982). In ADH and aspartate transcarbamylase, the zinc is fully coordinated tetrahedrally to four cysteines. The various cysteine ligands in ADH and aspartate transcarbamylase are separated by 2, 2 and 7 or 4, 22 and 2 amino acids, respectively (Vallee and Auld,

1990a,b; Vallee and Auld, 1992a). However the second zinc site observed in fibroblast collagenase (Lovejoy et al., 1994) may be the first non-cysteinyl structural zinc binding site. It is comprised of three histidine ligands and an aspartate tetrahedrally coordinated to the zinc with spacing intervals of 1, 12, and 12. As expected for a structural site there is no coordinated water molecule.

Synthetic Zn enzyme binding sites
Recognition, delineation and generalization of the scaffolding of the catalytic and structural sites of zinc enzymes provide a novel theoretical basis for the design and synthesis of model systems. Both the catalytic and structural potential of zinc in enzymes depend on the characteristics of the short and long spacers. For catalytic zinc atoms, one would expect that, minimally, models would have to mimic these features to achieve the potential of catalysis and specificity. Structural zinc atoms would be expected to induce and control folding of the peptide chain as well as local and overall conformation typical of a native enzyme.

A first step in the direction of synthesizing a peptide having the characteristics typical of a structural zinc atom has been taken with ADH. The structural zinc atom of mammalian alcohol ADHs is coordinated tetrahedrally by a short peptide segment containing four Cys ligands that constitute an isolated loop of the protein that contributes to subunit and substrate interactions. A 23-residue synthetic peptide which incorporates the loop stoichiometrically binds zinc or cobalt in a tetrahedral coordination geometry that mimics the metal-binding properties of intact ADH (Bergman et al., 1992). On the basis of absorption and magnetic circular dichroic spectra of the peptide, the product is virtually identical to that of the structural site of horse and human $(\beta_1\beta_1)$ liver ADHs.

The relationship of the overall catalytic potential of zinc enzymes to the nature of the amino acid spacers and the creation of an environment for metal ligands suitable to catalysis is relevant to the design of zinc enzyme models in general. The design of systems based on the above consideration should lead to the synthesis of molecules with both the specificity of and capacity for catalysis as well as the requisite primary structures and conformations characteristic of zinc enzymes (Vallee and Auld, 1992a).

Gene regulatory proteins

These are involved in DNA transcription and contain functionally essential zinc atoms. The involvement of zinc in the regulation of their activities has attracted much attention and generated a large body of data that promises to reveal how specific genes are expressed and what role zinc may have in this process. The structures of the DNA binding domain of three types of zinc activator proteins critical to replication and/or transcription have been determined by X-ray diffraction and NMR. As many as 500 (or more) of these molecules are presumed to contain zinc (Rhodes and Klug, 1993).

The critical role of zinc in DNA and RNA synthesis and cell division became apparent in the 1970s (Vallee, 1977; Auld, 1979; Vallee and Falchuk, 1981). In 1983, Wu and his collaborators found that *Xenopus* transcription factor IIIA (TFIIIA), which activates the transcription of the 5S RNA gene, contains 2–3 mol of zinc per mol of protein (Hanas et al., 1983), focusing attention on a specific role for zinc in transcription. "Zinc finger" has rapidly both become a part of the language and a concept conveying even more general ideas.

Zinc fingers
X-ray crystallography of a native zinc-finger protein (Pavletich and Pabo, 1991, 1993) together with NMR data on single or double synthetic "fingers" (Vallee and Auld, 1993c) provide direct evidence for the structure of zinc fingers that serve as a prediction of function. A definitive understanding of how these zinc-finger domains serve in site-specific recognition of a particular DNA-binding protein first came from the X-ray diffraction analysis of the three zinc finger DNA-binding domain derived from the mouse immediate early protein Zif 268 complexed with a consensus DNA-binding site (Pavletich and Pabo, 1991).

Each zinc-finger domain consists of an antiparallel β-sheet containing two Cys and an α-helix containing two His held together by coordination of the Cys and His residues to the central zinc ion and by a set of hydrophobic residues. The interatomic distance between Zn(1) and Zn(2) is 26.6 Å and that between Zn(2) and Zn(3) is 27.4 Å (Vallee and Auld, 1992b). Importantly, in each zinc finger one zinc atom is coordinated to two His and two Cys as its base.

Each of the three zinc fingers uses Arg, Asp, and His residues from the N-terminal portion of its α-helix to make contact with guanine base pairs in the major groove (Fig. 4A). These residues derive from the central peptide loop between the second Cys and first His of each "finger" and include the amino acid immediately preceding the α-helix and the second, third and sixth ones of the α-helix.

The results of these studies demonstrate how zinc makes a direct and unsuspected contribution to the overall binding energy. The tetrahedral geometry around the zinc atom orients the finger for site-specific interaction. Zinc also promotes a specific interaction of the first ligating His to the DNA backbone; the Nϵ of His coordinates the zinc while the Nδ hydrogen bonds to the phosphodiester oxygens.

The crystal structure of a complex of the five zinc fingers from human GLI oncogene and a high affinity DNA binding site has been deter-

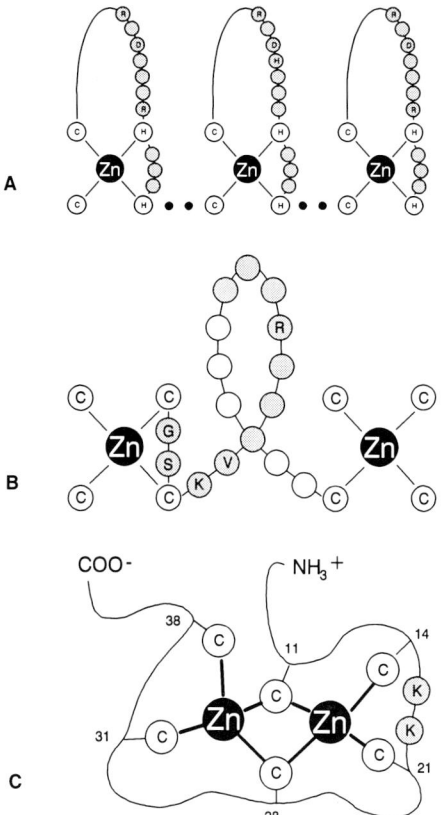

Figure 4. Schematic of (A) zinc fingers of Zif 268, (B) zinc twist of glucocorticoid receptor, and (C) zinc cluster of GAL4. The amino acids involved in DNA recognition are labeled in shaded circles.

mined at 2.6 Å resolution (Pavletich and Pabo, 1993). Recognition of the DNA still comes from the central peptide loop of individual fingers, but surprisingly not all the fingers are involved. Fingers 4 and 5 of GLI appear to be the most important for recognition of specific bases. Fingers 2 and 3 generally only contact the DNA backbone. Finger one does not contact DNA at all. In addition new amino acid residues, e.g., Tyr, Ala, and Ser in addition to Lys, Arg, and Asp are involved in binding interactions with specific bases. The prediction of how many zinc fingers and which central loop residues are important for recognition is thus still open to investigation.

Zinc twists
The glucocorticoid (GR) and estrogen (ER) receptors belong to the nuclear receptor class of proteins. Upon binding to a hormone, the

receptor translocates from the cytoplasm to the nucleus where it binds to a specific DNA sequence (glucocorticoid or estrogen responsive elements, GRE or ERE) and thereby modulates transcription. The ^{1}H and ^{113}Cd NMR solution structures of the 71 amino acid segment of the DNA-binding domain of GR (Härd et al., 1990; Pan et al., 1990; Kellenbach et al., 1991) and ^{1}H-NMR structure of a corresponding 84 amino acid fragment of ER (Schwabe et al., 1990) show two zinc atoms each coordinated to four cysteines in these particular receptors, separated by ~ 13 and 12 Å, respectively.

Remarkably, these zinc-binding sites are not independent zinc finger-like substructures, but rather fold to form a single structural domain (Härd et al., 1990; Schwabe et al., 1990). The DNA-binding recognition site derives from the α-helix that is part of the linking peptide region between the two zinc sites, accounting for our choice of calling this a zinc twist (Vallee et al., 1991) (Fig. 4B). A second α-helix which is anchored by the C-terminal coordinating cysteines of the second zinc site perpendicularly crosses the first helix near its midpoint, stabilizing it by hydrophobic interaction between the two helices. The globular peptide domain of the resultant zinc twists has a DNA recognition site between the two zinc atoms, each of which is coordinated to four Cys. This is in marked contrast to the arrangement encountered in zinc fingers where the DNA recognition site comes from within the zinc coordination site where zinc is coordinated to two His and two Cys (Fig. 4A).

X-ray crystallographic analysis of the DNA-binding domain of GR complexed to either a GRE consensus sequence containing a symmetric but abnormal spacing of four nucleotides, GRE_{4s} or a natural three nucleotide space GRE_{3s}, shows how the presence of DNA profoundly influences the structure of GR (Luisi et al., 1991). In both cases the presence of DNA causes the GR domain to dimerize, a situation which is not encountered in the NMR solution studies when GRE is not present, even when [GR] is 1 mM. In addition, if the three nucleotide spaced GRE_{3s} is used, each DNA recognition site of the dimer does interact properly with the major groove of the DNA, while one-half of the dimer is out of register when the abnormal GRE_{4s} is used to form the complex.

The mode of interaction of GR with DNA is in general agreement with the molecular models for dimer interaction with DNA as proposed on the basis of NMR studies of uncomplexed monomeric DNA-binding domains of GR and ER (Härd et al., 1990; Schwabe et al., 1990). The two zinc atoms are coordinated tetrahedrally to the expected eight cysteines with the central α-helix of the zinc twist forming a complex with the major groove of the DNA. These studies further stress the importance of Arg-466, Val-462, and Lys-461 in this interaction (Vallee and Auld, 1993c). The x-ray diffraction studies further reveal how the second zinc-binding site influences the dimerization process. Ala-477, Arg-479, and Asp-481 from the five amino acid loop between the first two cysteines and Ile-483, Ile-487, and Asn-491 from the nine amino acid loop between the second and third cysteines makes critical dimer interface contacts (Luisi et al., 1991).

The function of zinc in the nuclear receptor family is clearly twofold. It stabilizes the helix involved in DNA recognition and aids in orienting the peptide fold of the second zinc site which is critical to the dimerization process.

Metallothionein
The first zinc enzymes and metallothionein were discovered concurrently. Their metal content has long been known to be extraordinary (Margoshes and Vallee, 1957), but the function of this molecule has remained elusive much as it has received ever increasing attention (Kägi, 1991). An understanding of the relationship between zinc and the structures of this protein is essential to a comprehension of the metal's specific roles in the activities of this and other zinc-containing molecules and their overall biological importance.

Structural studies of native mammalian metallothionein demonstrate that the arrangement of their 7 g-at zinc and/or cadmium/mol is unusual. The metal atoms are present in the form of clusters in both their solution and crystal structures (Braun et al., 1992). Remarkably, such a cluster structure has not been described hitherto for any organic, naturally occurring zinc and/or cadmium coordination complex and is unique so far to biology. The zinc atoms of Zn_7-metallothionein are organized into two distinct metal clusters, Zn_4Cys_{11} (residues 33–60) and Zn_3Cys_9 (residues 5–29), with five and three cysteine residues acting as bridging ligands between two metal atoms in each cluster, respectively. The interatomic distance between the single thiolate-bridged zinc atoms in the Zn_2Cd_5 metallothionein crystal is 3.88 Å with clusters reminiscent of iron-sulfur in ferredoxin and rubredoxins. It is conceivable that the two metal thiolate clusters serve to regulate the metabolic roles of these metals as indicated by fluctuations in the clusters and consequent intramolecular metal exchange (Vasák et al., 1985). Spontaneous mutual exchange among binding sites occurs within minutes and seconds in cluster A and B, respectively (Otvos et al., 1989).

The metal-free form of the protein, thionein, can remove zinc from the gene activator protein TFIIIA (Zeng et al., 1991). This may bear on potential interprotein metal exchange and the physiological function of metallothionein, particularly in regard to the GAL4 DNA binding proteins.

Zinc clusters

The GAL4 transcription factor from *Saccharomyces cerevisiae*, an 881 amino acid transcription factor that in the presence of galactose regulates the expression of genes encoding the DNA-binding domain, has been expressed in *E. coli*. Various Gal 4 and Lac 9 expressed fragments contain 2 moles of zinc per mole of protein (Vallee and Auld, 1992a).

^{1}H and ^{113}Cd NMR studies of either two zinc or two cadmium GAL4(62*) (Pan and Coleman, 1991) and GAL4(7-49) (Gadhavi et al., 1991) have shown that the two metal atoms coordinate to the six cysteines in a binuclear zinc thiolate cluster (Fig. 4C). The interatomic zinc-zinc distance for this zinc cluster is ~ 3.5 Å, reflecting a structure of this zinc coordination site that is distinctly different from those occurring in zinc fingers and zinc twists (Vallee et al., 1991). Cys-11 and -28 are the bridging ligands while the Cys-14/21 and Cys-31/38 pairs bind as monodentate ligands to the first and second zinc atoms, respectively (Pan and Coleman, 1991; Gadhavi et al., 1991).

Much as the Cys spacings (2, 6, 6, and 2 amino acids) in the first zinc coordination site of GR and that of GAL4 are identical, the structures of their zinc complexes differ significantly (Figs 4B, 4C). Furthermore, it is apparent that some Cys (e.g., found in the GR and ER receptors are neither involved in binding zinc, nor do they affect the capacity of some Cys residues to form bridging ligands between two zinc atoms. The primary sequences of GAL4 and GR are necessary but not sufficient to permit definitive structural predictions. Considering the limited number of structure determinations of DNA-binding zinc proteins broad extrapolations from primary to tertiary structure and their relationship to function would seem premature.

The DNA-recognition sites of zinc clusters have been suggested to be in the region of residues 12 to 19 based on two findings (Vallee and Auld, 1992a): this region has a propensity to form an α-helix (Gadhavi et al., 1991) and in the GAL4 family of fungal transcription factors it contains several Lys and Arg residues (Vallee et al., 1991). The side chain of such amino acids thus could interact with DNA bases and/or the phosphate backbone.

The crystal structure of the yeast transcription factor fragment GAL4(1-65) bound to a 17 bp DNA fragment that is a consensus of 11 known GAL4 binding sites has recently been reported (Marmorstein et al., 1992). The GAL4(1-65) binds to the DNA as a symmetrical dimer. Each subunit folds into three distinct modules: the zinc cluster (8-40), extended linker (41-49), and an α-helical dimerization element (50-64). The zinc cluster contains two short α helices, residues 10-22 and 27-39, with the first and fourth being zinc binding Cys residues. The zinc cluster recognizes a conserved CCG triplet at each end of the site through direct contacts with the major groove.

Specific base interactions occur at Lys-17 and -18 (Fig. 4C). The carbonyl of the amide of Lys-17 accepts a hydrogen bond from the N4 nitrogen of cytosine 8 while the carbonyl of Lys 18 accepts H bonds from N4 nitrogens of both cytosines 7 and 8. The ϵ-amino group donates H-bonds to the O6 of guanine 6 and to N7 of guanine 7. Zinc stabilizes these interactions since it is coordinated to both Cys-14 and Cys-21, both part of this α-helical region and flanking the Lys residues critical to recognizing the DNA triplet.

^{1}H-^{113}Cd two-dimensional NMR spectra of the GAL4(1–65) peptide (Baleja et al., 1992) and the GAL4(7–49) peptide (Kraulis et al., 1992) binding to DNA generally show the same main features as are seen in the crystal structure (Marmorstein et al., 1992).

Clearly, the results on GAL4, GR, ER, Zif 268 and GLI extend to members of those protein families. Other proteins may yet be shown to contain multiple zinc atoms which might encompass variable numbers and types of ligands, e.g., glutamic or aspartic acid and histidine. When such proteins contain multiple zinc centers, their zinc-zinc distances may well prove important indices both of overall structure and function in any given zinc protein. As additional structures become available, the detection of yet further motifs revealing other permutations of zinc chemistry may be expected.

Acknowledgements
This research was supported in part by NIH Grants GM 47534 and 53265.

References

Auld, D.S. (1979) The role of zinc biochemical processes. *Adv. Chem. Ser.* 172: 112–133.
Auld, D.S. (1992) The astacin family. *Faraday Discuss.* 93: 117–120.
Baleja, J.D., Marmorstein, R., Harrison, S.C. and Wagner, G. (1992) Solution structure of the DNA-binding domain of Cd2-GAL4 from *S. cerevisiae*. *Nature* 356: 450–453.
Baumann, U., Wu, S., Flaherty, K.M. and McKay, D.B. (1993) Three-dimensional structure of the alkaline protease of *Pseudomonas aeruginosa*: a two-domain protein with a calcium binding parallel beta roll motif. *EMBO J.* 12: 3357–3364.
Becker, A.B. and Roth, R.A. (1992) An unusual active site identified in a family of zinc metalloendoproteases. *Proc. Natl. Acad. Sci. USA* 89: 3835–3840.
Bergman, T., Jörnvall, H., Härd, T., Holmquist, B. and Vallee, B.L. (1993) A synthetic approach to analysis of the structural zinc site of alcohol dehydrogenase. *Adv. Exp. Med. Biol.* 328: 419–428.
Betts, L., Xiang, S., Short, S.A., Wolfenden, R. and Carter, C.W. (1994) The 2.3 Å crystal structure of an enzyme: transition-state analog complex. *J. Mol. Biol.* 235: 635–656.
Bode, W., Gomis-Rüth, F.X. and Stöcker, W. (1993) Astacins, serralysins, snake venom and matrix metalloproteinases exhibit identical zinc-binding environments (HEXXHXXGXXH and Met-turns) and topologies and should be grouped into a common family, the "metzinins." *FEBS Lett.* 331: 134–140.
Brändén, C.I., Jörnvall, H., Eklund, M. and Furugren, B. (1975) Alcohol dehydrogenase. *Enzymes*: 11, Part A: 103–190.
Braun, W., Vasák, M., Robbins, A.H., Stout, C.D., Wagner, G., Kägi, J.H.R. and Wüthrich, K. (1992) Comparison of the NMR solution structure and the x-ray crystal structure of rat metallothionein-2. *Proc. Natl. Acad. Sci. USA* 89: 10124–10248.

Burley, S.K., David, P.R., Sweet, R.M., Taylor, A. and Lipscomb, W.N. (1992) Structure determination and refinement of bovine lens leucine aminopeptidase and its complex with bestatin. *J. Mol. Biol.* 224: 113–140.

Dideberg, O., Charlier, P., Dive, G., Joris, B., Frère, J.M. and Ghuysen, J.M. (1982) Structure of a Zn^{2+}-containing D-alanyl-D-alanine-cleaving carboxypeptidase at 2.5 Å resolution. *Nature* 299: 469–470.

Gadhavi, P.L., Davis, A.L., Povey, J.F., Keeler, J. and Laue, E.D. (1991) Polypeptide-metal cluster connectivities in Cd(II) GAL4. *FEBS Lett.* 281: 223–226.

Gomis-Rüth, F.X., Kress, L.F. and Bode, W. (1993a) First structure of a snake venom metalloproteinase: a prototype for metalloproteinases/collagenases. *EMBO J.* 12: 4151–4157.

Gomis-Rüth, F.X., Stöcker, W., Huber, R., Zwilling, R. and Bode, W. (1993b) Refined 1.8 Å X-ray crystal structure of astacin, a zinc-endopeptidase from the crayfish *Astacus astacus* L. Structure determination, refinement, molecular structure and comparison with thermolysin. *J. Mol. Biol.* 229: 945–968.

Haeggström, J.Z., Wetterholm, A., Vallee, B.L. and Samuelsson, B. (1990) Leukotriene A4 hydrolase: a zinc metalloenzyme. *Biochem. Biophys. Res. Commun.* 173: 431–437.

Hanas, J.S., Hazuda, D.J., Bogenhagen, D.F., Wu, F.Y.-H. and Wu, C.-W. (1983) *Xenopus* transcription factor A requires zinc for binding to the 5 S RNA gene. *J. Biol. Chem.* 258: 14120–14125.

Hansen, S., Hough, E., Svensson, L.A., Wong, Y.-L. and Martin, S.F. (1993) Crystal structure of phospholipase C from *Bacillus cereus* complexed with a substrate analog. *J. Mol. Biol.* 234: 179–187.

Härd, T., Kellenbach, E., Boelens, R., Kaptein, R., Dahlman, K., Carlstedt-Duke, J., Freedman, L.P., Maler, B.A., Hyde, E.I., Gustafsson, J.-Å. and Yamamoto, K. (1990) ^{1}H NMR studies of the glucocorticoid receptor DNA-binding domain: sequential assignments and identification of secondary structure elements. *Biochemistry* 29: 9015–9023.

Honzatko, R.B., Crawford, J.L., Monaco, H.L., Ladner, J.E., Edwards, B.F., Evans, D.R., Warren, S.G., Wiley, D.C., Ladner, R.C. and Lipscomb, W.N. (1982) Crystal and molecular structures of native and CTP-liganded aspartate carbamoyltransferase from *Escherichia coli*. *J. Mol. Biol.* 160: 219–263.

Kägi, J.H.R. (1991) Overview of metallothionein. *Meth. Enzymol.* 205: 613–626.

Kannan, K.K., Nostrand, B., Fridborg, K., Lövgren, S., Orlsson, A. and Petef, M. (1975) Crystal structure of human erythrocyte carbonic anhydrase B: Three-dimensional structure at a nominal 2.2 Å resolution. *Proc. Natl. Acad. Sci. USA* 72: 51–55.

Kallenbach, E., Maler, B.A., Yamamoto, K.R., Boelens, R. and Kaptein, R. (1991) Identification of the metal coordinating residues in the DNA binding domain of the glucocorticoid receptor by ^{113}Cd-^{1}H heteronuclear NMR spectroscopy. *FEBS Lett.* 291: 367–370.

Kim, E.E. and Wyckoff, H.W. (1991) Reaction mechanism of alkaline phosphatase based on crystal structures. Two-metal ion catalysis. *J. Mol. Biol.* 218: 449–464.

Kraulis, P.J., Raine, A.R.C., Gadhavi, P.L. and Laue, E.D. (1992) Structure of the DNA-binding domain of zinc GAL4. *Nature* 356: 448–450.

Liljas, A., Kannan, K.K., Bergstén, P.C., Waara, I., Fridborg, K., Strandberg, B., Carlbom, V., Järup, L., Lövgren, S. and Petef, M. (1972) Crystal structure of human carbonic anhydrase C. *Nature* 235: 131–137.

Lovejoy, B., Cleasby, A., Hassel, A.M., Longley, K., Luther, M.A., Weigl, D., McGeehan, G., McElroy, A.B., Drewry, D., Lambert, M.H. and Jordan, S.R. (1994) Structure of the catalytic domain of fibroblast collagenase complexed with an inhibitor. *Science* 263: 375–377.

Luisi, B.F., Xu, W.X., Otwinowski, Z., Freedman, L.P., Yamamoto, K.R. and Sigler, P.B. (1991) Crystallographic analysis of the interaction of the glucocorticoid receptor with DNA. *Nature* 352: 497–505.

Margoshes, M. and Vallee, B.L. (1957) A cadmium protein from equine kidney cortex. *J. Am. Chem. Soc.* 79: 4813.

Maret, W. and Vallee, B.L. (1993) Cobalt as a probe and label of proteins. *Meth. Enzymol.* 226: 52–71.

Marmorstein, R., Corey, M., Ptashne, M. and Harrison, S.H. (1992) DNA recognition by GAL4: structure of a protein-DNA complex. *Nature* 356: 408–414.

Matthews, B.W., Jansonius, J.N., Colman, P.M., Schoenborn, B.P. and Dupourque, D. (1972) Three-dimensional structure of thermolysin. *Nature New Biol.* 238: 37–41.

Medina, J.F., Wetterholm, A., Rådmark, O., Shapiro, R., Haeggström, J.Z., Vallee, B.L. and Samuelsson, B. (1991) Leukotriene A4 hydrolase: determination of the three zinc-binding ligands by site-directed mutagenesis and zinc analysis. *Proc. Natl. Acad. Sci. USA* 88: 7620–7624.

Murphy, J.E., Xu, X. and Krantrowitz, E.R. (1993) Conversion of a magnesium binding site into a zinc binding site by a single amino acid substitution in *Escherichia coli* alkaline phosphatase. *J. Biol. Chem.* 268: 21497–21500.

Otvos, J.D., Chen, S.-M. and Liu, X. (1989) NMR insights into the dynamics of metal interaction with metallothionein. *In*: D.H. Hamer and D.R. Winge (eds): *Metal Ion Homeostasis: Molecular Biology and Chemistry*. Alan R. Liss, New York, pp 197–206.

Pan, T., Freedman, L.P. and Coleman, J.E. (1990) Cadmium-113 NMR studies of the DNA binding domain of the mammalian glucocorticoid receptor. *Biochemistry* 29: 9218–9225.

Pan, T. and Coleman, J.E. (1991) Sequential assignments of the ^{1}H NMR resonances of Zn(II)2 and ^{113}Cd(II)2 derivatives of the DNA-binding domain of the GAL4 transcription factor reveal a novel structural motif for specific DNA recognition. *Biochemistry* 30: 4212–4222.

Pauptit, R.A., Karlsson, R., Picot, D., Jenkins, J.A., Niklaus-Reimer, A.-S. and Jansonius, J.N. (1988) Crystal structure of neutral protease from *Bacillus cereus* refined at 3.0 Å resolution and comparison with the homologous but more thermostable enzyme thermolysin. *J. Mol. Biol.* 168: 525–537.

Pavletich, N.P. and Pabo, C.O. (1991) Zinc finger-DNA recognition: crystal structure of Zif268-DNA complex at 2.1 Å. *Science* 252: 809–818.

Pavletich, N.P. and Pabo, C.O. (1993) Crystal structure of a five-finger GLI-DNA complex: new perspectives on zinc fingers. *Science* 261: 1701–1707.

Quiocho, F.A. and Lipscomb, W.N. (1971) Carboxypeptidase A: a protein and an enzyme. *Adv Prot. Chem.* 25: 1–58.

Rhodes, D. and Klug, A. (1993) Zinc fingers. *Sci. Amer.* 268: 56–65.

Schmid, M.F. and Herriott, J.R. (1976) Structure of carboxypeptidase B at 2.8 Å resolution. *J. Mol. Biol.* 103: 175–190.

Schwabe, J.W.R., Neuhaus, D. and Rhodes, D. (1990) Solution structure of the DNA-binding domain of the oestrogen receptor. *Nature* 348: 458–461.

Soler, D., Nominzu, T., Brown, W.E., Chen, M., Ye, Q.-Z., Van Wart, H.E. and Auld, D.S. (1994) Zinc content of promatrilysin, matrilysin, and the stromelysin catalytic domain. *Biochem. Biophys. Res. Commun.* 201: 917–923.

Stöcker, W., Ng, M. and Auld, D.S. (1990) Fluorescent oligopeptide substrates for kinetic characterization of the specificity of Astacus protease. *Biochemistry* 29: 10418–10425.

Sutton, B.J., Artymiuk, P.J., Cordero-Borboa, A.E., Little, C., Phillips, D.C. and Waley, S.G. (1987) An x-ray crystallographic study of beta-lactamase II from *Bacillus cereus* at 0.35 nm resolution. *Biochem. J.* 248: 181–188.

Thayer, M.M., Flaherty, K.M. and McKay, D.B. (1991) Three-dimensional structure of the elastase of *Pseudomonas aeruginosa* at 1.5 Å resolution. *J. Biol. Chem.* 266: 2864–2871.

Vallee, B.L. and Williams, R.J.P. (1968) Metalloenzymes: the entatic nature of their active sites. *Proc. Natl. Acad. Sci. USA* 59: 498–505.

Vallee, B.L. (1977) Zinc biochemistry in normal and neoplastic growth processes. *Experientia* 33: 600.

Vallee, B.L. and Falchuk, K.H. (1981) Zinc and gene expression. *Phil. Trans. R. Soc. London Ser. B* 294: 185–197.

Vallee, B.L. (1983) The biological roles of zinc. *In*: T.G. Spiro (ed.): *Zinc Enzymes*. Wiley, New York, pp 2–24.

Vallee, B.L. and Auld, D.S. (1989) Short and long spacer sequences and other structural features of zinc binding sites in zinc enzymes. *FEBS Lett.* 257: 138–140.

Vallee, B.L. and Auld, D.S. (1990a) Active-site zinc ligands and activated H_2O of zinc enzymes. *Proc. Natl. Acad. Sci. USA* 87: 220–224.

Vallee, B.L. and Auld, D.S. (1990b) Zinc coordination, function, and structure of zinc enzymes and other proteins. *Biochemistry* 29: 5647–5659.

Vallee, B.L., Coleman, J.E. and Auld, D.S. (1991) Zinc fingers, zinc clusters, and zinc twists in DNA-binding protein domains. *Proc. Natl. Acad. Sci. USA* 88: 999–1003.

Vallee, B.L. and Auld, D.S. (1992a) Functional zinc-binding motifs in enzymes and DNA-binding proteins. *Faraday Discuss.* 93: 47–65.

Vallee, B.L. and Auld, D.S. (1992b) Active zinc binding sites of zinc metalloenzymes. In: H. Birkedal-Hansen, Z. Werb, H. Welgus and H. Van Wart (eds): *Matrix Metalloproteinases and Inhibitors*. Gustav Fischer Verlag, Stuttgart, pp 5–19.

Vallee, B.L. and Auld, D.S. (1993a) Cocatalytic zinc motifs in enzyme catalysis. *Proc. Natl. Acad. Sci. USA* 90: 2715–2718.

Vallee, B.L. and Auld, D.S. (1993b) New perspective on zinc biochemistry: cocatalytic sites in multi-zinc enzymes. *Biochemistry* 32: 6493–6500.

Vallee, B.L. and Auld, D.S. (1993c) Zinc: Biological functions and coordination motifs. *Acc. Chem. Res.* 26: 543–551.

Vallee, B.L. and Falchuk, K.H. (1993) The biochemical basis of zinc physiology. *Physiol. Rev.* 73: 79–118.

Vâsák, M., Hawkes, G.E., Nicholson, J.K. and Sadler, P.J. (1985) ^{113}Cd NMR studies of reconstituted seven-cadmium metallothionein: evidence for structural flexibility. *Biochemistry* 24: 740–747.

Volbeda, A., Lahm, A., Sakiyama, F. and Suck, D. (1991) Crystal structure of Penicillium citrinum P1 nuclease at 2.8 A resolution. *EMBO J.* 10: 1607–1618.

Wilson, D.K. and Quiocho, F.A. (1993) A pre-transition-state mimic of an enzyme: X-ray structure of adenosine deaminase with bound 1-deazaadenosine and zinc-activated water. *Biochemistry* 32: 1689–1694.

Zeng, J., Vallee, B.L. and Kägi, J.H.R. (1991) Zinc transfer from transcription factor IIIA fingers to thionein clusters. *Proc. Natl. Acad. Sci. USA* 88: 9984–9988.

NMR Structural studies on the zinc finger domains of nuclear hormone receptors

R.M.A. Knegtel, M.A.A. van Tilborg, R. Boelens and R. Kaptein

Bijvoet Center for Biomolecular Research, Utrecht University, Padualaan 8, NL-3584 CH Utrecht, The Netherlands

Summary. This chapter presents an overview of the application of modern NMR methods in structural studies of the DNA binding domains (DBDs) of nuclear hormone receptors. The DBDs studied so far comprise those of the glucocorticoid, estrogen, retinoic acid and retinoid X receptors. NMR spectroscopy has allowed the elucidation of the first structures of this family of C_4-type zinc fingers, which led to a better understanding of their role in gene regulation. Crystallographic studies provided insight in protein-protein and protein-DNA interactions. Subsequent studies, applying NMR, have provided deeper insight into a diversity of issues concerning these proteins, ranging from backbone dynamics and metal coordination to the interaction of these domains with their DNA target sites. From this work a picture emerges of a class of closely related zinc-binding proteins which, despite their strong sequence homology, exhibit interesting structural and functional differences between members of different subfamilies.

Introduction

Since it was discovered in the 1980s that a number of transcription factors recognize DNA through a zinc binding domain, they have been the focus of a vast amount of biochemical and biophysical research (Kaptein, 1991; Rhodes and Klug, 1993). Especially the superfamily of nuclear hormone receptors (Palmer, 1991) has proved to be of great interest, for these proteins have been identified to be involved in a diversity of essential cellular processes (Evans, 1988). The nuclear hormone receptors consist of six domains (cf. Fig. 1), each fulfilling a distinct function in the process of ligand induced gene regulation. The unstructured A/B domain is a weakly conserved domain of variable length containing a transactivation function. The E/F-domain binds small hydrophobic ligands among which are the steroid hormones, thyroid hormone, vitamin D3 and several vitamin A derivatives. This domain is connected via a linker D to the highly conserved C-domain (cf. Fig. 2) which binds to specific DNA sequences, referred to as response elements for the respective ligands. The DNA binding domain only folds into a functionally active protein when two zinc ions are bound. Each of these two zinc ions is tetrahedrally coordinated by four conserved cystein residues, forming the so-called zinc fingers (Freedman et al., 1988). Three residues at the C-terminal end of the first finger (the

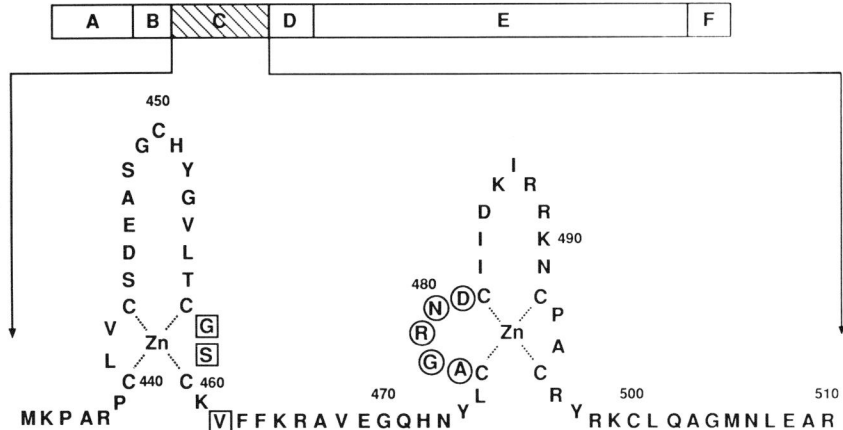

Figure 1. A schematic representation of the modular organization of the nuclear hormone receptors. Residues Cys440–Arg510 of the rat glucocorticoid receptor DNA binding domain are illustrated in detail. The boxed residues indicate amino acids that are essential for discrimination between the glucocorticoid and estrogen response elements. The circled residues are involved in protein-protein interactions in the dimeric GRDBD-GRE complex.

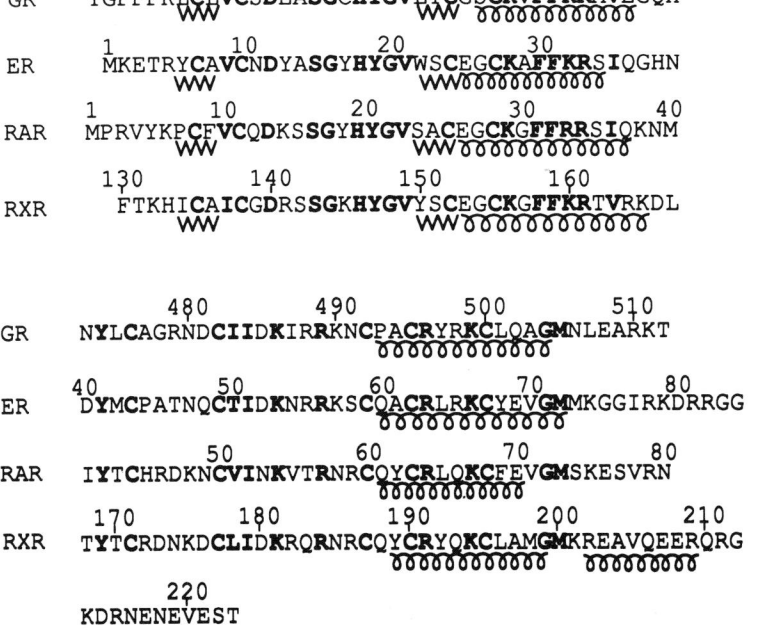

Figure 2. Sequences of the RAR, GR, ER and RXR-DBDs. Conserved residues are shown in bold and α-helical and β-sheet regions are indicated. The numbering of the residues which is used is the same as in the original papers.

boxed residues in Fig. 1) were found to determine the DNA binding specificity of the GR DBD and these residues are usually referred to as the P box (Umesuno et al., 1989; Danielsen et al., 1989). Five residues in the first N-terminal loop of the second zinc finger (circled in Fig. 1) were found to be responsible for protein-protein interactions determining the cooperative binding of the GR DBD homodimers. These amino acids are referred to as the D box (Dahlman-Wright et al., 1991). In this chapter we will focus on the structural properties of the DNA binding domains of several members belonging to the hormone receptor family which have been characterized using high-resolution NMR spectroscopy.

Within the superfamily of hormone receptors a distinction can be made between two subfamilies. The first subfamily comprises, among others, the receptors for glucocorticoid (GR), estrogen (ER) and progesterone (PR). In their inactive state these receptors are present in the cytosol, usually complexed with heat-shock factors. After binding of the ligand and dissociation of the receptor/heat-shock protein complexes the receptors are translocated in the cell nucleus where they bind to the DNA as homodimers. The vitamin D3 (VD3R), thyroid hormone (TR), retinoic acid (RAR) and retinoid X (RXR) receptors are members of the second subfamily and reside within the cell nucleus. After ligand binding they form heterodimers with the RXR and bind subsequently to the corresponding response elements.

The DNA target sites of the two subfamilies differ strikingly in both spacing and relative orientation of the two halfsites (cf. Fig. 3). GR and ER bind as homodimers to inverse repeats with a three basepair spacer.

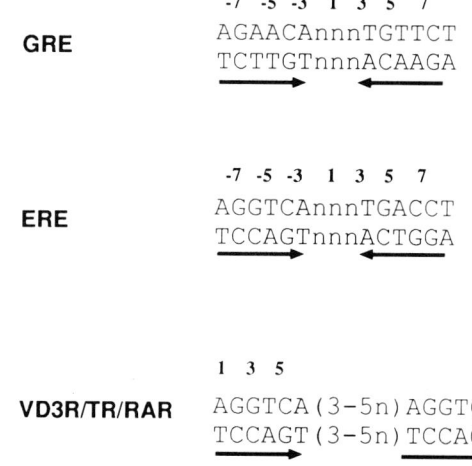

Figure 3. Sequences of the GRE, ERE, VD3R, TR and RAR response elements. The hexameric consensus response elements are underlined and the numbering scheme of the bases is indicated.

By contrast, VD3R, TR and RAR bind as heterodimers with RXR to direct repeats with spacers of three, four and five basepairs, respectively. The RXR is also capable of forming homodimeric complexes with response elements consisting of a direct repeat separated by a single basepair. The determinants of the specificity for these different relative orientations and spacings appear to reside in the DNA binding domains which makes structural studies of these proteins and their complexes with DNA essential in order to understand the basis of the molecular recognition process at an atomic level.

High resolution NMR spectroscopy has been applied quite successfully to the hormone receptor DNA binding domains, resulting in the structures of the GR and ER DBDs being elucidated in 1990 (Härd et al., 1990b; Schwabe et al., 1990). A few years later solution structures of the RAR and RXR DBDs were reported (Knegtel et al., 1993; Lee et al., 1993) and information was gained on the interaction of the GR DBD with its halfsite (Kellenbach et al., 1991; Remerowski et al., 1991). In addition, two complexes of the GR DBD and ER DBD homodimers complexed with DNA were solved by X-ray crystallography and yielded a wealth of information on the protein-DNA and protein-protein interactions (Luisi et al., 1991; Schwabe et al., 1993a). In the following sections we will discuss several features of the structure and dynamics of these domains in relation to their biological function. Special emphasis will be given to the NMR studies.

Overview

The glucocorticoid receptor and estrogen receptor DNA binding domains

The glucocorticoid receptor DNA binding domain
The DNA binding domain of the rat glucocorticoid receptor (GR DBD), consisting of residues Cys440-Arg510, was the first DNA binding domain in this superfamily which has been studied by high-resolution NMR spectroscopy. A low-resolution structure, providing a first view of this new class of zinc binding proteins was reported by Härd et al. (1990b). Several secondary structure elements could be identified based on characteristic patterns of NOE connectivities. These elements include two α-helical regions encompassing Ser459-Glu469 including the P-box and Pro493-Gly504 and a short stretch of anti parallel β-sheet involving the residues Cys440-Leu441 and Leu455-Cys457 with a hydrogen bond between the amide of Leu441 and the carbonyl of Leu455. The two zinc coordinating domains are linked to each other by a less well defined region comprising an extended structure from Gln471 to Leu475. The second zinc coordinating domain contains the D-box (Cys476-Cys482) which is disordered in solution. The original structure determination was

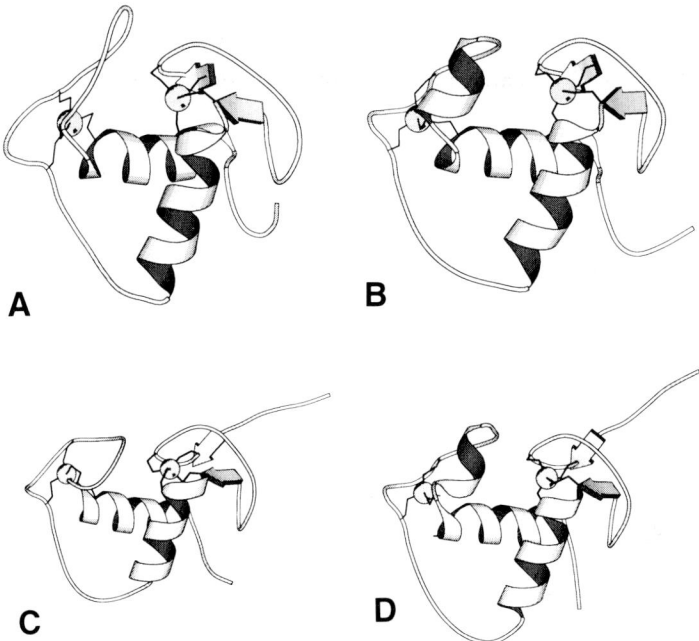

Figure 4. Ribbon representations (Kraulis, 1991) of: (a) the refined GRDBD solution structure (van Tilborg et al., manuscript in preparation), (b) the GRDBD-GRE$_{s4}$ crystal structure (Luisi et al., 1991), (c) the ERDBD refined NMR structure (Schwabe et al., 1993b) and the crystal structure of the ERDBD-ERE in the complex (Schwabe et al., 1993a).

based on a set of 470 (inter and intra) NOE connectivities. The metal-coordination residues in the DNA binding domain were identified by Kellenbach et al. (1991) using ^{113}Cd-^{1}H heteronuclear NMR spectroscopy on a GR DBD fragment with its two zinc ions replaced by ^{113}Cd ions. Residues Cys440, Cys443, Cys457 and Cys460, located in the first finger and residues Cys476, Cys482, Cys492 and Cys495 in the second finger were observed to coordinate the two metal ions as determined by 2D HSQC and HSQC-NOESY experiments. No functional role for the ninth conserved cysteine (Cys500) has been found yet.

Shortly after the solution structure was published, a crystallographic analysis of the interaction of the glucocorticoid receptor with DNA was reported by Luisi et al. (1991). Comparison of the initial NMR structure with the complex crystal structure of the GR DBD revealed some apparent differences within the second zinc coordinating domain, which is illustrated in Figure 4a and b. In the second finger of the crystal structure a third "distorted" helical segment was observed (Lys486-Asn491). This region makes both protein-protein and protein-DNA contacts in the crystal structure of the dimeric complex with DNA. There is a difference in opinion whether this region is disordered in the uncomplexed state and correctly folded only upon DNA binding or that it is already structured in the uncomplexed state (Baumann et al., 1993). Recently, two NMR structure refinements of GR DBD have been carried out (van Tilborg et al., manuscript in preparation; Baumann et al., 1993). A refinement of a GR DBD fragment consisting of 83 residues (GR DBD83) (Baumann et al., 1983), based on 906 distance constraints obtained from NOE intensities, had an overall backbone r.m.s.d. of 0.70 Å (calculated from 24 structures with respect to the average). The actual differences between the unbound and bound GR DBD were found to be small. The additional distorted helical region (K486-N491) was now also observed in the solution structure and the authors concluded that the distorted helix was not formed as a consequence of DNA binding, but is already present in the unbound state. The D-box, however, which also forms a part of the DBD dimerization surface, does seem to undergo a conformational change upon DNA binding.

Simultaneously, a longer fragment of GR DBD93 was refined (van Tilborg et al., manuscript in preparation) using relaxation matrix calculations (Bonvin et al., 1993, 1994). This structure was based on 1186 distance constraints, obtained from NOE buildup curves (Fig. 4a). The 14 refined structures have an overall backbone r.m.s.d. (with respect to the average structure) of 0.58 Å for the core of the protein (Cys440-Ser448, His451-Glu469, Pro493-Glu508). The second finger is followed by a second well defined helical region (Pro493-Ala503) and a short stretch of β-strand (Met505-Leu507), which was not observed in the unrefined structure (Härd et al. 1990a,b), but was also seen in the refined structure of Baumann et al. (1993). In this longer fragment the backbone of the second zinc coordinating domain is still disordered without a distinct helical structure. Specific helical NOE contracts were absent in the 2D and 3D NMR spectra for this region.

In order to establish whether this region in the second zinc coordinating domain displays enhanced local mobility, the extent of rapid backbone motion within the GR DBD has been investigated by Berglund et al. (1992). They applied proton detected heteronuclear NMR spectroscopy on ^{15}N-labeled DBD83. Averaged S^2 order parameters were found to be similar for various functional domains of the DBD ($\sim 0.86 \pm 0.07$) and in particular, no evidence was found for rapid backbone motions within the second domain. In addition, a comparison of ^{15}N-NMR relaxation measurements with a molecular dynamics simulation of the crystal structure in water was made (Eriksson et al., 1993). The authors concluded that uniform flexibility of the amide bonds throughout the backbone was observed for both the experimental and simulated order parameters and these results suggested that the second zinc coordinating domain is not disordered in the unbound state. However, both ^{15}N relaxation measurements and molecular dynamics simulations register only fast, short time-scale dynamics while the motions involving the second zinc finger domain could take place on a longer time scale inaccessible to current experimental or theoretical approaches. Moreover, thermodynamic studies by Spolar and Record (1994) have demonstrated the importance of a large negative heat capacity change in site-specific protein-DNA recognition by GR DBD. Based on experimental data by Härd et al. (1990c) these authors estimated that approximately 18 residues per monomer fold upon binding, a thermodynamic estimate which agrees with the size of the second zinc coordinating domain of GR DBD.

The estrogen receptor DNA binding domain
Related to the GR DBD, is the ER DBD of which the solution structure was reported by Schwabe et al. (1990). A refined solution structure was published a few years later (Schwabe et al., 1993b). The refined solution structure of the ER DBD is illustrated in Figure 4c. This structure determination was based on 351 distance constraints obtained from 2D ^{1}H homonuclear NMR data. An additional 23 χ_1 angle constraints, within a protein fragment comprising residues 1–75, were used in the structure calculations. For the last ten carboxyl terminal residues, only intra-residue and sequential NOE crosspeaks were observed and therefore these residues were not included in the reported structure calculation. A total of 49 structures was calculated. Superposition on the backbone of residues 4–35 and 59–74 results in a r.m.s.d. of 0.81 ± 0.15 Å with respect to the average structure. In each of the two zinc coordinating domains, the second pair of zinc binding cysteines forms the start of an α-helix. NOEs between side chains of the hydrophobic surface of these two helices (Phe30, Phe31, Ile35, Leu64, Cys67, Tyr68) define a precise relative perpendicular orientation and form a hydrophobic core with a backbone r.m.s.d. of 0.37 ± 0.07 Å (for the residues 24–35 and 59–71). Similar to what was observed in the refined GR DBD93 (van Tilborg et al., manuscript in preparation) the second zinc coordinating domain (residues 36–59) is also rather ill defined by the NMR data and no helical segment is observed. In fact, many of the ψ and ϕ angles in this region reflect an extended conformation. Residues Cys43 to Cys49, which correspond to the D-box, form the least well defined region of the solution structure.

The crystal structure of a dimeric DR DBB-DNA complex was reported by Schwabe et al. (1993a). The structure of ER DBD in this complex is illustrated in Figure 4d. The asymmetric unit contains two DNA duplexes, each bound by a protein dimer. If a comparison is made of the backbone conformations of the ordered residues (residues 4–35 and 59–74) between the averaged NMR structure and the crystal structure, a r.m.s.d. of 1.07 Å is obtained. Besides the positions of the C_α-atoms, also the conformations of many of the interior side chains are essentially identical. Surprisingly, the short helix in the second finger is present in both protein monomers of one dimer, but absent in the other.

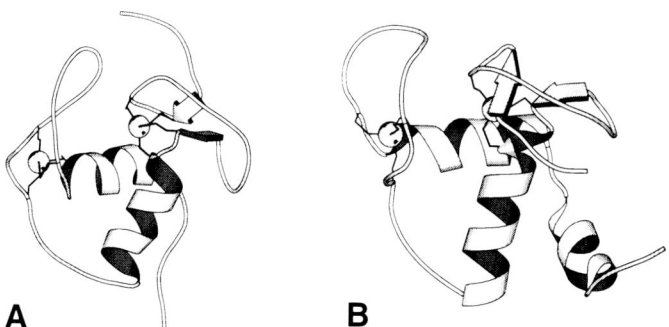

Figure 5. Ribbon representation of the refined NMR structure of RAR DBD (a) (Knegtel et al., 1993) and the NMR solution structure of the RXR DBD (b) (Lee et al. 1993).

The retinoic acid receptor-β and retinoid X receptor DNA binding domains

The retinoic acid receptor-β DNA binding domain

The human retinoic acid receptor-β was the first member of the second subfamily of nuclear hormone receptors of which the structure of its DNA binding domain was solved by NMR (Knegtel et al., 1993b). A total of 1244 distance restraints was collected from 2D NOE experiments, consisting of 448 intra-residue and 796 inter-residue NOEs. of 23 residues the χ-angle and of an additional 30 residues the φ-angle could be restrained, in both cases on the basis of short mixing time NOE and J-coupling data. Twenty-nine structures were generated with distance geometry and were refined by a combination of restrained molecular dynamics using GROMOS and relaxation matrix calculations using the "ensemble IRMA" protocol (Bonvin et al., 1993). The final 15 best structures had an r.m.s.d. from the average of 0.18 Å when only the α-helix backbones were superimposed and 0.37 Å when nine sidechains of core-residues were included in the superpositioning. The backbone of residues 5–80 was determined with a r.m.s.d. of 0.76 Å. The first zinc finger (residues 8–28) had a r.m.s.d. of 0.79 Å, while that of the second finger (residues 44–62) was 0.64 Å. It can be seen in Figure 5a that the RAR DBD has a global fold similar to that to the GR and ER DBDs. Again, two α-helices preceded by two zinc fingers cross each other perpendicularly, thus forming a hydrophobic core involving the conserved phenylalanines Phe31 and Phe32. An interesting difference with the previously studied DBDs was that the C-terminus was found to be in a structured, extended state, in contrast to the disorder that was observed in this region in the case of GR and ER DBDs. No distorted α-helical segment was observed in the second finger, as has been seen in the crystal structure of the GR and ER DBD-DNA complexes. In fact, several large $^3J_{H\alpha}$ couplings were observed in this region which implies an extended conformation and the availability of these scalar coupling constants allows a more precise determination of the second finger compared to the first finger.

The zinc coordinating cysteines were identified by replacing the zinc ions by ^{113}Cd and performing heteronuclear ^{113}Cd-^{1}H NMR experiments (Knegtel et al., 1993a). In 3D HSQC-HOHAHA spectra seven out of eight cysteines were observed to be J-coupled to the two cadmium ions. The missing Cys60 proton resonances were identified in 2D HSQC-HOHAHA spectra recorded with longer refocussing delays. In each zinc finger one cysteine exhibited a markedly smaller ^{113}Cd-^{1}H coupling constant (Cys8 in the first finger and Cys60 in the second) which could be due to an imperfect tetrahedral surrounding of the cadmium ions.

Retinoid X receptor DNA binding domain
A low-resolution structure of the 81-residue DNA binding domain of the RXR was determined from a set of 593 NOE derived inter-proton distances which included 197 intra-residue, 155 sequential, 82 medium-range, and 159 long-range NOEs (Lee et al., 1993). Out of 310 distance geometry structures a subset of 58 structures exhibiting low residual errors were selected for further optimization by means of restrained molecular dynamics calculations in the AMBER force field. The RXR DNA binding domain was found to contain an additional helical segment at the C-terminus of the protein, as shown in Figure 5b. The backbone r.m.s.d. with respect to the average for the entire sequence after superpositioning on the helices was 1.04 Å. The third helix observed in the RXR DBD is a secondary structure element previously unobserved in the nuclear hormone receptor It crosses the first helix while contacting the tip of the first zinc finger. Val205 in the C-terminal helix takes part in the hydrophobic core formed by helices 1 and 2 and some additional electrostatic interactions with charged residues in helix 1 are expected to take place. Interestingly, a valine is present at the same position in the sequence of the RAR DBD and this residue is also involved in hydrophobic contacts with apolar residues in the first helix. In the GR and ER DBD this valine is replaced by an arginine and an isoleucine, respectively, and the C-terminus is observed to be unstructured in solution. The conserved Gly-Met dipeptide which, by forming a distinct bend, separates the third helix from the second in the RXR DBD, is present in all DBDs studied. In the case of the RAR DBD a shorter fragment was studied than in the case of the RXR, with its sequence ending two residues before the end of the additional third α-helix. One might speculate that in a longer fragment the C-terminal stretch, that in the current RAR DNA binding domain structure is extended, might fold up to form an α-helix, similar to that observed in the RXR DBD.

DNA binding by nuclear hormone receptors

DNA binding by the glucocorticoid receptor DNA binding domain

A model for the GR DBD-GRE complex was first proposed by Härd et al. (1990b) based on the NMR structure and available genetic and biochemical data. A number of properties was considered. The GR DBD binds to the two GRE halfsites in a cooperative manner. Substitution of Ala477 to Asp481 (D-box) with the corresponding segment of another receptor yields recognition of the GRE but without cooperative binding. Residues that direct the discrimination between response elements are located in the N-terminus of the helical region of the first

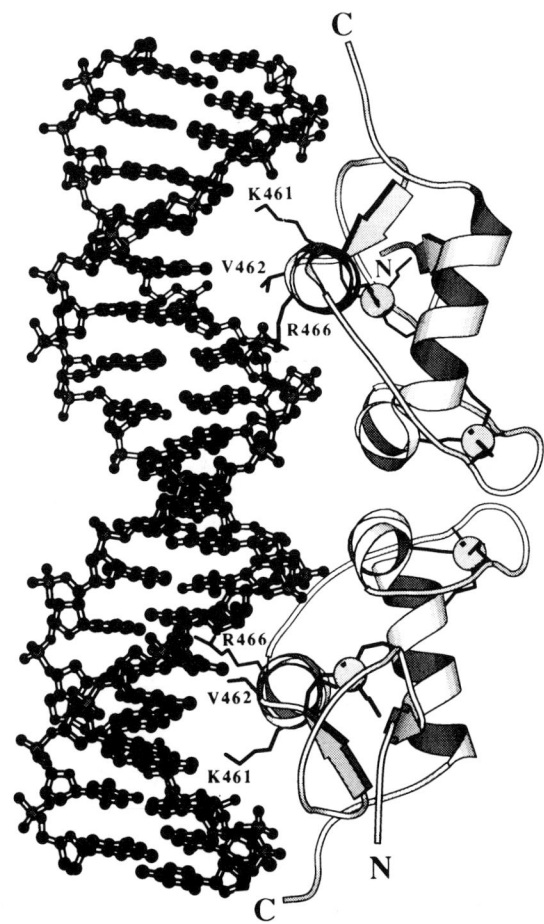

Figure 6. Schematic MOLSCRIPT (Kraulis, 1991) representation of the GRDBD-GRE$_{s4}$ complex with DNA (Luisi et al., 1991). The amino (N) and carboxyl (C) termini are indicated as well as the side chains of the three residues which make base specific contacts.

domain, the so-called P-box, corresponding to GR DBD residues 457 to 462. Mader et al. (1989) found the three discriminating amino acids to be Glu25, Gly26 and Ala29 for ER DBD and Gly458, Ser459 and Val462 for GR DBD. Missing base contact and phosphate ethylation interference analysis show that the two DBD molecules interact with the major groove of the GRE-half site. Figures 2 and 3 show the half-sites and protein sequences of four of the nuclear hormone receptors DNA binding domains.

In the proposed model, the first α-helix is the recognition helix which is located in the major groove of the DNA with residues Gly458, Ser459 and Asn462 contacting the DNA bases of the GRE while residues in the D-box are positioned such that they can form protein-protein contacts with the second monomer. The first NMR experiments on the complex were carried out by Kellenbach et al. (1991). A photo-CIDNP experiment was performed to study the interaction between the GR DBD93 and its corresponding GRE. Using a GRE halfsite, an unaffected polarization of Tyr452 and a suppression of the polarization of Tyr474 was observed upon DNA binding, suggesting that Tyr474 is involved in protein-DNA interactions while Tyr452 is not.

Later that year the crystal structure of two GR DBD-GRE complexes were reported by Luisi et al. (1991) (cf. Fig. 6). A ribbon representation of the complex is shown in Figure 6. In the crystallographic studies, the complexes contain DNA with a modification of the GRE halfsite spacing. The separation of the two GRE half-sites was increased by one base, from the natural three (GREs3, complex refined to a resolution of 4 Å) to an artificial four (GREs4, complex refined to a resolution of 2.9 Å). However, in this last complex, both DBD molecules did not bind equally to their specific half sites. One DBD is bound in the correct specific site while the second DBD is bound to the first DBD, but not sequence specifically to the DNA. Although the GR DBD is monomeric in solution (Freedman et al., 1988; Härd et al., 1990c) two molecules bind a GRE cooperatively. This cooperativity arises from favorable protein-protein contacts, and two effects may contribute to this interaction. First, the alignment of the protein subunits on the DNA, and secondly, conformational changes in the protein structure upon binding. It is intriguing that the protein-protein interactions overrule the specific interactions with the DNA on the GREs4. The crystal structure shows that the dimer is stabilized through inter-monomer salt-bridges, hydrophobic interactions, and hydrogen bonds between backbone and side chains. The contacts are made by a segment between the first and second coordinating cysteines (D-box, residues 477 to 481) of the second zinc-finger. These interactions include, Ile487–Leu475, Ile483–Ala477 and Asp481–Arg479.

In the crystal structure GRDBD is observed to interact with the DNA in the major groove by means of its first α-helix, as proposed in the model by Härd et al. (1990b) and three amino acids make specific base contacts. The methyl group of Val462, located in the P-box, contacts the methyl group of T_{+5} of the GRE. This contact is essential for the discrimination of the GR and ER response elements. The other contacts are hydrogen bonds made by Arg666 to Gua_{+4} and Lys461 to Gua_{-7}. The last two amino acids are conserved throughout the receptor superfamily. Besides these specific interactions a number of non-specific interactions with the DNA backbone is observed, involving side chains of residues His451, Tyr452, Arg489, Lys490 and Arg496 of the specific complex and Cys450, Tyr452, Arg466, Arg489 and Arg496 of the non-specific complex. Upon binding the DNA is deformed, as is manifested by an opening of the major groove by roughly 2 Å (as measured by cross-strand phosphate distances) in order to create an optimal match to the protein's recognition surface.

The complex of the GR DBD and half-site sequence of the consensus GRE was studied by two-dimensional NMR spectroscopy using 2D-NOESY and TOCSY experiments (Remerowski et al., 1991). A 10 base-pair oligonucleotide $5'd(GCT_2G_4T_5T_6C_7T_8GC)3'$. $5'd(GC-A_{-3}G_{-4}A_{-5}A_{-6}C_{-7}A_{-8}GC)3'$ containing the GRE half-site hexamer was used with GC base pairs at each end. Eleven NOE cross peaks between protein and DNA were identified between $Tyr452-Ade_{-8}$, $Val462-Thy_5$ and $Val462-Gua_4$. Using these protein-DNA contacts, it was concluded that the general features, i.e., orientation and placement of the recognition helix in the major groove, are the same as in the crystal.

Due to overlap of resonances of DNA thymine methyl groups and aliphatic protein side chains it is often difficult to observe NOEs between the two. A solution to this problem was the efficient stable isotope labeling of the major groove functional groups of DNA by Kellenbach et al. (1992). In combination with sensitive detection and editing methods, more simplified NMR spectra of protein-DNA complexes were obtained. Synthetic methods were reported for ^{15}N labeling of the exocyclic amino group of cytosine and ^{13}C-labeling of the methyl group of thymine in both GRE strands (Kellenbach et al., 1992). The reported NOE contacts observed in 2D heteronuclear experiments were in accordance with both the crystallographic (Luisi et al., 1991) and NMR studies (Remerowski et al., 1991). However, in contrast to the crystal structure, where only one of the two Val462 methyl groups contacts the Thy_5 methyl group, Kellenbach et al. (1992) suggested that in solution both methyls contact the Thy_5.

DNA binding by the estrogen receptor DNA-binding domain
Recently, the crystal structure of another DNA complex of a member of the nuclear receptor family, the estrogen receptor ERE complex, was reported by Schwabe et al. (1993a). Schwabe and co-workers determined the crystal structure at 2.4 Å resolution of a fully specific complex

between a dimer of the 84 residue ERDBD and its response element. The protein binds as a symmetrical dimer to its palindromic binding site consisting of two six base-pair consensus half sites with a three base-pair spacing. Like the GR DBD, the ER DBD is monomeric in solution, but two monomers bind cooperatively to DNA.

In the crystal structure Schwabe et al. (1993a) observed two different conformations for the ER DBD bound to DNA. In both ER DBD molecules of one of the two complexes in the asymmetric unit, residues Asn54 to Ser58 of the second finger form a third helix as also observed in the GR DBD-GRE complex (Luisi et al., 1991). This helix was not observed in the second complex where high temperature factors were found for the Cys43 to Cys59 fragment, which might be due to stabilization of an unfolded state by crystal packing. Although different conformations are found, all four ER DBD's show the same pattern of base-specific contacts consisting of hydrogen bonds to the base pairs. ER DBD interacts with the central four base pairs of the six base-pair half-site. Four specific contacts are made involving side chains on the surface of the recognition helix: Glu25 to Cyt_{+4}, Lys28 to Gua-7, Lys32 to Gua-6/Thy-5 and Arg33 to Gua_{+4}. Furthermore, three ordered water molecules form a network of hydrogen bonds between side chains and base pairs and four ordered water molecules extend the network to the phosphate backbone of DNA. The contacts to phosphates at the outer edge of the response element are made by residues Tyr17, His18 and Tyr19, and on the opposite side of the major groove by Lys57, Gln60, Arg33, Arg56 and Arg63.

As discussed in the GR DBD-GREs4 complex, three polar amino acids make base-specific contacts with the DNA, of which two are conserved throughout the superfamily of nuclear receptors namely Lys28, Arg33 and Lys461, Arg466 of the ER DBD and GR DBD, respectively. The contact made by Lys28 (Lys461 in GR DBD) is very similar in both GR and ER complexes. In the GR DBD, Arg466 makes a bidentate contact to a contact Gua_{+4}, while in the ER DBD Arg33 has a different conformation resulting in only one contact to Gua_{+4}, and a hydrogen bond with the phosphate of Thy_{-3}. Lys32 which is also a conserved residue, appears to play no role in the GR DBD complex, but is used in the ER structure to interact with the G-C and T-A base pairs, which are specific to the ERE. A hydrophobic contact is formed between the GR DBD specific Val462 to a thymine base that is differentiating the GRE from the ERE. No hydrophobic interactions appear to be involved in specific recognition by the ER DBD.

DNA binding by the retinoic acid- and retinoid X receptor
Until now no structural data have been published of complexes of the RAR DBD with its DNA target site. NMR studies of these complexes are hampered by extensive broadening of proton signals (M. Katahira

and R.M.A. Knegtel, unpublished results). The residues in the P-box of the RAR and RXR DBDs (amino acids Glu, Gly and Ala) which discriminate between different half-sites are the same as those of the ER P-box. In fact, from the 13 residues in the ER DBD-DNA complex (Schwabe et al., 1993) that have direct contacts to the DNA, only three residues are not conserved in the different DBD sequences. Using the ER DBD numbering, Tyr17 is replaced by Lys in the RXR, Lys32 is replaced by Arg in the RAR and Lys57 is replaced by Asn in both the RAR and RXR. This conservation of surface residues suggests that recognition of the AGGTCA half-site by the RAR and RXR DBDs is very similar to that of the ER DBD. The non-conserved residues Tyr17 and Lys57 make non-specific contacts to the phosphate backbone and Lys32 is functionally conserved in the RAR DBD through its replacement by an arginine. Preliminary NMR studies of the RAR DBD complexed with different DNA fragments containing the AGGTCA half-site (cf. Fig. 3) showed specific broadening of the AT_1, TA_4 and CG_5 basepair imino and adenine H8 protons (M. Katahira and R.M.A. Knegtel, unpublished results). Except for the AT_1 basepair, of which only the phosphate backbone of Ade_1 is directly contacted in the ER DBD complex, these bases are involved in specific hydrogen bonds with the protein and ordered water molecules.

DNA binding assays with truncated RXR DBD constructs (Lee et al., 1993) revealed that the C-terminal third helix is involved in protein-DNA recognition as well as protein-protein interactions in the homo-dimer DNA complex. On the basis of model building of two RXR DBDs bound to a DR-1 operator it was concluded that the third helix is in close proximity to the second zinc finger of the other RXR DBD. This configuration would also allow positively charged residues located in helix 3 to contact the DNA backbone and upon mutation of these residues lower binding affinities were observed, while dimerization remained unaffected.

On the basis of biochemical experiments (Perlmann et al., 1993; Zechel et al., 1994) it can be concluded that in the case of RXR heterodimeric complexes with DNA, the dimerization interface involves different regions of the DNA binding domain. In the heterodimer-DNA complex the RXR DBD binds near the 5' end of the DNA and the RAR DBD near the 3' end, thus allowing residues in the RAR DBDs first zinc finger to contact residues in the second finger of the RXR DBD.

Conclusions

In structural studies of the nuclear hormone receptor DNA binding domains, modern NMR spectroscopy has provided the first insights in the folding and dynamical behavior of this biologically important class of zinc proteins. The elucidation of the GR and ER DBD solution structures provided a first structural rationale for the wealth of biochemical and genetic studies performed on these proteins. Subsequent crystallographic and NMR studies of complexes between the GR DBD and DNA fragments confirmed the protein-DNA contacts observed in the crystal to be present also in solution. The comparison of solution and crystal structures of these domains revealed interesting conformational changes in the second zinc finger upon cooperative binding to DNA. Although analysis of backbone dynamics by measuring ^{15}N relaxation properties and the refined GR DBD of Baumann et al. (1993) suggest that an α-helical fold of the second zinc finger is already present in the unbound state, these studies cannot exclude flexibility on longer timescales. Furthermore, they appear to contradict the absence of secondary structure in this region in the refined structure of van Tilborg et al. (manuscript in preparation). NMR and crystallographic studies of the homologous ERDBD in the unbound and bound state demonstrated that also in that case the second finger folds upon cooperative binding to DNA. This model is also confirmed by the thermodynamic analysis of induced secondary structure in the GR DBD when bound to DNA (Spolar and Record, 1994). The true nature of the structural changes in the second zinc finger domain will be the subject of future investigations on complexes in solution.

Members of the subfamily of nuclear receptors which require dimerization with the RXR for efficient DNA binding have only very recently been investigated by NMR. Studies of the RAR and RXR DBDs revealed striking differences in the secondary structure of the C-terminal region compared to the previously studied GR and ER DBDs. The mechanism of heterodimerization and specific recognition of direct repeat response elements will certainly require further study.

Although several questions regarding the mechanisms of cooperative binding and specific DNA recognition have been answered by recent NMR and X-ray crystallography studies, many others still need to be solved. The combination of isotope labeling, protein mutagenesis and rapidly evolving NMR methodology might contribute to providing some of the missing links in the near future.

Acknowledgements
The authors thank Dr. P. Wright for providing us with the coordinates of the RXR-DBD, Dr. J. Schwabe for providing us with the crystal- and the refined NMR-structure coordinates of the ERDBD-ERE complex and ER-DBD, respectively, and Dr. P. Sigler for providing us with the GRDBD-GRE complex coordinates. This work was supported by the Netherlands Organization for Chemical Research (SON) with financial aid from the Netherlands Organization for Scientific Research (NWO).

References

Berglund, H., Kovács, H., Dahlman-Wright, K., Gustafsson, J-Å. and Härd, T. (1993) Backbone dynamics of the glucocorticoid receptor DNA-binding domain. *Biochemistry* 31: 12001–12011.

Bonvin, A.M.J.J., Rullmann, J.A.C., Lamerichs, R.M.J.N., Boelens, R. and Kaptein, R. (1993) "Ensemble" iterative relaxation matrix approach: a new NMR refinement protocol applied to the solution structure of Crambin. *Proteins* 15: 385–400.

Bonvin, A.M.J.J., Vis, H., Breg, J.N., Burgering, M.J.M., Boelens, R. and Kaptein, R. (1994) NMR solution structure of the Arc repressor using relaxation matrix calculation. *J. Mol. Biol.* 236: 328–341.

Dahlman-Wright, T., Wright, A., Gustafsson, J-Å. and Carlstedt-Duke, J. (1991) Interaction of the glucocorticoid receptor DNA-binding domain with DNA is mediated by a short segment of five amino acids. *J. Biol. Chem.* 266: 3107–3122.

Danielsen, M., Hinck, L. and Ringold, G.M. (1989) Two amino acids within the knuckle of the first zinc finger specify response element activation by the glucocorticoid receptor. *Cell* 57: 1131–1138.

Eriksson, M.A.L., Berglund, H., Härd, T. and Nilsson, L. (1993) A comparison of 15N relaxation measurements with a molecular dynamics simulation: backbone dynamics of the glucocorticoid receptor DNA-binding domain. *Proteins* 17: 375–390.

Evans, R.M. (1988) The steroid and thyroidhormone receptor super family. *Science* 240: 889–895.

Freedman, L.P., Luisi, B.F., Korszun, Z.R., Basavappa, R., Sigler, P.B. and Yamamoto, K.R. (1988) The function and structure of the metal coordination sites within the glucocorticoid receptor DNA binding domain. *Nature* 334: 543–546.

Härd, T., Kellenbach, E., Boelens, R., Kaptein, R., Dahlman, K., Carlstedt-Duke, J., Freedman, L.P., Maler, B.A., Hyde, E.I., Gustafsson, J-Å. and Yamamoto, K.R. (1990c) ^{1}H NMR studies of the glucocorticoid receptor DNA-binding domain: sequential assignments and identification of secondary structure elements. *Biochemistry* 29: 9015–9023.

Härd, T., Kellenbach, E., Boelens, R., Maler, B.A., Dahlman, K., Freedman, L.P., Carlstedt-Duke, J., Yamamoto, K.R., Gustafsson, J-Å. and Kaptein, R. (1990b) Solution structure of the glucocorticoid receptor DNA-binding domain. *Science* 249: 157–160.

Härd, T., Dahlman, K., Carlstedt-Duke, J., Gustafsson, J-Å. and Rigler, R. (1990a) Cooperativity and specificity in the interaction between DNA and the glucocorticoid receptor DNA-binding domain. *Biochemistry* 29: 5538–5364.

Kaptein, R. (1991) Distinguishing features. *Cur. Biol.* 1: 336–338.

Katahira, M., Knegtel, R.M.A., Boelens, R., Eib, D., Schilthuis, J.G., van der Saag, P.T. and Kaptein, R. (1992) Homo- and heteronuclear NMR studies of the human retinoic acid receptor β DNA-binding domain: sequential assignments and identification of secondary structure elements. *Biochemistry* 31: 6474–6480.

Kellenbach, E., Härd, T., Boelens, R., Dahlman, K., Carlstedt-Duke, J., Gustafsson, J-Å., van der Marel, G.A., van Boom, J.H., Maler, B.A., Yamamoto, K.R. and Kaptein, R. (1991) Photo-CIDNP study of the interaction between the glucocorticoid receptor DNA-binding domain and glucocorticoid response elements. *J. Biomol. NMR* 1: 105–110.

Kellenbach, E., Remerowski, M.L., Eib, D., Boelens, R., van der Marel, G.A., van den Elst, H., van Boom, J.H. and Kaptein, R. (1992) Synthesis of isotope labeled oligonucleotides and their use in an NMR study of a protein-DNA complex. *Nucleic Acids Res.* 20: 653–657.

Klug, A. and Rhodes, D. (1983) Zinc fingers. *Scientific American* 2: 32–39.

Knegtel, R.M.A., Boelens, R., Ganadu, M.L., George, A.V.E., van der Saag, P.T. and Kaptein, R. (1993a) Heteronuclear ^{113}Cd-^{1}H NMR study of metal coordination on the human retinoic acid receptor-β DNA binding domain. *Biochem. Biophys. Res. Comm.* 192: 492–498.

Knegtel, R.M.A., Katahira, M., Schilthuis, J.G., Bonvin, A.M.J.J., Boelens, R., Eib, D., van der Saag, P.T. and Kaptein, R. (1993b) The solution structure of the human retinoic acid receptor-β DNA-binding domain. *J. Biol. NMR* 3: 1–17.

Kraulis, P.J. (1991) MOLCRIPT: a program to produce both detailed and schematic plots of protein structures. *J. Appl. Cryst.* 24: 946–950.

Lee, M.S., Kliewer, S.A., Provencal, J., Wright, P.E. and Evans, R. (1993) Structure of the retinoid X receptor DNA binding domain: a helix required for homodimeric DNA binding. *Science* 260: 1117–1121.

Luisi, B.F., Xu, W.X., Otwinowski, Z., Freedman, L.P., Yamamoto, K.R. and Sigler, P.B. (1991) Crystallographic analysis of the interaction of the glucocorticoid receptor with DNA. *Nature* 352: 497–505.

Palmer, M.G. (1991) *Nuclear Hormone Receptors, Molecular Mechanisms, Cellular Functions, Clinical Abnormalities.* Academic Press, Inc., New York.

Perlmann, T., Rangarajan, P.N., Umesono, K. and Evans, R. (1993) Determinants for selective RAR and TR recognition of direct repeat HREs. *Genes Dev.* 7: 1411–1422.

Remerowski, M.L., Kellenbach, E., Boelens, R., van der Marel, G.A., van Boom, J.H., Maler, B.A., Yamamoto, K.R. and Kaptein, R. (1991) 1H NMR studies of DNA recognition by the glucocorticoid receptor: complex of the DNA-binding domain with a half-site response element. *Biochemistry* 30: 11620–11624.

Schwabe, J.W.R., Neuhaus, D. and Rhodes, D. (1990) Solution structure of the DNA-binding domain of the oestrogen receptor. *Nature* 348: 458–461.

Schwabe, J.W.R., Chapman, L., Finch, J.T. and Rhodes, D. (1993a) The crystal structure of the estrogen receptor DNA-binding domain bound to DNA: how receptors discriminate between their response elements. *Cell* 75: 567–578.

Schwabe, J.W.R., Chapman, L., Finch, J.T., Rhodes, D. and Neuhaus, D. (1993b) DNA recognition by the oestrogen receptor: from solution to the crystal. *Structure* 1: 187–204.

Spolar, S. and Record T., Jr., (1994) Coupling of local folding to site-specific binding proteins to DNA. *Science* 263: 777–784.

Umesuno, K. and Evans, R.M. (1989) Determinants of target gene specificity for steroid/thyroid hormone receptor. *Cell* 57: 1139–1146.

Zechel, C., Shen, X.Q., Chambon, P. and Gronemeyer, H. (1994) Dimerization interfaces formed between the DNA binding domains determine the cooperative binding of RXR/RAR and RXR/TR heterodimers to DR5 and DR4 elements. *EMBO J.* 13: 1414–1424.

Subject index

abzyme 79
activation interfacial 3
affinity labeling 236
  – group 229
allele-specific consensus motif 113
allyl carbamate 217
amino acid, bridging 266
amino acid compositional analysis 166
antibody 79
  – catalysis 121
apolipoprotein CII (apo CII) 14
attenuated total reflection (ATR) 27

bimolecular reaction 126
binding energy, programmable 132
bioanalysis 49
biocatalyst engineering 49
biotin 229

capillary column chromatography 146
cation-olefin cyclization 127
chaperone 68
circular dichroism (CD) 29
  – spectroscopy 27
consensus motif, allele-specific 113
cooperative binding 281
cyclobutane type dimer 174

deamination 179
destabilization 125
Dewar valence isomer 178
DNA
  – binding domain (DBD) 279
  – damage 173
  – photodamage 180
  – transcription 268

electrospray ionization 108
enzyme
  digestive – 3
  – catalysis 49
  repair – 184
estrogen receptor, DNA binding domain of 279
evolution 79, 121

fluorescence 229
Fourier transform infrared spectroscopy (FTIR) 27

glucocorticoid receptor, DNA binding domain of 279
N-glycopeptide 203
glycosyl azide 203
glycosyl fluoride 206
glycosyl triazole 206
glycosyltransferase 223

α-helix 30
  poly-valyl- – 36
heparin-binding 15
high performance liquid chromatography (HPLC)
  microcapillary – 111
  reversed-phase (RP) – 141
hormone receptor 281
  – DNA binding domain 282

in-gel proteolytic digestion 146
inactivation mechanism 53

Lewis^a 203
Lewis^x 201
lid domain 8
lipase 220
  colipase 6
  lipoprotein – 3
  pancreatic – 3
lipid/water interface 3
lipoprotein metabolism 3
low water system 49
low-density lipoprotein receptor-related protein 15

major histocompatibility complex (MHC) 106
β-mannoside 210
mass spectrometry 156
  tandem – 111
medium engineering 49
metalloendopeptidase 264
metallothionein 272
microbore column chromatography 146
microsequencing 150
mitochondrial targeting peptide 71
mutagenic property 173

nuclear hormone receptor 279
nuclear magnetic resonance (NMR) spectroscopy 27, 279

N magic angle spinning – 52
nuclear receptor family   290

O-glycopeptide   217
organic synthesis   49

peptidase   229
peptide   143
  – antigen   229
  – -antibody interaction   240
  – -binding site   106
  chloroplast transit –   72
  – conformation   233
  – hormone   229
  – hormone receptor   235
  – mapping   158
  – -mass fingerprinting   163
  phosphopeptide   241
  – receptor   230
  signal –   69
  – synthesis   230
  synthetic –   108
pH memory   49
phospholipase   5
  – A1   3
photoaffinity labeling   237
photolyases   173
ping-pong mechanism   134
protease   222, 229
protein   143
  – aggregation   53
  – dynamics   50
  – engineering   56
  – imprinting   49
  – ladder sequencing   157
  – -nucleic acid interaction   247
  surfactant-associated – SP-B   27
  surfactant-associated – SP-C   27
pulmonary surfactant   27
pyrimidine dimerization   173
pyrimidine (6-4) pyrimidine
  photoproduct   176

regioselective/enantioselective synthesis
  58

respiratory distress syndrome (RDS)   28
retinoic acid receptor DNA binding
  domain   279
retinoid X receptor DNA binding
  domain   279
ribozyme   79
RNA   79

sec-machinery   68
secondary structure   29
secretory pathway   68
sequence analysis
  C-terminal –   153
  N-terminal –   148
SH_2 domain   230
shuttle vector   184
sodium dodecyl sulfate polyacrylamide
  gel electrophoresis (SDS-PAGE)   143
solid phase synthesis   215
specificity   83
stop-transfer   71
structural analysis   143
structure-function relationship   22

thermostability   49
thioglycoside   219
transesterification   55
  – reaction   58
transition state   122
trichloroacetimidate   208
two-dimensional polyacrylamide gel
  electrophoresis (2-DE)   143

X-ray crystallography   3

zinc   259
zinc cluster   259
zinc finger   259, 279
zinc site
  catalytic –   259
  cocatalytic –   259
  structural –   259
zinc twist   259

EXPERIENTIA SUPPLEMENTUM

P. Jollès, *University of Paris V and C.R.N.S., Paris, France (Ed.)*

Proteoglycans

1994. 280 pages. Hardcover
ISBN 3-7643-2957-2 (EXS 70)

Proteoglycans are glycoconjugates which constitute a topic of current interest in biochemistry, molecular biology, cell biology, and medicine. They can no longer be considered, as they were a few years ago, as simple inert building blocks.

Proteoglycans belong to a versatile protein family whose potential functions arise either from their glycosaminoglycan chains or from specific regions of their protein cores. They help maintain the essential microenvironment for cell adhesion, migration, and proliferation through their ability to function as links between cells and the extracellular matrix, and as growth-factor binders.

Their importance is gaining recognition in a wide spectrum of research areas such as the brain and reproduction. In this book, various new research aspects of proteoglycans are reviewed in a series of chapters written by well-known scientists involved in this exciting field of research.

Birkhäuser Verlag • Basel • Boston • Berlin

MOLECULAR AND CELL BIOLOGY UPDATES

S. Papa, *Institute of Medical Biochemistry and Chemistry, University of Bari, Italy*
J.M. Tager, *E.C. Slater Institute, University of Amsterdam, The Netherlands (Eds)*

Biochemistry of Cell Membranes
A Compendium of Selected Topics

1995. 376 pages. Hardcover. ISBN 3-7643-5056-3 (MCBU)

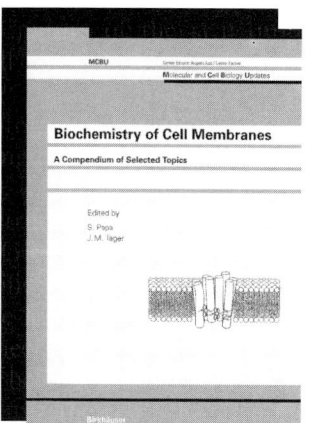

This book consists of a series of reviews on selected topics within the rapidly and vastly expanding field of membrane biology. Its aim is to highlight the most significant and important advances that have been made in recent years in understanding the structure, dynamics and functions of cell membranes.

Areas covered in this monograph include

- Signal Transduction
- Membrane Traffic: Protein and Lipids
- Bioenergetics: Energy Transfer and Membrane Transport
- Cellular Ion Homeostasis
- Growth Factors and Adhesion Molecules
- Structural Analysis of Membrane Proteins
- Membranes and Disease

Biochemistry of Cell Membranes should serve as a benchmark for indicating the most important lines for future research in these areas.

Birkhäuser Verlag • Basel • Boston • Berlin

BIOELECTROCHEMISTRY: PRINCIPLES AND PRACTICE

This book series provides a comprehensive compilation of all the physicochemical aspects of the different biochemical and physiological processes.

Bioelectrochemistry
General Introduction

Edited by
S.R. Caplan, I.R. Miller, *The Weizmann Institute of Science, Rehovot, Israel*
G. Milazzo †, *formerly Istituto Superiore di Sanità, Rome, Italy*

(BIOELECTROCHEMISTRY: PRINCIPLES AND PRACTICE: Vol. I)
1995. 384 pages. Hardcover. ISBN 3-7643-2687-5

This first of several planned volumes discusses nonequilibrium thermodynamics and kinetics, particularly enzyme catalysis, for processes and systems in the steady state. Methods of mathematical modeling by means of network simulations are also treated, since they serve to assess the transient behavior of a system on its way to a steady state. Water as a ubiquitous constituent plays an essential role in bioelectrochemical systems, hence its structure is carefully evaluated, both in the pure state and in the ionic hydration shell. Similarly, the interface between water and a membranous or biocolloidal phase is of major importance. The phenomena occurring at such interfaces, including diffuse double layers, as well as binding and adsorption of solutes, are extensively examined.

Further volumes will be added to the series which is intended as a set of source books for graduate and postgraduate students, as well as research workers, at all levels in bioelectrochemistry.

Birkhäuser Verlag • Basel • Boston • Berlin